DRONES ACROSS AMERICA

Unmanned Aircraft Systems (UAS) Regulation and State Laws

SARAH NILSSON

DRONES ACROSS AMERICA

Unmanned Aircraft Systems (UAS) Regulation and State Laws

AMERICAN BAR ASSOCIATION
Air & Space Law Forum

Cover design by Andrew Alcala / ABA Design

Printed in the United States of America.

20 19 18 17 5 4 3 2 1

ISBN: 978-1-63425-712-1

e-ISBN: 978-1-63425-713-8

Library of Congress Cataloging-in-Publication Data

Names: Nilsson, Sarah, author.
Title: Drones across America : unmanned aircraft systems (UAS), drone regulation, and state laws in the United States / Sarah Nilsson.
Description: Chicago : American Bar Association, 2017.
Identifiers: LCCN 2016049818 | ISBN 9781634257121
Subjects: LCSH: Drone aircraft—United States. | Aeronautics—United States—Safety measures. | Drone aircraft—United States—States. | Aeronautics, Commercial—Law and legislation—United States—States.
Classification: LCC KF2406 .N55 2017 | DDC 343.7309/75—dc23 LC record available at https://lccn.loc.gov/2016049818

CONTENTS

ACKNOWLEDGMENTS

I would like to thank the following people for their encouragement, advice, and support, without which I would not have undertaken and persevered with the writing of this text, not to mention inevitable subsequent updates.

Dr. Brent Bowen, Dean, College of Aviation, Embry-Riddle Aeronautical University, Prescott, Arizona, who was the first person to urge me to research "drone law." Within less than two weeks of me joining this college as an aviation law professor, he strongly suggested I write a white paper on Unmanned Aircraft law for an upcoming conference. This first paper sparked my interest in the topic, has led to the creation of three new courses, Global UAS Regulations, Global UAS Risk Management, and Unmanned Aircraft Pilot Ground School, and subsequently the production of this book.

Dr. Scott Hamilton, who has been my long-time mentor and friend, not to mention the reason so many opportunities in my aviation law profession came my way. From the start, he has been my biggest supporter. He had absolute faith in me when he offered me the task of writing a commercial space law portion of his textbook's 6th edition. I would never have believed I could accomplish my own book were it not for his initial belief in me.

Too many to mention, are all my other friends, colleagues, and students, who endure my constant conversations about aviation, the law, and especially commercial space travel. Thank you all for indulging me as I ran ideas past you, and thank you all for your kindness and respect.

Finally, I would like to thank the American Bar Association, Forum on Air & Space Law, for granting me this opportunity to bring this text to you, the reader. And, to you who purchased this book, I sincerely hope this provides some insight into this ever-changing area of the law. Trust me when I say, updates will most certainly be forthcoming!

INTRODUCTION

As you crack the spine of this book and enter the ever-dynamic world of unmanned aircraft systems (UAS), it is imperative to keep in mind that no sooner did I put fingers to keyboard to draft this text, the entire nation, much like the rest of the world, was still developing the regulations governing their use and operation. Hence, my words of caution are to read the footnotes and be sure to update your knowledge accordingly to ensure you are abreast of the most recent developments. For example, when the Federal Aviation Administration (FAA) updates its Orders and Circulars, the newer versions are followed by letters, A, B, C, etc. You can see that clearly in Advisory Circular (AC) 91-57A.

Furthermore, the FAA has just recently promulgated 14 CFR Part 107 for the purpose of regulating commercial UAS operations weighing less than 55 pounds. The FAA anticipates the development of further regulations to address beyond visual line of sight (BVLOS), heavier UAS, and more risky operations. The Remote Pilot in Command (PIC) certificate currently makes mention of a small UAS rating, meaning there may even be a large UAS rating in the future, or even a micro UAS category. Thus, this book is by no means a complete work.

It is also anticipated that more states and cities will develop rules to accommodate among others, privacy issues. And finally, as the technology unfolds, seemingly exponentially, these rules will have to adapt to keep pace. It is my sincere hope that this book will whet your appetite for more knowledge in this field, as it is just the beginning of an exciting chapter in aviation much like the early Wright Brother days when manned flight was becoming commercially viable.

Finally, I welcome your feedback and suggestions in anticipation of a second edition or online updates. Please do not hesitate to contact me! In my finest captain's voice, I now urge you to sit back, relax, and enjoy the flight (book)!

CHAPTER 1
Unmanned Aircraft System (UAS)

1. What's in a Name?

The term, *unmanned aircraft system (UAS)*, is but one term being used in the world today to describe an emerging technology that has entrepreneurs excited and lawmakers scrambling to enact laws to keep pace with the rapid growth. There is no consensus worldwide on what term should be used. Other terms include unmanned aircraft (UA), unmanned aerial vehicle (UAV), remotely piloted aircraft (RPA), remotely piloted aircraft system (RPAS), and the misconstrued term, drone. The Federal Aviation Administration (FAA) of the United States has chosen to use the term, UAS, despite the fact that the United States is one of 191 member states of the International Civil Aviation Organization (ICAO), and ICAO also uses the term, RPAS.[1] Thus, the term, *UAS*, shall be used throughout this book unless other terms are used by individual states within the United States.

The ICAO is a United Nations (UN) specialized agency, established by member states in 1944 to manage the administration and governance of the Convention on International Civil Aviation (Chicago Convention).[2] ICAO works with the Convention's member states and industry groups to reach consensus on international civil aviation Standards and Recommended Practices (SARPs) and policies in support of a safe, efficient, secure, economically sustainable, and environmentally responsible civil aviation sector. These SARPs and policies are used by ICAO member states to ensure that their local civil aviation operations and regulations conform to global norms.[3]

In addition to its core of work resolving consensus-driven international SARPs and policies among its member states and industry, and among many other priorities and

[1] *ICAO Member States*, http://www.icao.int/MemberStates/Member%20States.English.pdf.
[2] *About ICAO*, International Civil Aviation Organization (ICAO), http://www.icao.int/about-icao/Pages/default.aspx.
[3] *Id.*

programs, ICAO also (1) coordinates assistance and capacity building for states in support of numerous aviation development objectives; (2) produces global plans to coordinate multilateral strategic progress for safety and air navigation; (3) monitors and reports on numerous air transport sector performance metrics; and (4) audits states' civil aviation oversight capabilities in the areas of safety and security.[4]

The goal of ICAO in addressing unmanned aviation is to provide the fundamental international regulatory framework through SARPs, with supporting Procedures for Air Navigation Services (PANS) and guidance material, to underpin routine operation of UAS throughout the world in a safe, harmonized, and seamless manner comparable to that of manned operations. ICAO Circular 328, Unmanned Aircraft Systems (UAS) issued in 2011, was the first step in reaching that goal. It was the culmination of several meetings that began in 2005. The purpose of this circular was to: (1) apprise states of the emerging ICAO perspective on the integration of UAS into nonsegregated airspace and at aerodromes; (2) consider the fundamental differences from manned aviation that such integration will involve; and (3) encourage states to help with the development of ICAO policy on UAS by providing information on their own experiences associated with these aircraft. ICAO believes that unmanned aircraft are indeed aircraft, as does the United States, and thus existing SARPs apply to a very great extent. That said, the complete integration of UAS at aerodromes and in the various airspace classes does, however, necessitate the development of UAS-specific SARPs to supplement those already existing.[5]

Since at least 2007, ICAO has sought input on UAS matters from two other large organizations, namely, Radio Technical Commission for Aeronautics (RTCA) and European Organization for Civil Aviation Equipment (EUROCAE). RTCA is a private, not-for-profit association founded in 1935. They are the premier public–private partnership venue for developing consensus among diverse, competing interests on critical aviation modernization issues in an increasingly global enterprise.[6] RTCA has provided the foundation for virtually every modern technical advance in aviation. Their products serve as the basis for government certification of equipment used by the tens of thousands of aircraft flying daily through the world's airspace. Utilized as a Federal Advisory Committee, RTCA works in response to requests from the FAA to develop comprehensive, industry-vetted, and endorsed recommendations for the federal government on issues ranging from technical performance standards to operational concepts for air transportation. Their deliberations are open to the public. Aviation community volunteers function in a consensus-based, collaborative, peer-reviewed environment in which they develop their products such as training manuals and documents.[7]

---

[4] *Id.*
[5] *ICAO Circular 328*, https://www.trafikstyrelsen.dk/~/media/Dokumenter/05%20Luftfart/Forum/UAS%20-%20droner/ICAO%20Circular%20328%20Unmanned%20Aircraft%20Systems%20UAS.aspx.
[6] *About Us*, Radio Technical Commission for Aeronautics (RTCA), http://www.rtca.org/content.asp?pl=49&contentid=49.
[7] *Id.*

RTCA provides the venue for public–private collaboration that achieves the necessary improvements in the safety and efficiency of the air transportation system. The FAA has taken positive action on recommendations that it has sought from RTCA, leading to tangible outcomes such as timely availability of certified avionics and improvements in the performance of the air traffic management system. One publication that continues to serve as a reference is RTCA, DO-304, Guidance Material and Considerations for Unmanned Aircraft Systems.[8]

EUROCAE is a nonprofit organization dedicated to aviation standardization of both airborne and ground systems and equipment since 1963. It was formed in Lucerne, Switzerland, as a European forum focusing on electronic equipment for air transport. The organization is composed of members who specialized in one or several technical fields of aeronautics, and many of them are considered to be among the world's leaders in their domain. Members include equipment and airframe manufacturers, regulators, European and International civil aviation authorities, air navigation service providers (ANSP), and airports.[9]

According to ICAO, remotely piloted aircraft are but one type of unmanned aircraft. All unmanned aircraft, whether remotely piloted, fully autonomous, or combinations thereof, are subject to the provisions of Article 8 of the Convention on International Civil Aviation (Document 7300), signed in Chicago on December 7, 1944, and amended by the ICAO Assembly. The Chicago Convention of 1944 replaced the Paris Convention of 1919 that developed the legal framework for international civil aviation. Article 8 of the Chicago Convention entitled "pilotless aircraft" provides that:

> No aircraft capable of being flown without a pilot shall be flown without a pilot over the territory of a contracting state without special authorization by that state and in accordance with the terms of such authorization. Each contracting state undertakes to ensure that the flight of such aircraft without a pilot in regions open to civil aircraft shall be so controlled as to obviate danger to civil aircraft.[10]

It is worth further defining the ICAO definitions for the following three terms in order to compare them to the terms used by the FAA. First, remotely piloted aircraft is an aircraft where the flying pilot is not on board the aircraft. According to ICAO, this is a subcategory of unmanned aircraft. The FAA uses the term *unmanned aircraft (UA)*, and sees UA as a subcategory of the broader term *aircraft*. Second, remotely piloted aircraft

[8] *ICAO Circular 328*, https://www.trafikstyrelsen.dk/~/media/Dokumenter/05%20Luftfart/Forum/UAS%20-%20droner/ICAO%20Circular%20328%20Unmanned%20Aircraft%20Systems%20UAS.aspx.

[9] *About Us*, European Organization for Civil Aviation Equipment (EUROCAE), https://www.eurocae.net/organisation/about.

[10] *ICAO Manual on Remotely Piloted Aircraft (RPAS)*, http://www4.icao.int/demo/pdf/rpas/10019_cons_en%20-%20Secured.pdfhttp://www.wyvernltd.com/wp-content/uploads/2015/05/ICAO-10019-RPAS.pdf

system (RPAS) is a set of configurable elements consisting of a remotely piloted aircraft, its associated remote pilot station(s), the required command and control links, and any other system elements as may be required, at any point during flight operation. The FAA uses the term *unmanned aircraft system (UAS)*. Third, RPA observer is a remote crewmember that, by visual observation of the remotely piloted aircraft, assists the remote pilot in the safe conduct of the flight.[11] The FAA uses the term *visual observer (VO)*.

According to ICAO, in the broadest sense, the introduction of UAS does not change any existing distinctions between model aircraft and aircraft. Model aircraft, generally recognized as intended for recreational purposes only, fall outside the provisions of the Chicago Convention, being exclusively the subject of relevant national regulations, if any.[12]

In 2015, ICAO published the *Manual on Remotely Piloted Aircraft Systems (RPAS)*, ICAO document 10019 An/507. It primarily adds to the UAS glossary; defines the ICAO regulatory framework and scope of the manual; gives an introduction to RPAS; explains special authorization; describes type certification and airworthiness approvals; describes RPA registration; outlines responsibilities of the RPAS operator; defines safety management; details licensing and competencies; describes RPAS operations; explains Detect and Avoid (DAA); explains Command and Control (C2) link; defines air traffic control (ATC) communications; describes remote pilot station (RPS); defines the integration of RPAS operations into air traffic management (ATM) and ATM procedures; and explains the use of aerodromes.[13]

This 2015 manual addresses RPAS as one subset of a much broader collection, UAS. According to ICAO, RPAS are envisioned to be an equal partner in the civil aviation system, able to interact with ATC and other aircraft on a real-time basis. The scope of ICAO provisions in the next five to ten years will be to facilitate integration of RPAS operating in accordance with instrument flight rules (IFR) in controlled airspace and at controlled aerodromes. While not excluding visual line-of-sight (VLOS) operations from consideration, these are viewed to be a lower priority for global harmonization of international flights.[14]

ICAO's RPAS panel aims to deliver standards for unmanned aircraft to the organization's governing council in 2018. Once approved, the standards will guide ICAO's 191 member states in setting their own national regulations. The overall process of producing RPAS standards is expected to take a decade or longer.[15]

---

[11]*ICAO Circular 328*, https://www.trafikstyrelsen.dk/~/media/Dokumenter/05%20Luftfart/Forum/UAS%20-%20droner/ICAO%20Circular%20328%20Unmanned%20Aircraft%20Systems%20UAS.aspx

[12]*ICAO Circular 328*, https://www.trafikstyrelsen.dk/~/media/Dokumenter/05%20Luftfart/Forum/UAS%20-%20droner/ICAO%20Circular%20328%20Unmanned%20Aircraft%20Systems%20UAS.aspx.

[13]*ICAO Manual on Remotely Piloted Aircraft (RPAS)*, http://www4.icao.int/demo/pdf/rpas/10019_cons_en%20-%20Secured.pdfhttp://www.wyvernltd.com/wp-content/uploads/2015/05/ICAO-10019-RPAS.pdf.

[14]*Id.*

[15]Bill Carey, *ICAO Panel Will Recommend First UAV Standards in 2018*, Aviation International News (AIN) (Jan. 6, 2015, 9:04 AM), http://www.ainonline.com/aviation-news/aerospace/2015-01-06/icao-panel-will-recommend-first-uav-standards-2018.

Just as there is no worldwide consensus on the terms used to describe this emerging technology, there is also no standardization when it comes to regulating the operations. Each country's governing aviation authority is trying to regulate its UAS as best as it sees fit. And within the United States, each state seems to be regulating UAS activities within its borders as it too sees fit. Sometimes, individual cities have taken to enacting their own ordinances. This is causing a great deal of legal turmoil because of the FAA's field preemption over the National Airspace System (NAS), but more on this issue later.

2. Categories of Unmanned Aircraft

In order to make sense of the plethora of federal and state, and sometimes even city ordinances, pertaining to UAS in the United States, one must categorize UAS on two parallel levels. On the one level, the rules change depending on the weight of the aircraft, and on the other, and more critical level, the rules change based on anticipated use. As to weight categories, there are micro UAS (four proposed subcategories of small UAS, but more on this later), small UAS (aircraft weighing less than fifty-five pounds), and large UAS (aircraft weighing more than fifty-five pounds).[16] As to the three "use" categories, there are civil or commercial (nongovernmental), public or governmental, and hobby or recreational.[17]

To enable easy and efficient use, this book shall follow these three use categories. But first, the foundation must be laid, so the reader has an overview of the congressional mandate and federal landscape that has shaped our current processes. To that end, the next chapter, the federal legal environment, lays this foundation so the reader can gain an understanding of how and why these rules were developed. After that, the book is divided into three categories based on use, namely, commercial, governmental, and hobby or recreational.

---

[16] *Small UAS Notice of Proposed Rulemaking*, Federal Aviation Administration (FAA), https://www.faa.gov/regulations_policies/rulemaking/recently_published/ https://www.faa.gov/uas/nprm.

[17] *UAS FAQs*, Federal Aviation Administration (FAA), https://www.faa.gov/uas/faqs.

CHAPTER 2
The Federal Legal Environment

1. Federal Aviation Administration (FAA)

On May 21, 1958, Senator A. S. "Mike" Monroney (D-OK) introduced a bill to create an independent Federal Aviation Agency to provide for the safe and efficient use of national airspace. Two months later, on August 23, 1958, the president signed the Federal Aviation Act, which transferred the Civil Aeronautics Authority's functions to a new independent Federal Aviation Agency responsible for civil aviation safety.[1]

President Johnson, concerned about the lack of a coordinated transportation system, believed a single department was needed to develop and carry out comprehensive transportation policies and programs across all transportation modes. In 1966, Congress authorized the creation of a cabinet department that would combine major federal transportation responsibilities. This new Department of Transportation (DOT) began full operations on April 1, 1967. On that day, the Federal Aviation Agency became one of several modal organizations within DOT and received a new name, the Federal Aviation Administration (FAA). At the same time, the Civil Aeronautics Board (CAB)'s accident investigation function was transferred to the new National Transportation Safety Board.[2] This is how the FAA first came to be the governing aviation authority in the United States.

2. FAA Modernization and Reform Act (FMRA) of 2012

When the FAA Modernization and Reform Act (FMRA) was signed into law in 2012 (Public Law 112-95), Section 333 of this Act granted authority to the FAA to establish an interim policy that would bridge the gap between the current state and National Airspace

---

[1] *History*, Federal Aviation Administration (FAA), http://www.faa.gov/about/history/brief_history.
[2] *Id.*

System (NAS) operations as they would be once the small UAS rule was finalized. Current FAA policy is based on whether the unmanned aircraft is used as a public aircraft, civil aircraft, or as a model aircraft.[3] Additionally, in 2012, Congress confirmed that UAS, including those used for recreation or hobby purposes, are aircraft consistent with the statutory definition set forth in 49 United States Code § 40102(a)(6), namely, any contrivance invented, used, or designed to navigate, or fly in, the air.[4]

The parts of this act that are of significance to the UAS community in this country come from Subtitle B—Unmanned Aircraft Systems—and are as follows: Section 331—Definitions; Section 332—Integration of civil unmanned aircraft systems into national airspace system; Section 333—Special rules for certain unmanned aircraft systems; Section 334—Public unmanned aircraft systems; Section 335—Safety studies; and Section 336—Special rule for model aircraft.[5] Each of these parts shall now be described in turn.

SECTION 331—DEFINITIONS

In this subtitle, the following definitions apply:

1. ARCTIC: The term "Arctic" means the United States zone of the Chukchi Sea, Beaufort Sea, and Bering Sea north of the Aleutian chain.
2. CERTIFICATE OF WAIVER; CERTIFICATE OF AUTHORIZATION: The terms "certificate of waiver" and "certificate of authorization" mean a Federal Aviation Administration grant of approval for a specific flight operation.
3. PERMANENT AREAS: The term "permanent areas" means areas on land or water that provide for launch, recovery, and operation of small, unmanned aircraft.
4. PUBLIC UNMANNED AIRCRAFT SYSTEM: The term "public unmanned aircraft system" means an unmanned aircraft system that meets the qualifications and conditions required for operation of a public aircraft (as defined in section 40102 of title 49, United States Code).
5. SENSE AND AVOID CAPABILITY: The term "sense and avoid capability" means the capability of an unmanned aircraft to remain a safe distance from and to avoid collisions with other airborne aircraft.
6. SMALL UNMANNED AIRCRAFT: The term "small unmanned aircraft" means an unmanned aircraft weighing less than 55 pounds.
7. TEST RANGE: The term "test range" means a defined geographic area where research and development are conducted.

---

[3] *Frequently Asked Questions,* FEDERAL AVIATION ADMINISTRATION (FAA), http://www.faa.gov/uas/faqs.

[4] UAS RTF ARC, FEDERAL AVIATION ADMINISTRATION (FAA), http://www.faa.gov/regulations_policies/rulemaking/committees/documents/index.cfm/committee/browse/committeeID/537http://www.faa.gov/uas/publications/media/RTFARCFinalReport_11-21-15.pdf.

[5] *Public Law 112-95*, https://www.gpo.gov/fdsys/pkg/PLAW-112publ95/pdf/PLAW-112publ95.pdf.

8. UNMANNED AIRCRAFT: The term "unmanned aircraft" means an aircraft that is operated without the possibility of direct human intervention from within or on the aircraft.
9. UNMANNED AIRCRAFT SYSTEM: The term "unmanned aircraft system" means an unmanned aircraft and associated elements (including communication links and the components that control the unmanned aircraft) that are required for the pilot in command to operate safely and efficiently in the national airspace system.[6]

SECTION 332—INTEGRATION OF CIVIL UNMANNED AIRCRAFT SYSTEMS INTO NATIONAL AIRSPACE SYSTEM

a. REQUIRED PLANNING FOR INTEGRATION
   1. COMPREHENSIVE PLAN: Not later than 270 days after the date of enactment of this Act, the Secretary of Transportation, in consultation with representatives of the aviation industry, Federal agencies that employ unmanned aircraft systems technology in the national airspace system, and the unmanned aircraft systems industry, shall develop a comprehensive plan to safely accelerate the integration of civil unmanned aircraft systems into the national airspace system.
   2. CONTENTS OF PLAN: The plan required under paragraph (1) shall contain, at a minimum, recommendations or projections on:
      A. the rulemaking to be conducted under subsection (b), with specific recommendations on how the rulemaking will
         i. define the acceptable standards for operation and certification of civil unmanned aircraft systems;
         ii. ensure that any civil unmanned aircraft system includes a sense and avoid capability; and
         iii. establish standards and requirements for the operator and pilot of a civil unmanned aircraft system, including standards and requirements for registration and licensing;
      B. the best methods to enhance the technologies and subsystems necessary to achieve the safe and routine operation of civil unmanned aircraft systems in the national airspace system;
      C. a phased-in approach to the integration of civil unmanned aircraft systems into the national airspace system;
      D. a timeline for the phased-in approach described under subparagraph (C);
      E. creation of a safe
      F. airspace designation for cooperative manned and unmanned flight operations in the national airspace system;

---

[6] *Id.*

G. establishment of a process to develop certification, flight standards, and air traffic requirements for civil unmanned aircraft systems at test ranges where such systems are subject to testing;

H. the best methods to ensure the safe operation of civil unmanned aircraft systems and public unmanned aircraft systems simultaneously in the national airspace system; and

I. incorporation of the plan into the annual NextGen Implementation Plan document (or any successor document) of the Federal Aviation Administration.

3. DEADLINE: The plan required under paragraph (1) shall provide for the safe integration of civil unmanned aircraft systems into the national airspace system as soon as practicable, but not later than September 30, 2015.
4. REPORT TO CONGRESS: Not later than 1 year after the date of enactment of this Act, the Secretary shall submit to Congress a copy of the plan required under paragraph (1).
5. ROADMAP: Not later than 1 year after the date of enactment of this Act, the Secretary shall approve and make available in print and on the Administration's Internet Web site a 5-year roadmap for the introduction of civil unmanned aircraft systems into the national airspace system, as coordinated by the Unmanned Aircraft Program Office of the Administration. The Secretary shall update the roadmap annually.

b. RULEMAKING: Not later than 18 months after the date on which the plan required under subsection (a)(1) is submitted to Congress under subsection (a)(4), the Secretary shall publish in the Federal Register:
   1. a final rule on small unmanned aircraft systems that will allow for civil operation of such systems in the national airspace system, to the extent the systems do not meet the requirements for expedited operational authorization under section 333 of this Act;
   2. a notice of proposed rulemaking to implement the recommendations of the plan required under subsection (a)(1), with the final rule to be published not later than 16 months after the date of publication of the notice; and
   3. an update to the Administration's most recent policy statement on unmanned aircraft systems, contained in Docket No. FAA–2006–25714.

c. PILOT PROJECTS
   1. ESTABLISHMENT: Not later than 180 days after the date of enactment of this Act, the Administrator shall establish a program to integrate unmanned aircraft systems into the national airspace system at 6 test ranges. The program shall terminate 5 years after the date of enactment of this Act.
   2. PROGRAM REQUIREMENTS: In establishing the program under paragraph (1), the Administrator shall:

A. safely designate airspace for integrated manned and unmanned flight operations in the national airspace system;
B. develop certification standards and air traffic requirements for unmanned flight operations at test ranges;
C. coordinate with and leverage the resources of the National Aeronautics and Space Administration and the Department of Defense;
D. address both civil and public unmanned aircraft systems;
E. ensure that the program is coordinated with the Next Generation Air Transportation System; and
F. provide for verification of the safety of unmanned aircraft systems and related navigation procedures before integration into the national airspace system.

3. TEST RANGE LOCATIONS: In determining the location of the 6 test ranges of the program under paragraph (1), the Administrator shall:
A. take into consideration geographic and climatic diversity;
B. take into consideration the location of ground infrastructure and research needs; and
C. consult with the National Aeronautics and Space Administration and the Department of Defense.

4. TEST RANGE OPERATION: A project at a test range shall be operational not later than 180 days after the date on which the project is established.

5. REPORT TO CONGRESS
A. IN GENERAL: Not later than 90 days after the date of the termination of the program under paragraph (1), the Administrator shall submit to the Committee on Commerce, Science, and Transportation of the Senate and the Committee on Transportation and Infrastructure and the Committee on Science, Space, and Technology of the House of Representatives a report setting forth the Administrator's findings and conclusions concerning the projects.
B. ADDITIONAL CONTENTS: The report under subparagraph (A) shall include a description and assessment of the progress being made in establishing special use airspace to fill the immediate need of the Department of Defense:
i. to develop detection techniques for small unmanned aircraft systems; and
ii. to validate the sense and avoid capability and operation of unmanned aircraft systems.

d. EXPANDING USE OF UNMANNED AIRCRAFT SYSTEMS IN ARCTIC

1. IN GENERAL: Not later than 180 days after the date of enactment of this Act, the Secretary shall develop a plan and initiate a process to work with relevant

Federal agencies and national and international communities to designate permanent areas in the Arctic where small unmanned aircraft may operate 24 hours per day for research and commercial purposes. The plan for operations in these permanent areas shall include the development of processes to facilitate the safe operation of unmanned aircraft beyond line of sight. Such areas shall enable over-water flights from the surface to at least 2,000 feet in altitude, with ingress and egress routes from selected coastal launch sites.

2. AGREEMENTS: To implement the plan under paragraph (1), the Secretary may enter into an agreement with relevant national and international communities.
3. AIRCRAFT APPROVAL: Not later than 1 year after the entry into force of an agreement necessary to effectuate the purposes of this subsection, the Secretary shall work with relevant national and international communities to establish and implement a process, or may apply an applicable process already established, for approving the use of unmanned aircraft in the designated permanent areas in the Arctic without regard to whether an unmanned aircraft is used as a public aircraft, a civil aircraft, or a model aircraft.[7]

SECTION 333—SPECIAL RULES FOR CERTAIN UNMANNED AIRCRAFT SYSTEMS

a. IN GENERAL: Notwithstanding any other requirement of this subtitle, and not later than 180 days after the date of enactment of this Act, the Secretary of Transportation shall determine if certain unmanned aircraft systems may operate safely in the national airspace system before completion of the plan and rulemaking required by section 332 of this Act or the guidance required by section 334 of this Act.
b. ASSESSMENT OF UNMANNED AIRCRAFT SYSTEMS: In making the determination under subsection (a), the Secretary shall determine, at a minimum:
    1. which types of unmanned aircraft systems, if any, as a result of their size, weight, speed, operational capability, proximity to airports and populated areas, and operation within visual line of sight do not create a hazard to users of the national airspace system or the public or pose a threat to national security; and
    2. whether a certificate of waiver, certificate of authorization, or airworthiness certification under section 44704 of title 49, United States Code, is required for the operation of unmanned aircraft systems identified under paragraph (1).

---

[7] *Id.*

c. REQUIREMENTS FOR SAFE OPERATION: If the Secretary determines under this section that certain unmanned aircraft systems may operate safely in the national airspace system, the Secretary shall establish requirements for the safe operation of such aircraft systems in the national airspace system.[8]

SECTION 334—PUBLIC UNMANNED AIRCRAFT SYSTEMS

a. GUIDANCE: Not later than 270 days after the date of enactment of this Act, the Secretary of Transportation shall issue guidance regarding the operation of public unmanned aircraft systems to:
    1. expedite the issuance of a certificate of authorization process;
    2. provide for a collaborative process with public agencies to allow for an incremental expansion of access to the national airspace system as technology matures and the necessary safety analysis and data become available, and until standards are completed and technology issues are resolved;
    3. facilitate the capability of public agencies to develop and use test ranges, subject to operating restrictions required by the Federal Aviation Administration, to test and operate unmanned aircraft systems; and
    4. provide guidance on a public entity's responsibility when operating an unmanned aircraft without a civil airworthiness certificate issued by the Administration.

b. STANDARDS FOR OPERATION AND CERTIFICATION: Not later than December 31, 2015, the Administrator shall develop and implement operational and certification requirements for the operation of public unmanned aircraft systems in the national airspace system.

c. AGREEMENTS WITH GOVERNMENT AGENCIES
    1. IN GENERAL: Not later than 90 days after the date of enactment of this Act, the Secretary shall enter into agreements with appropriate government agencies to simplify the process for issuing certificates of waiver or authorization with respect to applications seeking authorization to operate public unmanned aircraft systems in the national airspace system.
    2. CONTENTS: The agreements shall:
        A. with respect to an application described in paragraph (1):
            i. provide for an expedited review of the application;
            ii. require a decision by the Administrator on approval or disapproval within 60 business days of the date of submission of the application; and
            iii. allow for an expedited appeal if the application is disapproved;

[8] *Id.*

B. allow for a one-time approval of similar operations carried out during a fixed period of time; and

C. allow a government public safety agency to operate unmanned aircraft weighing 4.4 pounds or less, if operated:

   i. within the line of sight of the operator;
   ii. less than 400 feet above the ground;
   iii. during daylight conditions;
   iv. within Class G airspace; and
   v. outside of 5 statute miles from any airport, heliport, seaplane base, spaceport, or other location with aviation activities.[9]

SECTION 335—SAFETY STUDIES

The Administrator of the Federal Aviation Administration shall carry out all safety studies necessary to support the integration of unmanned aircraft systems into the national airspace system. [10]

SECTION 336—SPECIAL RULE FOR MODEL AIRCRAFT

a. IN GENERAL: Notwithstanding any other provision of law relating to the incorporation of unmanned aircraft systems into Federal Aviation Administration plans and policies, including this subtitle, the Administrator of the Federal Aviation Administration may not promulgate any rule or regulation regarding a model aircraft, or an aircraft being developed as a model aircraft, if:

   1. the aircraft is flown strictly for hobby or recreational use;
   2. the aircraft is operated in accordance with a community-based set of safety guidelines and within the programming of a nationwide community-based organization;
   3. the aircraft is limited to not more than 55 pounds unless otherwise certified through a design, construction, inspection, flight test, and operational safety program administered by a community-based organization;
   4. the aircraft is operated in a manner that does not interfere with and gives way to any manned aircraft; and
   5. when flown within 5 miles of an airport, the operator of the aircraft provides the airport operator and the airport air traffic control tower (when an air traffic facility is located at the airport) with prior notice of the operation (model aircraft operators flying from a permanent location within 5 miles of an airport should establish a mutually-agreed upon operating procedure

---

[9] *Id.*
[10] *Id.*

with the airport operator and the airport air traffic control tower (when an air traffic facility is located at the airport)).

b. STATUTORY CONSTRUCTION: Nothing in this section shall be construed to limit the authority of the Administrator to pursue enforcement action against persons operating model aircraft who endanger the safety of the national airspace system.

c. MODEL AIRCRAFT DEFINED: In this section, the term "model aircraft" means an unmanned aircraft that is:
   1. capable of sustained flight in the atmosphere;
   2. flown within visual line of sight of the person operating the aircraft; and
   3. flown for hobby or recreational purposes.[11]

In the aftermath of FMRA of 2012, some citizens have seen the need to amend and add to this. Two noteworthy ones are House Resolution (HR) 4432 and Senate Bill (SB) 1314, both entitled Commercial UAS Modernization Act.

3. House Resolution (HR) 4432 and Senate Bill (SB) 1314

On February 2, 2016, Representative Earl Blumenauer introduced House Resolution (HR) 4432—Commercial UAS Modernization Act.[12] This bill proposes to establish an interim rule for the operation of small, unmanned aircraft for commercial purposes, and for other purposes.

HR 4432 proposes to amend the FAA Modernization and Reform Act of 2012 (PL 112-95) by: (1) adding Section 337—Operation of small unmanned aircraft for commercial purposes—dealing with lack of airworthiness certificates, liability insurance, registration, testing requirements now to include proficiency test as well as the original knowledge test, accident reporting now to be accomplished within 2 days of the accident; (2) adding Section 338—Micro UAS Operations—for UAS weighing up to 4.4 lbs.—limit of 400 feet AGL, visual line of sight (VLOS), daylight, 5 statute miles of an airport, exemption from airmen certification and airworthiness standards; (3) adding Section 339—Deputy Associate Administrator for Unmanned Aircraft whose duties it shall be to create an achievable comprehensive research and development program for the safe integration of UAS in the National Airspace System (NAS); (4) adding Section 4—Joint Aircraft System Research and Development Data Collection and Analysis Program—at the William J. Hughes Technical Center: (a) for research on safety standards for detect and avoid, command and control, autonomous aircraft systems, and air traffic management for beyond line of sight (BLOS); (b) use of Test Sites; and (c) Air Traffic Management Pilot Program.[13]

---

[11] *Id.*

[12] H.R. 4432, https://www.congress.gov/bill/114th-congress/house-bill/4432/text?q=%7B%22search%22:[%22commercial+uas%22]%7D&resultIndex=1.

[13] *Id.*

HR 4432, at the time of writing, had been referred to the House Science, Space, and Technology Committee, and reads as follows:

Be it enacted by the Senate and House of Representatives of the United States of America in Congress assembled,

SECTION 1. SHORT TITLE: This Act may be cited as the "Commercial UAS Modernization Act".

SECTION 2. INTERIM RULE FOR THE OPERATION OF SMALL UNMANNED AIRCRAFT FOR COMMERCIAL PURPOSES:

a. In General: Subtitle B of title III of the FAA Modernization and Reform Act of 2012 (Public Law 112-95) is amended by adding at the end the following:

SECTION 337. OPERATION OF SMALL UNMANNED AIRCRAFT FOR COMMERCIAL PURPOSES

a. In General: An individual may operate a small unmanned aircraft for commercial purposes without an airworthiness certificate within the United States, subject to the requirements under subsection (b) and the operating restrictions under subsection (c) during the period beginning on the date of the enactment of the Commercial UAS Modernization Act and ending on the effective date of a final rule based on the notice of proposed rulemaking issued on February 23, 2015, entitled 'Operation and Certification of Small Unmanned Aircraft Systems' (80 Fed. Reg. 9544).

b. General Requirements
    1. LIABILITY INSURANCE: A small unmanned aircraft may not be operated for commercial purposes during the period set forth in subsection (a) unless the Administrator receives an attestation that the owner of such aircraft has a liability insurance policy covering the operation of such aircraft.
    2. REGISTRATION: A small unmanned aircraft may not be operated for commercial purposes unless the owner has registered the aircraft as required by the Federal Aviation Administration.
    3. TESTING REQUIREMENTS
        A. EXAM DEVELOPMENT: Not later than 30 days after the date of the enactment of the Commercial UAS Modernization Act, the Administrator of the Federal Aviation Administration shall develop an initial aeronautical knowledge test that meets the requirements set forth in the notice of proposed rulemaking referred to in subsection (a).
        B. REQUIREMENTS: An individual may not operate a small unmanned aircraft for commercial purposes unless such individual has:

i. received a passing grade on the test developed under subparagraph (A);
ii. passed a proficiency test administered by a test site selected pursuant to section 332(c); and
iii. demonstrated the ability to fly the aircraft in accordance with the operating restrictions set forth in subsection (c).

4. CERTIFICATION: A small unmanned aircraft may not be operated for commercial purposes until the operator of a test site selected pursuant to section 332(c), in collaboration with a designated airworthiness representative, certifies that the small unmanned aircraft:
   A. meets the requirements for small unmanned aircraft set forth in the notice of proposed rulemaking referred to in subsection (a); and
   B. is capable of operating within the limits described in subsection (c).

c. Operating Restrictions: During the period set forth in subsection (a), small unmanned aircraft operated for commercial purposes:
   1. may only be operated under visual line of sight rules;
   2. may not be operated higher than 500 feet above ground level;
   3. may not be operated unless the operator has prior authorization from the air traffic control facility having jurisdiction over that airspace:
      A. in Class B, Class C, or Class D airspace; or
      B. within the lateral boundaries of the surface area of Class E airspace designated for an airport;
   4. may only be operated in daylight conditions;
   5. shall yield right of way to all other users of the National Airspace System;
   6. may not be operated by any individual with any physical or mental condition that the individual knows, or has reason to know, would interfere with the safe operation of the aircraft; and
   7. may only be operated after a preflight inspection (as described in the notice of proposed rulemaking referred to in subsection (a)).

d. Accident Reporting: The operator of a small unmanned aircraft that is involved in any accident causing personal injury or property damage, other than to the small unmanned aircraft, shall report such accident to the Federal Aviation Administration not later than 2 days after such accident.

SECTION 338. MICRO UAS OPERATIONS

a. Micro UAS Classification: The Administrator of the Federal Aviation Administration shall provide for a micro UAS classification of unmanned aircraft systems, the aircraft component of which may not weigh more than 4.4 pounds, including payload.

b. Micro UAS Operational Limitations: The operation of a micro UAS shall be subject to the exemptions under subsection (c) only if the micro UAS is operated:
   1. less than 400 feet above ground level;
   2. at an airspeed of not greater than 40 knots;
   3. within the visual line of sight of the operator;
   4. during daylight; and
   5. at least 5 statute miles from the geographic center of an airport as denoted on a current aeronautical chart published by the Federal Aviation Administration, except that upon notice to the airport operator and air traffic control tower, such airport operator may allow an individual to operate a micro UAS within 5 statute miles of a tower-controlled airport.

c. Micro UAS Exemptions
   1. An operator of a micro UAS that complies with the limitations of operation under subsection (b) shall not be required to pass any aeronautical knowledge test or meet any age or experience requirement, including any requirements under section 44703 of title 49, United States Code, part 61 of title 14, Code of Federal Regulations, and any other rule or regulation pertaining to airman certification.
   2. A micro UAS and the component parts and equipment of such micro UAS shall not be required to meet airworthiness certification standards or to obtain certificates of airworthiness.

d. Clerical Amendment: The table of contents in section 1(b) of the FAA Modernization and Reform Act of 2012 is amended by inserting after the item relating to section 336 the following:

Section 337. Operation of small unmanned aircraft for commercial purposes.
Section 338. Micro UAS operations.

SECTION 3. DEPUTY ASSOCIATE ADMINISTRATOR FOR UNMANNED AIRCRAFT

a. In General: Subtitle B of title III of the FAA Modernization and Reform Act of 2012 (Public Law 112-95), as amended by section 2(a), is further amended by adding at the end the following:

SECTION 339. DEPUTY ASSOCIATE ADMINISTRATOR FOR UNMANNED AIRCRAFT

a. Appointment: The Administrator of the Federal Aviation Administration (referred to in this section as the 'Administrator') shall appoint a Deputy Associate Administrator for Unmanned Aircraft (referred to in this section as the 'Deputy Associate Administrator'), who shall report to the Administrator and to the Secretary of Transportation.

b. Principal Duties: The Deputy Associate Administrator shall create an achievable comprehensive research and development plan for the safe integration of unmanned aircraft into the National Airspace System that:
   1. takes into account work being done at other Federal agencies, in conjunction with their industry collaborators;
   2. is based on an initial audit of current unmanned aircraft activity across the Federal Government in order to identify gaps and overlaps; and
   3. allows for programmatic exemptions based on previous analysis.

c. Other Issues: The Deputy Associate Administrator, in consultation with the Administrator, shall develop strategies for resolving:
   1. unmanned aircraft spectrum issues;
   2. barriers to unmanned aircraft operating beyond line of sight;
   3. barriers to allowing payload carriage and the feasibility of developing a classification of small UAS air carriers; and
   4. barriers to utilizing automated unmanned aircraft systems.

d. Exemptions
   1. IN GENERAL: Not later than 90 days after the date of the enactment of the Commercial UAS Modernization Act, the Deputy Associate Administrator, in consultation with the Administrator, shall expedite and expand exemptions from the interim operating restrictions otherwise applicable to unmanned aircraft under section 337.
   2. EXEMPTIONS: The exemptions authorized under paragraph (1) may include:
      A. beyond line of sight operations;
      B. programmatic exemptions based on previous analysis;
      C. extended visual line of sight and marginal visual flight rules weather conditions; and
      D. heavier unmanned vehicles."

e. Clerical Amendment: The table of contents in section 1(b) of the FAA Modernization and Reform Act of 2012 is amended by inserting after the item relating to section 338, as added by section 2(b), the following:

Section 339. Deputy Associate Administrator for Unmanned Aircraft.

SECTION 4. JOINT AIRCRAFT SYSTEM RESEARCH AND DEVELOPMENT DATA COLLECTION AND ANALYSIS PROGRAM

a. Establishment: The Administrator of the Federal Aviation Administration shall establish a joint aircraft system research and development data collection and analysis program at the William J. Hughes Technical Center (referred to in this section as the "Center").

b. Research And Development Priorities: The Director of the Center shall set priorities for data collection, analysis, and research under the program established under subsection (a), including identifying safety standards for detect and avoid, command and control, autonomous aircraft systems, and air traffic management for beyond visual line of sight operations for such aircraft.

c. Use of Test Sites: The program established under subsection (a) shall utilize the 6 unmanned aircraft system test sites of the Federal Aviation Administration to:
   1. conduct research;
   2. collect data;
   3. develop quarterly milestones to expedite commercial unmanned aircraft system operations; and
   4. work with other Federal agencies, the Center of Excellence for Unmanned Aircraft Systems, federally funded research and development centers, industry, academia, and others, as appropriate, to implement commercial unmanned aircraft system operations.
d. Air Traffic Management Pilot Program
   1. IMPLEMENTATION: The Administrator of the Federal Aviation Administration, acting through the Center, and the Administrator of the National Aeronautics and Space Administration, shall implement an air traffic management pilot program to research and test a new regulatory structure for commercial and other operations of small unmanned aircraft in controlled and uncontrolled airspace below 1,200 feet above ground level.
   2. MANAGEMENT TESTING: The Center shall partner with a neutral third party to test the management of small unmanned aircraft in the airspace described in paragraph (1).
e. Report: Not later than 180 days after the date of the enactment of this Act, and every 180 days thereafter, the Director of the Center shall submit a report that summarizes the actions taken under subsections (b), (c), and (d) to:
   1. the Committee on Commerce, Science, and Transportation of the Senate;
   2. the Committee on Appropriations of the Senate;
   3. the Committee on Transportation and Infrastructure of the House of Representatives; and
   4. the Committee on Appropriations of the House of Representatives.[14]

On May 13, 2015, Senator Cory Booker introduced a similar bill, SB 1314, to the Committee on Commerce, Science, and Transportation.[15] The Committee on Small Business and Entrepreneurship later reviewed this bill on March 10, 2016.[16] The bill reads as follows:

To establish an interim rule for the operation of small unmanned aircraft for commercial purposes and their safe integration into the national airspace system.

---

[14]*Id.*
[15]SB 1314, https://www.congress.gov/bill/114th-congress/senate-bill/1314/related-bills.
[16]*Id.*

Be it enacted by the Senate and House of Representatives of the United States of America in Congress assembled,

SECTION 1. SHORT TITLE

This Act may be cited as the "Commercial UAS Modernization Act".

SECTION 2. INTERIM RULE FOR THE OPERATION OF SMALL UNMANNED AIRCRAFT FOR COMMERCIAL PURPOSES

a. IN GENERAL—Subtitle B of title III of the FAA Modernization and Reform Act of 2012 (Public Law 112–95) is amended by adding at the end the following:

SEC. 337. OPERATION OF SMALL UNMANNED AIRCRAFT FOR COMMERCIAL PURPOSES

a. IN GENERAL—A person may operate a small unmanned aircraft for commercial purposes without an airworthiness certificate within the United States, subject to the requirements under subsection (b) and the operating restrictions under subsection (c) during the period beginning on the date of the enactment of this Act and ending on the effective date of a final rule based on the Notice of Proposed Rulemaking Operation and Certification of Small Unmanned Aircraft Systems (80 Fed. Reg. 9544, February 23, 2015).

b. GENERAL REQUIREMENTS –
    1. LIABILITY INSURANCE—A small unmanned aircraft may not be operated for commercial purposes during the period set forth in subsection (a) unless the Administrator receives an attestation that the owner of such aircraft has a liability insurance policy covering the operation of such aircraft.
    2. REGISTRATION—A small unmanned aircraft may not be operated for commercial purposes unless the owner has registered the aircraft under section 3(a) of the Commercial UAS Modernization Act.
    3. TESTING REQUIREMENTS –
        A. EXAM DEVELOPMENT—Not later than 30 days after the date of the enactment of the Commercial UAS Modernization Act, the Administrator of the Federal Aviation Administration shall develop an initial aeronautical knowledge test that meets the requirements set forth in the notice referred to in subsection (a).
        B. REQUIREMENTS—An individual may not operate a small unmanned aircraft for commercial purposes unless he or she –
            i. has received a passing grade on the test developed under subparagraph (A);
            ii. passed a proficiency test administered by a test site selected pursuant to section 332(c); and

iii. has demonstrated the ability to fly the aircraft in accordance with the operating restrictions set forth in subsection (c).

4. CERTIFICATION—A small unmanned aircraft may not be operated for commercial purposes until the operator of a test site selected pursuant to section 332(c), in collaboration with a designated airworthiness representative, certifies that the small unmanned aircraft –
   - A. meets the requirements for small unmanned aircraft set forth in the notice referred to in subsection (a); and
   - B. is capable of operating within the limits described in subsection (c).

c. OPERATING RESTRICTIONS—During the period set forth in subsection (a), small unmanned aircraft operated for commercial purposes –
1. may only be operated under visual line of sight rules;
2. may not be operated higher than 500 feet above ground level;
3. may not be operated, unless the operator has prior authorization from the air traffic control facility having jurisdiction over that airspace –
   - A. in Class B, Class C, or Class D airspace; or
   - B. within the lateral boundaries of the surface area of Class E airspace designated for an airport;
4. shall comply with model aircraft operating standards set forth in Advisory Circular 91–57, which was issued by the Federal Aviation Administration on June 9, 1981, or the current revision of such standards;
5. may only be operated in daylight conditions;
6. shall yield right of way to all other users of the National Airspace System;
7. may not be operated by any individual with any physical or mental condition that the individual knows, or has reason to know, would interfere with the safe operation of the aircraft; and
8. may only be operated after a preflight inspection (as described in the notice referred to in subsection (a)).

d. ENFORCEMENT—The Secretary of Transportation, or designee, may bring a civil action against a person in a district court of the United States to enforce this section or a requirement or regulation prescribed, or an order or any term of a certificate or permit issued, under this section. The action may be brought in the judicial district in which the person does business or the violation occurred.

e. ACCIDENT REPORTING—The owner or operator of a small unmanned aircraft that is involved in any accident causing personal injury or property damage, other than to the small unmanned aircraft, shall report such accident to the Federal Aviation Administration not later than 2 days after such accident.

f. CLERICAL AMENDMENT—The table of contents in section 1(b) of the FAA Modernization and Reform Act of 2012 is amended by inserting after the item relating to section 336 the following:

Sec. 337. Operation of small unmanned aircraft for commercial purposes.

SECTION 3. DEPUTY ASSOCIATE ADMINISTRATOR FOR UNMANNED AIRCRAFT

a. IN GENERAL—Subtitle B of title III of the FAA Modernization and Reform Act of 2012 (Public Law 112–95), as amended by section 2(a), is further amended by adding at the end the following:

SECTION 338. DEPUTY ASSOCIATE ADMINISTRATOR FOR UNMANNED AIRCRAFT

a. APPOINTMENT—The Administrator of the Federal Aviation Administration (referred to in this section as the 'Administrator') shall appoint a Deputy Associate Administrator for Unmanned Aircraft (referred to in this section as the 'Deputy Associate Administrator'), who shall report to the Administrator and to the Secretary of Transportation.

b. REGISTRATION –
   1. PROCEDURES—Not later than 30 days after the date of the enactment of the Commercial UAS Modernization Act, the Deputy Associate Administrator, in consultation with the Administrator, shall develop procedures for registering small unmanned aircraft.
   2. FEES—The Deputy Associate Administrator is authorized to collect reasonable fees, in an amount to be determined by the Deputy Associate Administrator, from the owner or operator of the small unmanned aircraft as part of the registration process.

c. PRINCIPAL DUTIES—The Deputy Associate Administrator shall –
   1. create an achievable comprehensive research and development plan for the safe integration of unmanned aircraft into the National Airspace System, which –
      A. takes into account work being done at other Federal agencies, in conjunction with their industry collaborators;
      B. is based on an initial audit of current unmanned aircraft activity across the Federal Government in order to identify gaps and overlaps; and
      C. allows for programmatic exemptions based on previous analysis.

d. OTHER ISSUES—The Deputy Associate Administrator, in consultation with the Administrator, shall develop strategies for resolving –

1. unmanned aircraft spectrum issues;
2. barriers to unmanned aircraft operating beyond line of sight;
3. barriers to allowing payload carriage; and
4. barriers to utilizing automated unmanned aircraft systems.

e. EXEMPTIONS –
   1. IN GENERAL—Not later than 90 days after the date of the enactment of this Act, the Deputy Associate Administrator, in consultation with the Administrator, shall expedite and expand exemptions from the interim operating restrictions otherwise applicable to unmanned aircraft under section 337.
   2. EXEMPTIONS—The exemptions authorized under paragraph (1) may include –
      A. beyond line of sight operations;
      B. programmatic exemptions based on previous analysis;
      C. extended visual line of sight and marginal visual flight rules weather conditions; and
      D. heavier unmanned vehicles.

f. CLERICAL AMENDMENT—The table of contents in section 1(b) of the FAA Modernization and Reform Act of 2012 is amended by inserting after the item relating to section 337, as added by section 2(b), the following:

Sec. 338. Deputy Associate Administrator for Unmanned Aircraft.

SECTION 4. JOINT AIRCRAFT SYSTEM RESEARCH AND DEVELOPMENT DATA COLLECTION AND ANALYSIS PROGRAM

a. ESTABLISHMENT—The Administrator of the Federal Aviation Administration shall establish a joint aircraft system research and development data collection and analysis program at the William J. Hughes Technical Center (referred to in this section as the "Center").

b. RESEARCH AND DEVELOPMENT PRIORITIES –

   The Director of the Center shall set priorities for data collection, analysis, and research under the program established under subsection (a), including identifying safety standards for detect and avoid, command and control, autonomous aircraft systems, and air traffic management for beyond-visual-line of sight operations for such aircraft.

c. USE OF TEST SITES—The program established under subsection (a) shall utilize the 6 unmanned aircraft system test sites of the Federal Aviation Administration –
   1. to conduct research;
   2. to collect data;
   3. to develop quarterly milestones to expedite commercial unmanned aircraft system operations; and

4. to work with other Federal agencies, the Center of Excellence for Unmanned Aircraft Systems, Federally funded research and development centers, industry, academia, and others, as appropriate, to implement the plan referred to in paragraph (3).

d. AIR TRAFFIC MANAGEMENT PILOT PROGRAM –
   1. IMPLEMENTATION—The Administrator of the Federal Aviation Administration, acting through the Center, and the Administrator of the National Aeronautics and Space Administration, shall implement an air traffic management pilot program to research and test a new regulatory structure for commercial and other operations of small unmanned aircraft in controlled and uncontrolled airspace below 1,200 feet above ground level.
   2. MANAGEMENT TESTING—The Center shall partner with a neutral third party to test the management of small unmanned aircraft in the airspace described in paragraph (1).

e. REPORT—Not later than 180 days after the date of the enactment of this Act, and every 180 days thereafter, the Director of the Center shall submit a report that summarizes the actions taken under subsections (b), (c), and (d) to –
   1. the Committee on Commerce, Science, and Transportation of the Senate;
   2. the Committee on Appropriations of the Senate;
   3. the Committee on Transportation and Infrastructure of the House of Representatives; and
   4. the Committee on Appropriations of the House of Representatives.[17]

4. Roadmap

In the section above dealing with the FMRA of 2012, it was mentioned that the FAA had been tasked with creating a roadmap. In 2013, the FAA Administrator, Michael P. Huerta, presented the FAA's Roadmap for Integration of Civil Unmanned Aircraft Systems (UAS) in the National Airspace System (NAS). The FAA and the UAS Aviation Rulemaking Committee (ARC) worked together for about a year, between 2012 and 2013, to produce this roadmap. It was recognized that unmanned aircraft offer new ways for commercial enterprises and public operators to increase operational efficiency, decrease costs, and enhance safety. This roadmap was to allow the FAA to safely and efficiently integrate them into the NAS. To this end, the roadmap outlined the actions and considerations needed to enable UAS integration into the NAS, and aligned proposed FAA actions with Congressional mandates from the FAA Modernization and Reform Act of 2012. This plan also provided goals, metrics, and target dates for the FAA and its government and industry partners to use in planning key activities for UAS integration. As previously

[17] *Id.*

mentioned, the FAA Modernization and Reform Act of 2012 (Public Law 112-95) mandated the creation and publication of this five-year roadmap for the FAA's process of developing regulations, policy, procedures, guidance material, and training requirements to support safe and efficient UAS operations in the NAS, while coordinating with relevant departments and agencies to address related key policy areas of concern such as privacy and national security.[18]

5. Notice of Proposed Rulemaking (NPRM)

In the section above dealing with the FMRA of 2012, it was mentioned that the FAA had been tasked with developing regulations pertaining to Section 333, governing commercial UAS operations in the NAS. The Administrative Procedure Act outlines the rulemaking process. First, a Notice of Proposed Rulemaking (NPRM) is published, followed by a period open to public comment that typically lasts for sixty days, then a period whereby the various FAA committees review and/or address these comments. Ultimately, a final rule is published in the Federal Register, and thirty days thereafter it becomes law.

On February 15, 2015, the FAA published its NPRM proposing regulations for the commercial use of UAS weighing less than fifty-five pounds. The sixty-day window for public comment closed on April 24, 2015. Thereafter, numerous committees at the FAA reviewed the 4,567 comments received, after which the FAA published their final rule. This final rule becomes law thirty days after publication in the Federal Register. As of the time of writing, the FAA has already published the final rule, also known as 14 CFR Part 107, effective August 29, 2016.[19]

Major provisions of the proposed 14 CFR Part 107 for small UAS fall into four main categories: (1) operational limitations; (2) operator certification and responsibilities; (3) aircraft requirements; and (4) model aircraft.[20]

1. *Operational Limitations*

   Unmanned aircraft must weigh less than fifty-five pounds. Visual line-of-sight is permitted only, meaning that the unmanned aircraft must remain within VLOS of the operator or visual observer. At all times, the small UAS must remain close enough to the operator for the operator to be capable of seeing the aircraft with vision unaided by any device other than corrective lenses. Small UAS may not operate over any persons not directly involved in the operation. Daylight-only operations, meaning from official sunrise to official sunset, local time, are permitted. UAS must yield right-of-way to other aircraft, manned or unmanned. UAS may use a visual observer (VO), but it is not required. First-person view camera

---

[18] *Roadmap*, Federal Aviation Administration (FAA), https://www.faa.gov/uas/legislative_programs/uas_roadmap/media/UAS_Roadmap_2013.pdf.
[19] *UAS*, Federal Aviation Administration (FAA), http://www.faa.gov/uas/nprm/.
[20] *Id.*

cannot satisfy the "see-and-avoid" requirement but can be used as long as the requirement is satisfied in other ways. The maximum airspeed allowed is 100 mph (87 knots). The maximum altitude allowed is 500 feet above ground level (AGL). The minimum weather visibility is three miles from the control station. No operations are allowed in Class A airspace, which is from 18,000 feet to 60,000 feet. Operations in Class B, C, D, and E airspace are allowed with the required ATC permission. Operations in Class G airspace are allowed without ATC permission. No person may act as an operator or VO for more than one unmanned aircraft operation at one time. Careless or reckless operations are prohibited. The operator must perform a preflight inspection. A person may not operate a small UAS if he or she knows or has reason to know of any physical or mental condition that would interfere with the safe operation of a small UAS. Finally, the rule proposes a micro UAS option that would allow operations in Class G airspace, over people not involved in the operation, provided the operator certifies he or she has the requisite aeronautical knowledge to perform the operation.[21]

2. *Operator Certification and Responsibilities*

Pilots of a small UAS would be considered "operators." Operators would be required to: (1) pass an initial aeronautical knowledge test at an FAA-approved knowledge testing center; (2) be vetted by the Transportation Safety Administration (TSA); (3) obtain a UAS operator certificate with a small UAS rating (which, like current pilot airman certificates, will not expire); (4) pass a recurrent aeronautical knowledge test every twenty-four calendar months; (5) be at least seventeen years old; (6) make available to the FAA, upon request, the small UAS for inspection or testing, and any associated documents/records required to be kept under the proposed rule; (7) report an accident to the FAA within ten days of any operation that results in injury or property damage; and (8) conduct a preflight inspection, to include specific aircraft and control station system checks to ensure the small UAS is safe for operation.[22]

3. *Aircraft Requirements*

FAA airworthiness certification is not required. However, the operator must maintain a small UAS in a condition for safe operation and prior to flight must inspect the UAS to ensure that it is in a condition for safe operation. Aircraft registration is required, and it is essentially the same requirements that apply to all other aircraft. Aircraft markings are required and they are essentially the same requirements that apply to all other aircraft. If the aircraft is too small to display

---

[21] *Id.*
[22] *Id.*

markings in standard size, then the aircraft simply needs to display markings in the largest practicable manner.[23]

4. *Model Aircraft*

The proposed rule would not apply to model aircraft that satisfy all of the criteria specified in Section 336 of Public Law 112-95. The proposed rule would codify the FAA's enforcement authority in part 101 by prohibiting model aircraft operators from endangering the safety of the NAS.[24]

6. Micro UAS Aviation Rulemaking Committee Recommendations

On April 1, 2016, the Aviation Rulemaking Committee (ARC) gave the FAA their final recommendations on micro UAS.[25] The FAA had chartered the Micro UAS ARC to provide recommendations to the FAA Administrator on a regulatory framework for the classification and operation of micro UAS. Recall that the FAA had contemplated a micro UAS classification in the small UAS NPRM mentioned above. The stated objective of the ARC was to consider recommendations for a performance-based standard that would allow for micro UAS to be operated over people who are not directly participating in the operation of the UAS or under a covered structure, which would ultimately contribute to an enforceable rule imposed by the FAA. The ARC was composed of members representing a diverse set of aviation stakeholders, including the UAS industry with members like 3D Robotics, GoPro Inc., GoogleX, DJI, etc. to name a few, and associations like Academy of Model Aeronautics (AMA), Aircraft Owners and Pilots Association (AOPA), Association for Unmanned Vehicle Systems International (AUVSI), among many others.[26]

The ARC was focused on flight over people and, in furtherance of that goal, identified four small UAS categories, defined primarily by level of risk of injury posed for operations over people. For each category, the ARC recommended a risk threshold that correlates to either weight or impact energy equivalent and, to the extent necessary to minimize the risks associated with that category, additional performance standards and operational restrictions. Each category is presumed to be subject to the restrictions of proposed part 107 (other than the restriction related to flight over people), including but not limited to altitude and time of operations restrictions.

1. *Category 1:* Under Category 1, a small UAS may operate over people if the weight (including accessories/payload, e.g., cameras) is 250 grams or less. Based on the data the ARC received, the ARC believes that the level of risk of injury posed by this category of UAS is so low that no performance standards and no operational

---

[23] *Id.*

[24] *Id.*

[25] *Micro UAS ARC Recommendations*, Federal Aviation Administration (FAA), https://www.faa.gov/uas/publications/media/Micro-UAS-ARC-FINAL-Report.pdf.

[26] *Id.*

restrictions beyond those imposed by the proposed part 107 are necessary. To demonstrate that a small UAS qualifies for Category 1 operations over people, the manufacturer of the UAS must either: (1) label the product retail packaging of the small UAS with the actual weight of the aircraft or a general statement that the aircraft weighs 250 grams or less; or (2) declare that the aircraft weighs 250 grams or less, and submit that declaration to the FAA in a form and manner acceptable to the FAA.

Under Categories 2, 3, and 4, a small UAS may operate over people if it does not exceed the impact energy threshold specified for each category, as certified by the manufacturer using industry-consensus test methods, and if its operator complies with operational restrictions specified for each category. Because the level of risk increases between Categories 2, 3, and 4, the performance-based standards and operational restrictions are scaled-up in each category to mitigate the increased risks.

2. *Category 2:* Under Category 2, a small UAS may operate over people if the manufacturer of the UAS certifies to the FAA that the UAS does not, in the most probable failure modes, exceed the typical or likely impact energy threshold, and if it complies with industry consensus performance standards. The operator must also comply with the operator instruction manual, must maintain minimum set-off distances of 20 feet above people's heads, or 10 feet laterally away from people, and may not operate so close to people as to create an undue hazard to those people. The ARC was presented with information from Canada and the commercial space industry suggesting that the risk calculation would result in a value of $12J/cm^2$ and that a quadcopter UAS weighing in the range of 4 to 5 pounds would qualify.
3. *Category 3:* Under Category 3, a small UAS may operate over people if the manufacturer of the UAS certifies to the FAA that the UAS does not, in the most probable failure modes, exceed the typical or likely impact energy threshold, and if it complies with industry-consensus performance standards. Flight over crowds or dense concentrations of people is never permitted under this category. In addition to that restriction, Category 3 UAS may only operate over people if: (1) the operation is conducted over a closed- or restricted-access work site with the permission of the site's owner or operator; or (2) overflight of people is limited to those who are transient or incidental to the operation, i.e., the overflight of people is incidental to the operation and is not sustained. Additionally, the performance standards and operational restrictions that apply to Category 2 operations also apply to Category 3.
4. *Category 4:* Under Category 4, a small UAS may operate over people, including flights over crowds or dense concentrations of people prohibited in Category 3,

if the manufacturer of the UAS certifies that the UAS does not, in the most probable failure modes, exceed the typical or likely impact energy threshold, if the UAS complies with industry consensus performance standards, and if the operation is conducted in compliance with a documented, risk mitigation plan, which was developed and adopted in accordance with industry consensus standards for conducting risk mitigation. The performance standards and operational restrictions that apply to Category 2 operations also apply to Category 4.

To demonstrate that a small UAS qualifies for Categories 2, 3, or 4 operations over people, the manufacturer of the UAS must: (1) declare that the small UAS meets industry consensus standards applicable to the category; (2) submit that declaration to the FAA in a form and manner acceptable to the FAA; (3) label the product or product retail packaging in accordance with industry consensus standards; and (4) provide an operating manual to the operator that includes operator instructions for flight over people. The operator is responsible for knowing what category of operations his or her UAS qualifies for, and what operational limitations he or she must follow.

7. UAS Test Sites

In the section above dealing with the FMRA of 2012, it was mentioned that the FAA had been tasked with selecting six test sites to allow the agency development of research findings and operational experiences to help ensure the safe integration of UAS into the NAS together with a system featuring NextGen technologies and procedures. As of 2013, the FAA had selected the following six UAS test sites: University of Alaska; State of Nevada; New York's Griffiss International Airport; North Dakota Department of Commerce; Texas A&M University; and Virginia Polytechnic Institute and State University (Virginia Tech).[27]

1. *University of Alaska:* The University of Alaska proposal contained a diverse set of test site range locations in seven climatic zones as well as geographic diversity with test site range locations in Hawaii and Oregon. The research plan included the development of a set of standards for UAS categories, state monitoring, and navigation. Alaska also plans to work on safety standards for UAS operations.[28]
2. *State of Nevada:* Nevada's project objectives concentrated on UAS standards and operations as well as operator standards and certification requirements. Their research will also include a concentrated look at how air traffic control procedures will evolve with the introduction of UAS into the civil environment and how

---

[27] *UAS Test Sites*, Federal Aviation Administration (FAA), https://www.faa.gov/news/fact_sheets/news_story.cfm?newsid=15575.
[28] *Id.*

these aircraft will be integrated with NextGen. Nevada's selection contributes to geographic and climatic diversity.[29]

3. *New York's* Griffiss *International Airport:* Griffiss International plans to work on developing test and evaluation as well as verification and validation processes under FAA safety oversight. They also plan to focus their research on sense and avoid capabilities for UAS and their sites will aid in researching the complexities of integrating UAS into the congested, northeast airspace.
4. *North Dakota Department of Commerce:* North Dakota plans to develop UAS airworthiness essential data and validate high reliability link technology. They will also conduct human factors research. North Dakota's application was the only one to offer a test range in the temperate (continental) climate zone and included a variety of different airspace that will benefit multiple users.
5. *Texas A&M University:* Located in Corpus Christi, Texas A&M plans to develop system safety requirements for UAS vehicles and operations with a goal of protocols and procedures for airworthiness testing. The selection of Texas A&M contributes to geographic and climatic diversity.
6. *Virginia Polytechnic Institute and State University (Virginia Tech):* Virginia Tech plans to conduct UAS failure mode testing and identify and evaluate operational and technical risk areas. This proposal includes test site range locations in both Virginia and New Jersey.

Data and other information related to the operation of UAS that is generated by the six test site operators will help the FAA answer key research questions such as solutions for "sense and avoid," command and control, ground control station standards, and human factors, airworthiness, lost link procedures, and the interface with the air traffic control system. This data will help the FAA to develop regulations and operational procedures for future commercial and civil use of the NAS.[30]

8. Fact Sheet to States

Between 2012 when FMRA was put forth, and the time of writing, in 2016, the United States has been awaiting FAA regulations on UAS. States and cities have not been patient. More than half of this nation's states have enacted their own laws, sometimes rightly so, other times in direct conflict with the federal field preemption of the FAA. By the end of 2015, the FAA decided it was high time to release a Fact Sheet to States, educating them on these matters by defining their jurisdiction and authority.[31] Here is the entire document entitled

---

[29] *Id.*

[30] *Id.*

[31] *Fact Sheet to States,* FEDERAL AVIATION ADMINISTRATION (FAA), http://www.faa.gov/uas/regulations_policies/media/UAS_Fact_Sheet_Final.pdf

State and Local Regulation of Unmanned Aircraft Systems (UAS) Fact Sheet, by Federal Aviation Administration, Office of the Chief Counsel, published on December 17, 2015.

BACKGROUND

Unmanned aircraft systems (UAS) are aircraft subject to regulation by the FAA to ensure safety of flight, and safety of people and property on the ground. States and local jurisdictions are increasingly exploring regulation of UAS or proceeding to enact legislation relating to UAS operations. In 2015, approximately 45 states have considered restrictions on UAS. In addition, public comments on the Federal Aviation Administration's (FAA) proposed rule, "Operation and Certification of Small Unmanned Aircraft Systems" (Docket No. FAA-2015-0150), expressed concern about the possible impact of state and local laws on UAS operations.

Incidents involving unauthorized and unsafe use of small, remote-controlled aircraft have risen dramatically. Pilot reports of interactions with suspected unmanned aircraft have increased from 238 sightings in all of 2014 to 780 through August of this year. During this past summer, the presence of multiple UAS in the vicinity of wild fires in the western U.S. prompted firefighters to ground their aircraft on several occasions.

This fact sheet is intended to provide basic information about the federal regulatory framework for use by states and localities when considering laws affecting UAS. State and local restrictions affecting UAS operations should be consistent with the extensive federal statutory and regulatory framework pertaining to control of the airspace, flight management and efficiency, air traffic control, aviation safety, navigational facilities, and the regulation of aircraft noise at its source.

Presented below are general principles of federal law as they relate to aviation safety, and examples of state and local laws that should be carefully considered prior to any legislative action to ensure that they are consistent with applicable federal safety regulations. The FAA's Office of the Chief Counsel is available for consultation on specific questions.

WHY THE FEDERAL FRAMEWORK

Congress has vested the FAA with authority to regulate the areas of airspace use, management and efficiency, air traffic control, safety, navigational facilities, and aircraft noise at its source. 49 U.S.C. §§ 40103, 44502, and 44701-44735. Congress has directed the FAA to "develop plans and policy for the use of the navigable airspace and assign by regulation or order the use of the airspace necessary to ensure the safety of aircraft and the efficient use of airspace." 49 U.S.C. § 40103(b)(1). Congress has further directed the FAA to "prescribe air traffic regulations on the flight of aircraft (including regulations on safe altitudes)" for navigating, protecting, and identifying aircraft; protecting individuals and property on the ground; using the navigable airspace efficiently; and preventing collision between aircraft, between aircraft and land or water vehicles, and between aircraft and airborne objects. 49 U.S.C. § 40103(b)(2).

A consistent regulatory system for aircraft and use of airspace has the broader effect of ensuring the highest level of safety for all aviation operations. To ensure the maintenance of a safe and sound air transportation system and of navigable airspace free from inconsistent restrictions, FAA has regulatory authority over matters pertaining to aviation safety.

REGULATING UAS OPERATIONS

In § 333 of the FAA Modernization and Reform Act of 2012 (Public Law No. 112-95), Congress directed the Secretary to determine whether UAS operations posing the least amount of public risk and no threat to national security could safely be operated in the national airspace system (NAS) and if so, to establish requirements for the safe operation of these systems in the NAS.

On February 15, 2015, the FAA proposed a framework of regulations that would allow routine commercial use of certain small UAS in today's aviation system, while maintaining flexibility to accommodate future technological innovations. The FAA's Notice of Proposed Rulemaking offered safety rules for small UAS (under 55 pounds) conducting nonrecreational or nonhobby operations. The proposed rule defines permissible hours of flight, line-of-sight observation, altitude, operator certification, optional use of visual observers, aircraft registration and marking, and operational limits.

Consistent with its statutory authority, the FAA is requiring Federal registration of UAS in order to operate a UAS. Registering UAS will help protect public safety in the air and on the ground, aid the FAA in the enforcement of safety-related requirements for the operation of UAS, and build a culture of accountability and responsibility among users operating in U.S. airspace. No state or local UAS registration law may relieve a UAS owner or operator from complying with the Federal UAS registration requirements. Because Federal registration is the exclusive means for registering UAS for purposes of operating an aircraft in navigable airspace, no state or local government may impose an additional registration requirement on the operation of UAS in navigable airspace without first obtaining FAA approval.

Substantial air safety issues are raised when state or local governments attempt to regulate the operation or flight of aircraft. If one or two municipalities enacted ordinances regulating UAS in the navigable airspace and a significant number of municipalities followed suit, fractionalized control of the navigable airspace could result. In turn, this 'patchwork quilt' of differing restrictions could severely limit the flexibility of FAA in controlling the airspace and flight patterns, and ensuring safety and an efficient air traffic flow. A navigable airspace free from inconsistent state and local restrictions is essential to the maintenance of a safe and sound air transportation system. See *Montalvo v. Spirit Airlines*, 508 F.3d 464 (9th Cir. 2007), and *French v. Pan Am Express, Inc.*, 869 F.2d 1 (1st Cir. 1989); see also *Arizona v. U.S., 567 U.S.* ___, 132.

S.Ct. 2492, 2502 (2012) ("Where Congress occupies an entire field . . . even complimentary state regulation is impermissible. Field preemption reflects a congressional decision to foreclose any state regulation in the area, even if it is parallel to federal standards."), and *Morales v. Trans World Airlines, Inc.,* 504 U.S. 374, 386-87 (1992).

EXAMPLES OF STATE AND LOCAL LAWS FOR WHICH CONSULTATION WITH THE FAA IS RECOMMENDED

- Operational UAS restrictions on flight altitude, flight paths; operational bans; any regulation of the navigable airspace. For example—a city ordinance banning anyone from operating UAS within the city limits, within the airspace of the city, or within certain distances of landmarks. Federal courts strictly scrutinize state and local regulation of overflight. *City of Burbank v. Lockheed Air Terminal*, 411 U.S. 624 (1973); *Skysign International, Inc. v. City and County of Honolulu*, 276 F.3d 1109, 1117 (9th Cir. 2002); *American Airlines v. Town of Hempstead*, 398 F.2d 369 (2d Cir. 1968); *American Airlines v. City of Audubon Park*, 407 F.2d 1306 (6th Cir. 1969).
- Mandating equipment or training for UAS related to aviation safety such as geofencing would likely be preempted. Courts have found that state regulation pertaining to mandatory training and equipment requirements related to aviation safety is not consistent with the federal regulatory framework. *Med-Trans Corp. v. Benton*, 581 F. Supp. 2d 721, 740 (E.D.N.C. 2008); *Air Evac EMS, Inc. v. Robinson*, 486 F. Supp. 2d 713, 722 (M.D. Tenn. 2007).

EXAMPLES OF STATE AND LOCAL LAWS WITHIN STATE AND LOCAL GOVERNMENT POLICE POWER

Laws traditionally related to state and local police power—including land use, zoning, privacy, trespass, and law enforcement operations—generally are not subject to federal regulation. *Skysign International, Inc. v. City and County of Honolulu*, 276 F.3d 1109, 1115 (9th Cir. 2002).

Examples include:

- Requirement for police to obtain a warrant prior to using a UAS for surveillance.
- Specifying that UAS may not be used for voyeurism.
- Prohibitions on using UAS for hunting or fishing, or to interfere with or harass an individual who is hunting or fishing.
- Prohibitions on attaching firearms or similar weapons to UAS.

CONTACT INFORMATION FOR QUESTIONS

The FAA's Office of the Chief Counsel is available to answer questions about the principles set forth in this fact sheet and to consult with you about the intersection of federal, state, and local regulation of aviation, generally, and UAS operations, specifically. You may contact the Office of Chief Counsel in Washington, D.C. or any of the following Regional Counsels:

FAA Office of the Chief Counsel, Regulations Division (AGC-200), 800 Independence Ave. SW, Washington, DC 20591, (202) 267-3073

Alaskan Region, Office of the Regional Counsel, 222 West 7th Ave., Anchorage, AK 99513, (909) 271-5269, (AK)

Central Region, Office of the Regional Counsel, 901 Locust St., Room 506, Kansas City, MO 61406-2641, (816) 329-3760, (IA, KS, MO, NE)

Eastern Region, Office of the Regional Counsel, 1 Aviation Plaza, Room 561, Jamaica, NY 11434-4848, (718) 553-3285, (DC, DE, MD, NJ, NY, PA, VA, WV)

Great Lakes Region, Office of the Regional Counsel, O'Hare Lake Office Center, 2300 East Devon Ave., Des Plaines, IL 60018, (847) 294-7313, (IL, IN, MI, MN, ND, OH, SD, WI)

New England Region, Office of the Regional Counsel, 12 New England Executive Park, Burlington, MA 01803, (781) 238-7040, (CT, ME, MA, NH, RI, VT)

Northwest Mountain Region, Office of the Regional Counsel, 1601 Lind Ave. SW, Renton, WA 98055-4056, (425) 227-2007, (CO, ID, MT, OR, UT, WA, WY)

Southern Region, Office of the Regional Counsel, 1701 Columbia Ave., Suite 530, College Park, GA 30337, (404) 305-5200, (AL, FL, GA, KY, MS, NC, SC, TN)

Southwest Region, Office of the Regional Counsel, 6N-300, 10101 Hillwood Parkway Dr., Fort Worth, TX 76177, (817) 222-5099, (AR, LA, NM, OK, TX)

Western-Pacific Region, Office of the Regional Counsel, P.O. Box 92007, Los Angeles, CA 90009, (310) 725-7100, (AZ, CA, HI, NV)

APPENDIX—LIST OF AUTHORITIES

Federal Statutes

- 49 U.S.C. §§ 40103, 44502, and 44701-44735 (former Federal Aviation Act of 1958, as amended and recodified).
- FAA Modernization and Reform Act of 2012, Public Law No. 112-95 (Feb. 14, 2012),Subtitle B, "Unmanned Aircraft Systems."

Federal Regulations

- Title 14 of the Code of Federal Regulations, Chapter 1.

The U.S. Supreme Court

- "Congress has recognized the national responsibility for regulating air commerce. Federal control is intensive and exclusive. Planes do not wander about in the sky like vagrant clouds. They move only by federal permission, subject to federal inspection, in the hands of federally certified personnel and under an intricate system of federal commands. The moment a ship taxies onto a runway it is caught up in an elaborate and detailed system of controls. It takes off only by instruction from the control tower, it travels on prescribed beams, it may be diverted from its intended landing, and it obeys signals and orders. Its privileges, rights, and protection, so far as transit is concerned, it owes to the Federal Government alone and not to any state government." *Northwest Airlines v. State of Minnesota*, 322 U.S. 292, 303 (1944) (Jackson, R., concurring).

- "If we were to uphold the Burbank ordinance [which placed an 11 p.m. to 7 a.m. curfew on jet flights from the Burbank Airport] and a significant number of municipalities followed suit, it is obvious that fractionalized control of the timing of takeoffs and landings would severely limit the flexibility of FAA in controlling air traffic flow. The difficulties of scheduling flights to avoid congestion and the concomitant decrease in safety would be compounded." *Burbank v. Lockheed Air Terminal Inc.*, 411 U.S. 624, 639 (1973).
- "The Federal Aviation Act requires a delicate balance between safety and efficiency, and the protection of persons on the ground. . . . The interdependence of these factors requires a uniform and exclusive system of federal regulation if the congressional objectives underlying the Federal Aviation Act are to be fulfilled." *Burbank* at 638-639.
- "The paramount substantive concerns of Congress [in enacting the FAA Act] were to regulate federally all aspects of air safety . . . and, once aircraft were in 'flight,' airspace management . . . " *Burbank* at 644 (Rehnquist, J. dissenting).

U.S. Courts of Appeals

- "Air traffic must be regulated at the national level. Without uniform equipment specifications, takeoff and landing rules, and safety standards, it would be impossible to operate a national air transportation system." *Gustafson v. City of Lake Angeles*, 76 F.3d 778, 792-793 (6th Cir. 1996)(Jones, N., concurring).
- "The purpose, history, and language of the FAA [Act] lead us to conclude that Congress intended to have a single, uniform system for regulating aviation safety. The catalytic events leading to the enactment of the FAA [Act] helped generate this intent. The FAA [Act] was drafted in response to a series of fatal air crashes between civil and military aircraft operating under separate flight rules . . . In discussing the impetus for the FAA [Act], the Supreme Court has also noted that regulating the aviation industry requires a delicate balance between safety and efficiency. It is precisely because of 'the interdependence of these factors' that Congress enacted 'a uniform and exclusive system of federal regulation.'" *Montalvo v. Spirit Airlines*, 508 F.3d 464, 471 (9th Cir. 2007), citing *City of Burbank v. Lockheed Air Terminal Inc.*, 411 U.S. 624, 638-39 (1973).

- "[W]hen we look to the historical impetus for the FAA, its legislative history, and the language of the [FAA] Act, it is clear that Congress intended to invest the Administrator of the Federal Aviation Administration with the authority to enact exclusive air safety standards. Moreover, the Administrator has chosen to exercise this authority by issuing such pervasive regulations that we can infer a preemptive intent to displace all state law on the subject of air safety." *Montalvo* at 472.
- "We similarly hold that federal law occupies the entire field of aviation safety. Congress' intent to displace state law is implicit in the pervasiveness of the federal regulations, the dominance of the federal interest in this area, and the legislative goal of establishing a single, uniform system of control over air safety. This holding is fully consistent with our decision in *Skysign International, Inc. v. Honolulu*, 276 F.3d 1109 (9th Cir. 2002), where we considered whether federal law preempted state regulation of aerial advertising that was distracting and potentially dangerous to persons on the ground. In upholding the state regulations, we held that federal law has not 'preempt[ed] altogether any state regulation purporting to reach into the navigable airspace.' *Skysign* at 1116. While Congress may not have acted to occupy exclusively all of air commerce, it has clearly indicated its intent to be the sole regulator of aviation safety. The FAA, together with federal air safety regulations, establish complete and thorough safety standards for interstate and international air transportation that are not subject to supplementation by, or variation among, states." *Montalvo* at 473-474.
- "[W]e remark the Supreme Court's reasoning regarding the need for uniformity [concerning] the regulation of aviation noise, see *City of Burbank v. Lockheed Air Terminal*, 411 U.S. 624 (1973), and suggest that the same rationale applies here. In Burbank, the Court struck down a municipal anti-noise ordinance placing a curfew on jet flights from a regional airport. Citing the 'pervasive nature of the scheme of federal regulation,' the majority ruled that aircraft noise was wholly subject to federal hegemony, thereby preempting state or local enactments in the field. In our view, the pervasiveness of the federal web is as apparent in the matter of pilot qualification as in the matter of aircraft noise. If we upheld the Rhode Island statute as applied to airline pilots, 'and a significant number of [states] followed suit, it is obvious that fractionalized control . . . would severely limit the flexibility of the F.A.A . . . .' [citing *Burbank*] Moreover, a patchwork of state laws in this airspace, some in conflict with each other, would create a crazyquilt effect . . . The regulation of interstate flight-and flyers-must of necessity be monolithic. Its very nature permits no other conclusion. In the area of pilot fitness as in the area of aviation noise, the [FAA] Act as we read it 'leave[s] no room for . . . local controls.' [citing *Burbank*]. *French v. Pan Am Express, Inc.*, 869 F.2d 1, 6 (1st Cir. 1989).[32]

---

[32]*Id.*

9. Boggs v. Merideth

Compounding the issues mentioned in the Fact Sheet to States, in the previous section, is the issue between a person's private property rights and the FAA's right to govern said navigable airspace. As this issue has arisen many times throughout history, as far back as the 1940s in the *Causby* case, federal courts now have to revisit these issues. This next section details *Boggs v. Merideth,* the first one to enter the dockets.

On January 4, 2016, Kentucky resident, David Boggs, filed suit in the United States District Court of Western Kentucky asking the court to define clearly the rights of aircraft operators and property owners as they relate to unmanned aircraft. The lawsuit stems from an incident last year that gained national media attention in which a Hillview, Kentucky, resident, Merideth, shot down an unmanned aircraft being flown by Boggs, claiming that the UAS had trespassed and invaded his privacy. Although the shooter initially faced criminal charges, a state court judge eventually dismissed those charges. On October 26, 2015, Kentucky District Court Judge Rebecca Ward dismissed the criminal charges, saying that "he had a right to shoot" at the aircraft. Boggs, on the other hand, claimed that he was approximately 200 feet above the property at the time his UAS was shot down and did not view or record the defendant's property.[33]

James Mackler, Boggs's legal counsel said, "The tension between private property rights and the freedom to use the national airspace is important to both the unmanned aircraft industry and the general public. Property owners deserve to be free from harassment and invasion of their privacy. Likewise, aircraft operators need to know the boundaries in which they can legally operate without risk of being shot down. This lawsuit will give clarity to everyone."[34]

Among the factual allegations in the *Boggs v. Merideth* complaint are the following. (1) The FAA has the exclusive authority to govern airspace within the United States and the operation of aircraft. The federal government's interest in regulating aviation is paramount. Federal statutes and regulations, including the Transportation Laws of the United States, 49 U.S.C.A. § 40101 et seq., and the Airline Deregulation Act, 49 U.S.C.A. § 41713, and the regulations promulgated by the FAA pursuant to those laws, preempt state law in that area. (2) The FAA's airspace designations state that Class G airspace includes uncontrolled airspace that does not fall within any other classification between the surface and any overlying Class E airspace. Class G airspace is part of the navigable airspace the regulation of which substantially affects interstate commerce. (3) The FAA defines an "aircraft" as "a device that is used or intended to be used for flight in the air." 14 C.F.R. § 1.1. (4) On July 26, 2015, Boggs was operating by wireless controller an unmanned aircraft (also referred

---

[33] *First Lawsuit to Clarify the Rights of Drone Operators and Property Owners Filed in Federal Court,* Frost, Brown, Todd, LLC, http://www.frostbrowntodd.com/newsroom-press-first-lawsuit-to-clarify-the-rights-of-drone-operators-and-property-owners-filed-in-federal-court.html.

[34] *Id.*

to as a "drone") at an altitude of approximately 200 feet above ground level in Class G airspace over Bullitt County, Kentucky. (5) Boggs's aircraft contained an onboard camera capable of recording video and still photographs. His aircraft recorded video of the horizon, woods, and the rooftops of various houses. At no time was he capturing video or still images of Merideth or anyone on his property. Mackler included in the complaint the last image recorded by the aircraft prior to being shot by defendant. (6) After approximately two minutes of flight, Merideth shot Boggs's unmanned aircraft down with a shotgun, resulting in damage to Boggs's property. Merideth later alleged that Boggs's unmanned aircraft may have been taking video or still images of his daughter while hovering over his property, thus Merideth asserts he was protecting his family's privacy rights and preventing further trespass. (7) Merideth was charged by Kentucky authorities with felony wanton endangerment and criminal mischief. On October 26, 2015, Kentucky District Court Judge Rebecca Ward dismissed the criminal charges against Defendant saying that Defendant "had a right to shoot" at the aircraft. (8) Merideth, using the nickname "The Drone Slayer," continues to assert that he was justified in shooting Boggs's aircraft and vows to do it again, as evidenced by his Facebook page, which states, "Not only did I do it, but I meant to do it. And I'd do it again." (9) Indeed, Merideth has implicitly encouraged others to engage in the same conduct by selling shirts that have the following words printed on them, "Droneslayer. . . . We the people . . . have had enough!"

Given the foregoing, Boggs requested a declaratory judgment, as set forth in more detail below.

FIRST CLAIM FOR RELIEF

(Declaratory Judgment)

Plaintiff incorporates by reference the allegations in the foregoing paragraphs as though fully set forth. An actual controversy has arisen and now exists between Plaintiff and Defendant concerning their respective rights and obligations with respect to the damage caused to Plaintiff's aircraft by Defendant. The United States Government has exclusive sovereignty over airspace of the United States pursuant to 49 U.S.C.A. § 40103. The airspace, therefore, is not subject to private ownership nor can the flight of an aircraft within the navigable airspace of the United States constitute a trespass. Unmanned aircraft are aircraft consistent with Subtitle B of Public Law 112-95 and the existing definition of aircraft in Title 49 of the United States Code, 49 U.S.C. 40102. In addition, even if Plaintiff had viewed the defendant's property from the air, which he did not, such viewing would not violate the defendant's reasonable expectation of privacy according to well-established federal law. The U.S. Supreme Court has determined, as a general rule, that there is no reasonable expectation of privacy in the area surrounding a home in plain view from above. *California v. Ciraolo*, 476 U.S. 207, 106 S.Ct. 1809, 90 L.Ed.2d 210 (1986); *Florida v. Riley*, 488 U.S. 445, 109 S.Ct. 693, 102 L.Ed.2d 835 (1989) (plurality opinion). Resolution of the current dispute between Plaintiff and Defendant requires the application of this existing federal case law to an as yet unexamined technology—unmanned aircraft. Further, Congress has indicated its unambiguous intent to ensure the safety of aircraft. Pursuant to 18 U.S.C.A. § 32, whoever "sets fire to, damages, destroys, disables, or wrecks any aircraft in the special aircraft jurisdiction of the United States or any civil aircraft used, operated, or employed in interstate, overseas, or foreign air commerce" commits a felony. Although this statute may not create a private right of action, the interpretation of the statute is critical to the determination of the claims asserted herein. Should the court determine that this statute applies to unmanned aircraft, as it should, that would leave no room for Defendant's assertion of the right to self-help or the Kentucky District Court Judge's ruling that Defendant was "within his rights" to shoot the aircraft. Conversely, Kentucky law regarding trespass does not specifically address the rights of unmanned aircraft to traverse the skies above private property. It defines a trespasser as "any person who enters or goes upon the real estate of another without any right, lawful authority or invitation, either expressed or implied." Ky. Rev. Stat. Ann. § 381.231. A trespasser may be subject to civil suit and/or criminal prosecution. Kentucky law also permits resort to self-help in response to trespass. A landowner may use physical force "upon another person when the person believes that such force is immediately necessary to prevent the commission of criminal trespass." Ky. Rev. Stat. Ann. § 503.080. Given the clear conflict of federal and state laws, as applied to the facts of this action, Plaintiff desires a judicial determination of the respective rights and duties of Plaintiff and Defendant with respect to Plaintiff's rights to operate an aircraft within Class G airspace and recover damages for trespass to chattel caused by Defendant's intentional shooting of that aircraft. Plaintiff is, herein, asserting a claim for trespass to chattels as a result of the damages to his aircraft. The ruling of the Kentucky District Court and assertions made by Defendant regarding his belief that his actions were justified because Plaintiff was engaged in trespass and invasion of privacy are in direct conflict with established federal law governing the regulation of manned aircraft and airways and cannot be resolved without addressing how this law applies to unmanned aircraft. For those reasons, Plaintiff seeks the following declaratory judgment:

A. An unmanned aircraft is an "aircraft" according to Federal law.

B. An unmanned aircraft operating in Class G airspace in the manner alleged above is operating in "navigable airspace" within the exclusive jurisdiction of the United States.

C. That Plaintiff was operating his unmanned aircraft in the navigable airspace within the exclusive jurisdiction of the United States and not within Defendant's property;

D. That the operation of his unmanned aircraft in in the manner alleged above did not violate the defendant's reasonable expectation of privacy; and

E. That a property owner cannot shoot at an unmanned aircraft operating in navigable airspace within the exclusive jurisdiction of the United States when that aircraft is operating in the manner alleged above.

SECOND CLAIM FOR RELIEF

(Trespass to Chattels)

Plaintiff incorporates by reference the allegations in the foregoing paragraphs as though fully set forth. Defendant intentionally intermeddled with personal property in the possession of Plaintiff, specifically, his unmanned aircraft. Defendant impaired the property as to its condition, quality, to value. Plaintiff's property was damaged by the reduced value, condition and quality of his aircraft in an amount of approximately $1,500.00.

WHEREFORE, premises considered, Plaintiff demands the following relief:

1. That the Defendant be served with process and answer the allegations and claims set forth above;
2. That the court enter the declaratory judgment requested in the First Cause of Action, above; and
3. That the Court award to the Plaintiff such other legal and equitable relief as it deems appropriate, including monetary damages, prejudgment interest, and the costs of filing this action.[35]

At the time of writing, the case was referred to Judge David J. Hale.[36] On March 3, 2016, Merideth's attorney filed a Memorandum in Support of Motion to Dismiss.[37] In this memo the defendant alleged that the case presents a small claims court case packaged with a tenuous at best declaratory judgment action in order to invoke federal court jurisdiction. Furthermore, he alleged that subject matter jurisdiction is lacking and hence the matter should be dismissed.[38]

10. Privacy Issues

On February 15, 2015, the same day the FAA issued a NPRM to fulfill their duty, under the FAA Modernization and Reform Act of 2012, to integrate small UAS weighing less than 55 pounds into the NAS, the President of the United States issued a presidential memorandum promoting economic competitiveness while safeguarding privacy, civil rights, and civil liberties in domestic use of unmanned aircraft systems.[39] This memorandum

---

[35] *Id.*

[36] *Boggs v. Merideth,* 3:16-cv-00006, PACERMONITOR, https://www.pacermonitor.com/public/case/10270285/Boggs_v_Merideth.

[37] *Memorandum in Support of Motion to Dismiss,* https://assets.documentcloud.org/documents/2752801/14-1.pdf

[38] *Id.*

[39] *Presidential Memorandum,* https://www.whitehouse.gov/the-press-office/2015/02/15/presidential-memorandum-promoting-economic-competitiveness-while-safegua.

outlines (1) UAS policies and procedures for federal government use including privacy protections, collection and use, retention and dissemination of data, civil rights and civil liberties protections, accountability, transparency, and reports; (2) multi-stakeholder engagement process; (3) definitions; and (4) general provisions.[40]

In this Memorandum, the President made it clear that where UAS is the platform for collection, information must be collected, used, retained, and disseminated consistent with the Constitution, Federal law, and other applicable regulations and policies. Agencies must, for example, comply with the Privacy Act of 1974 (5 U.S.C. 552(a), also known as the Privacy Act), which, among other things, restricts the collection and dissemination of individuals' information that is maintained in systems of records, including personally identifiable information (PII), and permits individuals to seek access to and amendment of records. Additionally, to keep pace with the expected advancements in UAS technologies, agencies shall, prior to deployment of new UAS technology and at least every three years, examine and update their existing policies.[41]

On March 3, 2015, Senator Edward Markey introduced SB 635, entitled the Drone Aircraft Privacy and Transparency Act of 2015, to the Senate Committee on Commerce, Science, and Transportation.[42] As of March 10, 2016, the Committee on Small Business and Entrepreneurship was hearing the bill. This legislation is to amend the FAA Modernization and Reform Act of 2012 to direct the Secretary of Transportation to study and identify any potential threats to privacy protections posed by the integration of unmanned aircraft (drone) systems into the national airspace system, including any potential violations of privacy principles.

The bill directs the Secretary to establish certain procedures to allow for civil operation in the national airspace system of small drone systems that do not meet expedited operational authorization requirements. It requires such procedures to ensure that the integration of drone systems into the national airspace system is done in compliance with privacy principles. It prohibits the Secretary from approving, issuing, or awarding any certificate, license, or other grant of authority to operate a drone system in the national airspace system unless the application for it includes a data collection statement, meeting certain requirements, that provides reasonable assurance that the applicant will operate the drone system in accordance with privacy principles. It applies the same privacy principles requirement to any drone system to be operated by a law enforcement agency or a law enforcement agency contractor or subcontractor, except that the application for it shall include a data minimization statement, meeting certain requirements, instead of a data collection statement.

---

[40]*Id.*
[41]*Id.*
[42]Drone Aircraft Privacy and Transparency Act of 2015, https://www.congress.gov/bill/114th-congress/senate-bill/635.

Additionally, the bill directs the Administrator of the Federal Aviation Administration (FAA) to make any approved certificate, license, or other grant of authority, plus other specified information, available in a searchable format on the public FAA website. It prohibits a governmental entity from using a drone system, or requesting information or data collected by another entity through use of a drone system, for protective activities, or for law enforcement or intelligence purposes, except pursuant to a warrant issued using the procedures described in the Federal Rules of Criminal Procedure (or, in the case of a state court, issued using state warrant procedures) by a court of competent jurisdiction, or as permitted under the Foreign Intelligence Surveillance Act of 1978. It makes an exception to this prohibition in exigent circumstances when a law enforcement entity reasonably believes there is: (1) an imminent danger of death or serious physical injury; or (2) a high risk of an imminent terrorist attack by a specific individual or organization, according to the Secretary of Homeland Security.

Furthermore, the bill makes it unlawful to operate a drone system in a manner that is not in accordance with the terms of a data collection statement or in a manner violating any portion of the final rule for the procedures for civil operation of small drone systems required by this Act. Grants enforcement authority to the Federal Trade Commission. It also authorizes a state to bring a civil action on behalf of state residents in state or U.S. district court for injunctive relief against violations of this Act or related regulations if the state attorney general has reason to believe that an interest of state residents has been or is threatened or adversely affected by a prohibited act or practice.

Creates a private right of action in state or U.S. district court for persons injured by a prohibited act. Additionally, it allows a suit against a governmental entity. The bill requires the FAA to revoke the certificate, license, or other grant of authority to operate a drone system operated in a prohibited manner. Finally, it declares that nothing in this Act may be construed to apply to model aircraft flown strictly for hobby or recreational purposes and meeting certain other criteria.[43]

A similar House Resolution (HR) 1229, with the same title, Drone Aircraft Privacy and Transparency Act of 2015, was introduced on March 3, 2015, by Representative Peter Welch to the House—Transportation and Infrastructure; Energy and Commerce, and three days later, referred to the Subcommittee on Commerce, Manufacturing, and Trade.[44]

11. FAA Order: 8130.34C—Airworthiness for UAS

In 2005, the Associate Administrator for Aviation Safety determined that UAS could be given limited access to the NAS (refer to Title 14 CFR part 21, Certification Procedures for Products and Parts, 21.191, Experimental certificates). The FAA's Office of Aviation

[43]*Id.*

[44]Drone Aircraft Privacy and Transparency Act of 2015, https://www.congress.gov/bill/114th-congress/house-bill/1229/related-bills.

Safety (AVS) is an organization responsible for the certification, production approval, and continued airworthiness of aircraft; and certification of pilots, mechanics, and others in safety-related positions. Aviation Safety is also responsible for: (1) Certification of all operational and maintenance enterprises in domestic civil aviation; (2) Certification and safety oversight of approximately 7,300 U.S. commercial airlines and air operators; (3) Civil flight operations; and (4) Developing regulations.[45]

The Director of the Aircraft Certification Service, with concurrence from the Director of the Flight Standards Service, stipulated that this process be managed by the office of primary responsibility for § 21.195 (Experimental certificates: Aircraft to be used for market surveys, sales demonstrations, and customer crew training). The Aircraft Certification Service, Production and Airworthiness Division, AIR-200, is tasked with coordinating all aspects of issuing an experimental certificate or a special flight permit to a UAS, OPA, or OPA/UAS applicant. AIR-200 issued this document on August 2, 2013.[46]

This order establishes procedures for issuing special airworthiness certificates in the experimental category or special flight permits to UAS, optionally piloted aircraft (OPA), and aircraft intended to be flown as either a UAS or an OPA, under the designation "OPA/UAS." The procedures in this order apply to FAA manufacturing aviation safety inspectors (ASI), to FAA airworthiness ASIs, and to private persons or organizations delegated authority to issue special flight permits for production flight-testing new production aircraft.[47]

12. FAA Flight Standards Information Management Systems (FSIMS) for UAS

The FAA's Flight Standards National Field Office is responsible for ensuring standard application of national certification and surveillance programs, policies, and procedures. The office, which is located at Dulles, Virginia, provides services in all regions. The division is affiliated with the Eastern Region, which serves as the host region. The Eastern Region's services to this division include, but are not limited to, legal, personnel, payroll, acquisition, budgeting, and security.[48]

The Flight Standards National Field Office is responsible for field implementation of national programs, including oversight of the certification and surveillance of airmen, air operators, and air agencies engaged in air transportation under 14 CFR Part 121. These responsibilities include development of national certification and surveillance programs, policies, procedures, and comprehensive national surveillance plans; standardizing and continuously improving Air Transportation Oversight System (ATOS) processes, providing

---

[45] *Aviation Safety,* FAA AVS, http://www.faa.gov/about/office_org/headquarters_offices/.

[46] *FAA Order 8130.34C, Airworthiness for UAS,* FAA, http://www.faa.gov/regulations_policies/orders_notices/index.cfm/go/document.information/documentID/1021706.

[47] *Id.*

[48] *FAA FSIMS, Flight Standards National Field Office, FAA,* http://www.faa.gov/about/office_org/headquarters_offices/avs/offices/afs/afs900/.

assistance and support to ATOS Certificate Management Teams; the creation and deployment of the Safety Assurance System (SAS); and, promoting System Safety and Risk Management in Flight Standards.[49] The following part of FSIMS Volume 16 pertains to UAS and reads as follows:

16-1-1-1 PURPOSE. This volume provides a means by which prospective air operators, air agencies, or government flight operatives are authorized to conduct business in a manner which complies with all applicable regulations, the Federal Aviation Act of 1958 (FA Act), and Federal Aviation Administration (FAA) directives. The process is designed to preclude the certification and/or approval of applicants who are unwilling or unable to comply with the regulations or conform to safe operating practices unique to Unmanned Aircraft Systems (UAS) operations. This volume also provides policies necessary for reviewing and evaluating the safety and interoperability of proposed UAS flight operations conducted within the U.S. National Airspace System (NAS) for the Flight Standards Service Unmanned Aircraft Systems (UAS) Integration Office (AFS-80) when assessing applications for a Certificate of Waiver or Authorization (COA) or Special Airworthiness Certificate and incorporates FAA Notice N 8900.227, Unmanned Aircraft Systems (UAS) Operational Approval, dated 7/30/13.[50] *(Author note: FAA Notice N 8900.227 was cancelled on July 30, 2014, despite being referenced in this document that is still current as of the printing of this book.)*

16-1-1-3 AUDIENCE. The primary audience for this volume is Flight Standards Service (AFS) aviation safety inspectors (ASI), their managers and supervisors, and other operational and administrative employees. The aviation industry may use this volume as a reference only, and the general public may find it helpful for informational and planning purposes.[51]

13. FAA Enforcement Action

The FAA Office of the Chief Counsel supports the Agency's mission by furnishing timely and responsive legal services to the FAA Administrator and all Agency organizations at the Headquarters, Regional and Center levels. The principal areas of the Chief Counsel's legal practice include: (1) Aircraft Registration—Proposed Policy Clarification; (2) Airports and Environmental Law; (3) Enforcement and Compliance; (4) Litigation and General Law; (5) Office of Dispute Resolution for Acquisition (ODRA); (6) Employment and Labor Law; (7) Acquisition and Fiscal Law; and (8) Regulations.[52]

Components of the Office also serve as the FAA Administrator's adjudicative forums for civil penalty and acquisition disputes. Attorneys represent the agency before a variety

---

[49] *Id.*

[50] *UAS*, Flight Standards Information Management System (FSIMS), http://fsims.faa.gov/PICResults.aspx?mode=EBookContents&restricttocategory=all~menu.

[51] *Id.*

[52] *FAA Office of the Chief Counsel, Enforcement and Compliance*, FAA, https://www.faa.gov/about/office_org/headquarters_offices/agc/.

of forums, including: (1) National Transportation Safety Board (NTSB); (2) Merit Systems Protection Board (MSPB); (3) Equal Employment Opportunity Commission (EEOC); (4) FAA's Office of Dispute Resolution for Acquisition (ODRA); and (5) United States federal courts. The Counsel's office also works closely with the Office of the General Counsel of the Department of Transportation on issues that are common to modal administrations or that are of national significance to the aviation industry.[53]

The Enforcement Division develops and reviews agency enforcement policies and programs. It provides guidance to agency personnel on enforcement matters and reviews enforcement actions of national significance. The Division also represents the FAA in enforcement actions before the National Transportation Safety Board, the FAA Decisionmaker, and the United States courts of appeals.[54]

The following section details civil penalties. These two sections together will be further explained in the section dealing with *Administrator v. Pirker*.

14. Civil Penalties

Title 49 United States Code Chapter 463 entitled Penalties outlines Sections 46301 through 46319 dealing with the following topics pertinent to unmanned operators: (1) Section 46301—Civil Penalties; (2) Section 46302—False information; (3) Section 46303—Carrying a weapon; (4) Section 46304—Liens on aircraft; (5) Section 46305—Actions to recover civil penalties; (6) Section 46306—Registration violations involving aircraft not providing air transportation; (7) Section 46307—Violation of national defense airspace; (8) Section 46308—Interference with air navigation; (9) Section 46309—Concession and price violations; (10) Section 46310—Reporting and record keeping violations; (11) Section 46311—Unlawful disclosure of information; (12) Section 46312—Transporting hazardous material; (13) Section 46313—Refusing to appear or produce records; (14) Section 46314—Entering aircraft or airport area in violation of security requirements; (15) Section 46315—Lighting violations involving transporting controlled substances by aircraft not providing air transportation; (16) Section 46316—General criminal penalty when specific penalty not provided; (17) Section 46317—Criminal penalty for pilots operating in air transportation without an airman's certificate; (18) Section 46318—Interference with cabin or flight crew; and (19) Section 46319—Permanent closure of an airport without providing sufficient notice.[55]

Title 49 United States Code § 46301 outlines the civil penalties that have been applied to UAS violators, and reads as follows:

---

[53] *Id.*
[54] *Id.*
[55] 49 USC Chapter 463, http://www.gpo.gov/fdsys/granule/USCODE-2011-title49/USCODE-2011-title49-subtitleVII-partA-subpartiv-chap463-sec46301.

a. General Penalty. (1) A person is liable to the United States Government for a civil penalty of not more than $25,000 (or $1,100 if the person is an individual or small business concern) for violating –

A. Chapter 401 (except sections 40103(a) and (d), 40105, 40116, and 40117), chapter 411, chapter 413 (except sections 41307 and 41310(b)-(f)), chapter 415 (except sections 41502, 41505, and 41507-41509), chapter 417 (except sections 41703, 41704, 41710, 41713, and 41714), chapter 419, subchapter II or III of chapter 421, chapter 441 (except section 44109), section 44502(b) or (c), chapter 447 (except sections 44717 and 44719-44723), chapter 449 (except sections 44902, 44903(d), 44904, 44907(a)-(d)(1)(A) and (d)(1)(C)-(f), and 44908), section 47107(b) (including any assurance made under such section), or section 47133 of this title;

B. A regulation prescribed or order issued under any provision to which clause (A) of this paragraph applies;

C. Any term of a certificate or permit issued under section 41102, 41103, or 41302 of this title; or

D. A regulation of the United States Postal Service under this part.[56]

15. Administrator v. Pirker

On or about October 17, 2011, a Swiss national, Raphael Pirker, flew his Ritewing Zephyr powered glider aircraft in the vicinity of the University of Virginia, Charlottesville, Virginia, for monetary compensation from Lewis Communications for video and photographs taken during that flight.[57] A few months later, on April 13, 2012, the FAA advised him through a Notice of Proposed Assessment that he was to receive a civil penalty amounting to $10,000 pursuant to 49 USC §§ 46301(a)(1) and (d)(2) and 46301(a)(5).[58]

The FAA alleged that Pirker was the pilot in command of the Ritewing Zephyr powered glider, or UAS as they classified it, although he did not possess a pilot certificate issued by the FAA. Furthermore, Pirker was alleged to have deliberately operated the UAS at extremely low altitudes over vehicles, buildings, people, streets, and structures. More specifically, Pirker allegedly operated the UAS at altitudes of approximately 10 feet to approximately 1,500 feet over the University of Virginia in a careless or reckless manner so as to endanger the life or property of another in violation of 14 CFR § 91.13.[59] Additionally, Pirker was alleged to have operated the UAS directly towards an individual standing

---

[56] 49 USC § 46301, http://www.gpo.gov/fdsys/granule/USCODE-2011-title49/USCODE-2011-title49-subtitleVII-partA-subpartiv-chap463-sec46301.

[57] *Administrator v. Raphael Pirker*, National Transportation Safety Board (NTSB) Office of Administrative Law Judges, Docket CP-217, http://www.ntsb.gov/legal/alj/Documents/Pirker-CP-217.pdf.

[58] *Id.*

[59] 14 C.F.R. § 91.13, Federal Aviation Administration (FAA), http://www.ecfr.gov/cgi-bin/text-idx?rgn=div8&node=14:2.0.1.3.10.1.4.7.

on the sidewalk causing the individual to take immediate and evasive maneuvers so as to avoid being struck by the UAS. Among other allegations, Pirker operated the UAS within approximately 100 feet of an active heliport at the University.[60]

Upon receipt of the FAA's Notice of Proposed Assessment, Pirker retained an attorney from the firm of Kramer, Levin, Naftalis & Frankel, LLP and appealed to the NTSB Administrative Law Judge (ALJ).[61] On March 6, 2014, the opinion came down from Judge Patrick G. Geraghty, that Pirker's motion to dismiss should be granted and the Order of Assessment be dismissed. The reason was simply because his UAS was not an "aircraft" but instead a model aircraft subject only to the voluntary compliance with safety guidelines stated in Advisory Circular 91-57. Additionally, Judge Geraghty was of the opinion that the Policy Notices 05-01 and 08-01 were issued and intended for internal guidance for FAA personnel and were not a jurisdictional basis for asserting 14 CFR § 91 enforcement authority on model aircraft operations. Neither did Policy Notice 07-01 establish a jurisdictional basis for asserting 14 CFR § 91.13 enforcement on Pirker's model aircraft operation. More importantly, out of this ruling came this statement: "Specifically, that at the time of Respondent's model aircraft operation, as alleged herein, there was no enforceable FAA rule or FAR Regulation [sic] applicable to model aircraft or for classifying model aircraft as an UAS."[62]

Naturally, the FAA attorneys were displeased at Judge Geraghty's ruling and appealed, this time as procedure in an enforcement action dictates, to the full board of the NTSB. On November 18, 2014, the full board ruled that the term "aircraft" for the purposes of 14 CFR § 91.13 means "any device used for flight in the air" and therefore includes any aircraft, manned or unmanned, large or small, thereby including Pirker's Ritewing Zephyr powered glider. Pirker's case was thus remanded to the ALJ for a full factual hearing to determine whether he had operated his UAS in a careless or reckless manner so as to endanger the life or property of another, contrary to 14 CFR § 91.13.[63]

On January 22, 2015, Pirker announced that rather than go to trial once again, he would settle the charges with the FAA for $1,100, which was a fraction of the original $10,000 penalty. The two-page settlement agreement signed by both Pirker's attorney and the FAA's supervisory attorney stated that Pirker did not admit to any allegation of fact or law contained in the FAA's assessment order and that he settled to avoid the expense of litigation.[64]

---

[60] *Administrator v. Raphael Pirker,* National Transportation Safety Board (NTSB) Office of Administrative Law Judges, Docket CP-217, http://www.ntsb.gov/legal/alj/Documents/Pirker-CP-217.pdf.
[61] *Id.*
[62] Nilsson, S., *In the Aftermath of Pirker: Questions of Policy versus Law on UAS*, http://www.sarahnilsson.org/aviation/uas-uav-drone-news-1/.
[63] *Id.*
[64] *Id.*

16. Memo to Law Enforcement from FAA

According to the FAA, there is evidence of a considerable increase in the unauthorized use of small, inexpensive UAS by individuals and organizations, including companies. The FAA retains the responsibility for enforcing Federal Aviation Regulations, including those applicable to the use of UAS. The agency recognizes though that State and local Law Enforcement Agencies (LEA) are often in the best position to deter, detect, immediately investigate, and, as appropriate, pursue enforcement actions to stop unauthorized or unsafe UAS operations. The information provided in this memo was intended to support the partnership between the FAA and LEAs in addressing these activities. This document is intended to assist LEAs in understanding the legal framework that serves as the basis for FAA legal enforcement action against UAS operators for unauthorized and/or unsafe UAS operations (Section 1) and to provide guidance regarding the role of LEAs in deterring, detecting, and investigating unauthorized and/or unsafe UAS operations (Section 2).[65]

17. VOLUNTARY BEST PRACTICES FOR UAS PRIVACY, TRANSPARENCY, AND ACCOUNTABILITY

On June 10, 2016, the FAA release the final rule under 14 CFR Part 107 and together with this rule, they provided the public with a document to help guide best practices with UAS and privacy issues. This document, entitled Voluntary Best Practices for UAS Privacy, Transparency, and Accountability, is meant to ensure that UAS and the exciting possibilities that come with them live up to their full potential, and operators use this technology in a responsible, ethical, and respectful way, which includes a commitment to transparency, privacy, and accountability.

UAS operators may implement these Best Practices in a variety of ways, depending on their circumstances and technology uses, and evolving privacy expectations. In some cases, these Best Practices are meant to go beyond existing law and they do not—and are not meant to—create a legal standard of care by which the activities of any particular UAS operator should be judged. These Best Practices are also not intended to serve as a template for future statutory or regulatory obligations, in part because doing so would make these standards mandatory (not voluntary) and could therefore raise First Amendment concerns.

These voluntary Best Practices may be summed up as follows: (1) Inform others of your use of UAS; (2) Show care when operating UAS or collecting and storing covered data; (3) Limit the use and sharing of covered data; (4) Secure covered data; and (5) Monitor and comply with evolving federal, state, and local UAS laws. When it comes to Best Practices for newsgatherers and news reporting organizations, it is important

[65] *Memo to Law Enforcement*, FEDERAL AVIATION ADMINISTRATION (FAA), https://www.faa.gov/uas/regulations_policies/media/FAA_UAS-PO_LEA_Guidance.pdf.

to emphasize that newsgathering and news reporting are strongly protected by US law, including the First Amendment to the Constitution. The public relies on an independent press to gather and report the news and ensure an informed public. For this reason these Best Practices do not apply to newsgatherers and news reporting organizations. These entities may use UAS in the same manner as any other comparable technology to capture, store, retain and use data or images in public spaces. They should still operate under the ethics rules and standards of their organization, and according to existing federal and state laws.[66]

18. FAA JO7210.891—Unmanned Aircraft (UA) Operations in NAS

On November 25, 2015, the FAA published this notice, JO7210.891, to replace JO7210.889, which had been issued on October 27, 2015, but then cancelled two days later. On October 27, 2015, JO7210.889 had replaced JO7210.882 issued on July 9, 2015. What follows are excerpts of JO7219.891, which is scheduled to be cancelled on November 24, 2016.[67]

This notice provides information and guidance on air traffic policies and prescribes procedures for the planning, coordination, and services involving the operation of Unmanned Aircraft Systems (UAS) in the NAS. This notice applies to the following Air Traffic Organization (ATO) service units: Air Traffic Services, Mission Support, and System Operations; and all associated air traffic control facilities, the David J. Hurley Air Traffic Control System Command Center (ATCSCC); and the Flight Service's divisions at Federal Aviation Administration (FAA) Washington headquarters.

Unmanned aircraft (UA) activities must be provided air traffic services following the policy, criteria, and procedures in this notice and other air traffic publications.

The introduction and evolution of UAS into the NAS is a dynamic process. Numerous airframes, missions, and flight characteristics create situations not common to manned flight. In the application of Air Traffic Control (ATC) procedures and techniques by ATC personnel, consideration to the type of unmanned aircraft being flown is required. Numerous UA platforms are similar to rotorcraft type aircraft, while others are similar to fixed wing aircraft, but with limited and at times different performance characteristics than manned aircraft. With a vast array of proponents finding new technological opportunities for UAs, the challenge of integrating these aircraft into the NAS is large. As the usage of UA operations expands, the areas of operation will reach all classes of airspace and altitudes. Open and efficient exchange of information between impacted ATC facilities and the UA proponents about the UA, its mission, contingencies and communication methods, are a fundamental requirement for safe operation in the NAS.

---

[66] *Voluntary Best Practices,* Federal Aviation Administration (FAA), http://www.ntia.doc.gov/files/ntia/publications/voluntary_best_practices_for_uas_privacy_transparency_and_accountability_0.pdf

[67] *FAA Notice 7210-891,* Federal Aviation Administration (FAA), http://www.faa.gov/documentLibrary/media/Notice/Notice_UAS_7210.891.pdf.

This notice covers UAS operations in the NAS. For the purposes of this notice, a small UA is defined as weighing less than 55 pounds; a UA weighing 55 pounds or more is considered large.

FAA policy for UAS operations is that no person may operate a UAS, including tethered UAS, outside of active restricted, prohibited or warning areas in the NAS without specific authority.

UAS Operations Types and Authority.

a. Public.
    1. Public aircraft operations are defined in Title 14, Code of Federal Regulations (CFR), Part 1, Definitions & Abbreviations.
    2. For UAS operating as public aircraft, the authority is the Certificate of Waiver or Authorization (COA).

b. Civil.
    1. Any operation that does not meet the statutory criteria for a public aircraft operation is considered a civil aircraft operation and must be conducted in accordance with all FAA regulations applicable to the operation.
    2. For UAS operating as civil aircraft, the authority is a special airworthiness certificate or FAA Modernization and Reform Act of 2012 (FMRA) Section 333 exemption. In addition, a COA is also required.
        a. FMRA Section 333 Exemption—a grant of exemption in accordance with Section 333 and a COA may be used to perform commercial or other operations in low-risk, controlled environments. When the Section 333 exemption is granted, the petitioner will be issued a Blanket COA. If the requested operation does not meet the requirements of a Blanket COA, the proponent must apply for a Standard COA.
        b. Blanket COAs. When the Section 333 exemption is granted, a "Blanket COA" is automatically issued to the proponent; this COA allows small UAS (less than 55 pounds) operations during daytime Visual Flight Rule (VFR) conditions at specific altitudes and outside of certain distances from airports and heliports. The Blanket COA allows the proponent to operate:
            i. 5 nautical miles away from an airport with an operating control tower.
            ii. 3 nautical miles from an uncontrolled airport with an instrument approach procedure.
            iii. 2 nautical miles from all other airports, heliports and seaports.
            iv. At or below 200 feet AGL.
        c. Standard COAs. For any Civil UAS operation that does not fit into the parameters of the Blanket COA, the proponent must complete

the process to obtain a Standard COA. A waiver request to a Blanket COA will not be approved.

c. Model Aircraft (Hobbyist).
   1. A model aircraft flown strictly for hobby or recreational use.
   2. Most small UAS are owned by individuals and cannot be used for commercial purposes. Individuals flying for hobby or recreation should follow safety guidelines outlined in AC 91-57A which include:
      a. Fly at or below 400 feet and remain clear of surrounding obstacles.
      b. Keep the aircraft within visual line of sight at all times.
      c. Remain well clear of and do not interfere with manned aircraft operations.
      d. Don't fly within 5 miles of an airport unless the hobbyist contacts the airport operator and control tower (if operating at a tower controlled airport) before flying.
      e. Don't fly near people or stadiums.
      f. Don't fly an aircraft that weighs 55 lbs. or more.
      g. Don't be careless or reckless with the unmanned aircraft—hobbyists could be fined for endangering people or other aircraft.
      h. Model aircraft operators flying from a permanent location within 5 miles of an airport should establish a mutually agreed upon operating procedure with the airport operator and the airport air traffic control tower (if operating at a tower controlled airport).
      i. Model aircraft must not operate in Temporary Flight Restriction Areas (TFR), Prohibited Areas, Restricted Areas, Special Flight Rules Areas, or the Washington National Capital Region Flight Restricted Zone, without specific authorization.
   3. AC 91-57A advises that hobbyist UAS operators provide notice to the airport operator and the airport air traffic control tower (when an air traffic facility is located at the airport) if the operation will take place within 5 miles of an airport. However, if the airport operator or the air traffic control facility believes the operation could impact safety, the facility may deny the operation and notify the UAS operator of the specific objection.

NOTE—*The FAA recognizes that people and companies other than model aircraft operators might be flying UAS with the misunderstanding they are legally operating under the authority of AC 91–57A. AC 91–57A provides guidance to model aircraft operators and does not apply to use by persons or companies for business purposes.*

d. Emergency COA (ECOA) application requests.
   1. The FAA must ensure procedures are available to accommodate real-time application requests that will directly support emergency and law

enforcement-type operations. An ECOA may be considered when the following conditions apply:

- There is a situation of distress or urgency
- The proponent is operating under a currently approved COA
- The operation does not introduce any new risk in the airspace where the operation is proposed (determined by FAA Headquarters)

2. Emergency UAS COAs will not be considered for demonstration flights, flights to test capabilities or training.

UAS Operations and Guidance for the ATO (Not applicable to model aircraft operations).

a. General UAS Operational Information.

1. The UAS Pilot in Command (PIC) is to give way to all manned aircraft. UAs flying under IFR should be handled the same as manned IFR aircraft, however consideration should be given to the UA performance characteristics.
2. UAS operations should not impede, delay, or divert manned aircraft operations.
3. Lost Link Procedures will vary based upon the type UAS and must be included in the COA. ATC specialists must have this information available in its simplest form, to determine the actions a UA will take in these scenarios. Controllers in Charge (CIC)/Front Line Managers (FLM) should have a method of contacting the appropriate UA PIC.
4. Some UAS receive their communications through a satellite, which may result in latency issues.
5. The UA PIC cannot accept visual separation or visual approach clearances.
6. Operational communications with any UA PIC/proponent must be on a recorded line, when available.
7. In the event of a UAS emergency, procedures outlined in JO 7110.65, Air Traffic Control, Chapter 10, will be followed.
8. Strict UA proponent compliance with the provisions of the COA or using agency/FAA Memorandum of Agreement (MOA) is required.
9. UA PICs cannot accept Special VFR clearances.
10. UA operations conducted during nighttime hours must comply with all applicable CFRs unless a waiver of those CFRs is approved.
11. Flights outside of Class A airspace require that a NOTAM be issued by the proponent.
12. Air Traffic Facility Management at facilities where UAS operations are being conducted, are required to ensure air traffic controllers are familiar with the contents of each COA and any applicable LOAs impacting their area of specialization.

13. Flights below FL180 must have a dedicated observer. These duties may be performed by a ground based observer or chase plane.
    a. Daisy chaining of observers or observers on a moving platform may be authorized on a case by case basis.
    b. When a ground based/chase plane observer is required a pilot may not perform concurrent duties as the pilot and an observer.
    c. UA pilots and observers must be responsible for only one UA at a time, unless authorized in the COA.
14. All categories of Department of Defense (DOD) UAS operations that have a DOD MOA for Class D COA or notification and will be conducted wholly within Class D airspace that has an associated DOD-controlled non-joint-use airfield must follow uniform air traffic control procedures at all locations.
15. Procedures for non-joint-use DOD airfield operations will be specified by the DOD.
16. UAS that have been authorized by the Class G section of a using agency/ FAA MOA must remain within visual line of sight of the pilot or a certified observer in ready contact with the pilot.
17. Operations that cannot be contained wholly within active restricted areas or warning areas must be conducted following procedures outlined in the issued COA.

b. UAS operations in Class A airspace.
    1. UAS must operate on an IFR flight plan and a standard or emergency COA.
    2. UAS must comply with provision of 14 CFR 91.135.
    3. ATC must provide separation and ATC services per 7110.65 requirements with consideration given to UA performance characteristics and potential latency issues.

c. UAS operations within Class B airspace.
    1. Operations in Class B will be considered on a case-by-case basis and will be flown under a standard or emergency COA.
    2. UAS must comply with the provisions of 14 CFR 91.131, unless otherwise authorized by the jurisdictional ATC facility.
    3. ATC must apply Class B services and procedures in accordance with 7110.65, Chapter 7, Section 9.
    4. Alternate methods to ensure separation may be established by an approved Letter of Agreement to supplement the COA.

d. UAS operations within Class C airspace.
    1. UAS must comply with the provisions of 14 CFR 91.130 unless otherwise authorized by the jurisdictional ATC facility.

   2. ATC must apply Class C services and procedures within the designated Class C airspace and the associated outer area in accordance with 7110.65, Chapter 7, Section 8.
   3. Alternate methods to ensure separation may be established by an approved Letter of Agreement to supplement the COA.
   4. UAS must operate on a standard or emergency COA.

e. UAS operations within Class D airspace.
   1. UAS must comply with the provisions of 14 CFR 91.129, unless otherwise authorized by the jurisdictional ATC facility.
   2. UAS pilots may not conduct concurrent or simultaneous operations in the presence of manned aircraft unless approved procedures have been established by Letter of Agreement to supplement the COA.
   3. UAS must operate on a standard or emergency COA.

f. UAS Operations in Class E.

   UAS must comply with provisions of 14 CFR 91.127, unless otherwise authorized by the jurisdictional ATC facility. UAS must operate on a standard, blanket or emergency COA.

g. UAS Operations in Class G.

   UAS must comply with provisions of 14 CFR 91.126, unless otherwise authorized by the jurisdictional ATC facility. UAS must operate on a standard, blanket, emergency COA or a Class G notification in accordance with a using agency/FAA memorandum of agreement.

h. Letters of Agreement (LOA) between ATC Facilities and UA Proponents.
   1. Facility managers have the option of developing LOA's with UA proponents operating under a COA or a notification process under a using agency/FAA memorandum of agreement. It is recommended that AT facilities develop LOAs with UAS proponents that have any of the following:
      a. Recurring operations in the facility's airspace or movement areas.
      b. Operations that are complex.
      c. Operations in any class of airspace other than Class E and G.
      d. Operations that involve multiple A T facilities.
      e. Operations transitioning to and from Special Use Airspace (SUA) airspace.

NOTE—*LOAs are not required for ECOA operations.*

   2. Facility managers must develop an LOA with UA proponents transitioning to and from Class A airspace in their area of jurisdiction and in Class D, or Class E or G Airports with ATCTs where concurrent or simultaneous operations with manned aircraft occur.

3. If a facility determines an LOA would be beneficial, consider these items to be included in the LOA:
    a. Communications must be on a recorded line, if available, and the method of contact specifically described, i.e. phone numbers, frequencies.
    b. All references to a geographic position from any aerodrome originate at the published Airport Reference Point and are described via a fix/radial/distance from that point.
    c. Any specific altitude limitations, geographic boundary limitations, preferred route assignments, and periods of operations. This information must be provided to the ATC facility involved in the LOA via graphical depiction along with any preprogrammed lost link procedures, including routes, holding areas and altitudes. If there are multiple lost link procedures, they should be illustrated to show where they may change as the operation progresses along its route of flight from its origination to its planned destination point and a depiction of each, inclusive of altitudes. Only the geographic portion impacting the facility involved in the LOA must be available to that facility.
    d. Weather requirements for operations.
    e. The COA number(s) and the COA date(s) of expiration for any operation(s).
    f. IFR routings for non-radar environments.
4. When a facility manager decides to develop an LOA, in addition to coordinating with the appropriate OSG; all UAS LOAs, which are developed, must be forwarded to Air Traffic Procedures and the Emerging Technologies Office at 9-AJV-8-HQ-Correspondence@faa.gov for review before signing.[68]

19. Airspace Classification

Since the previous section refers to airspace, what follows below is a short classification of United States airspace. In the United States, there are two categories of airspace: (1) regulatory and (2) nonregulatory. The regulatory airspace consists of Class A, B, C, D, and E airspace, and restricted and prohibited areas. The nonregulatory airspace consists of military operations areas (MOAs), warning areas, alert areas, and controlled firing areas (CFAs). Within these two categories, there are four types of airspace: (1) controlled; (2) uncontrolled; (3) special use; and (4) other airspace. These aforementioned categories and types of airspace are dictated by the complexity or density of aircraft movements;

---

[68] *Id.*

the nature of the operations conducted within the airspace; the level of safety required; and lastly, the national and public interest.[69]

Controlled airspace is a generic term that covers the different classification of airspace (Class A, B, C, D, and E) and defined dimensions within which air traffic control service is provided to Instrument Flight Rules (IFR) flights and Visual Flight Rules (VFR) flights in accordance with the airspace classification.[70]

Special use airspace consists of that airspace wherein activities must be confined because of their nature, or wherein limitations are imposed upon aircraft operations that are not a part of those activities, or both. Prohibited and restricted areas are regulatory special use airspace. Warning areas, MOAs, alert areas, and CFAs are nonregulatory special use airspace.[71]

Other airspace areas include airport advisory/information services, military training routes, temporary flight restrictions (TFRs), parachute jump aircraft operations, published VFR routes, and terminal radar service area (TRSA).[72]

Class A Airspace: Class A Airspace is, generally, that airspace from 18,000 feet mean sea level (MSL) up to and including flight level (FL) 600, which is 60,000 feet MSL, including the airspace overlying the waters within 12 nautical miles (NM) off the coast of the 48 contiguous states and Alaska; and designated international airspace beyond 12 NM off the coast of the 48 contiguous states and Alaska within areas of domestic radio navigational signal or ATC radar coverage, and within which domestic procedures are applied. Unless otherwise authorized, within Class A, all persons must operate their aircraft under IFR.[73]

Class B Airspace: Class B Airspace is, generally, that airspace from the surface to 10,000 feet MSL surrounding the nation's busiest airports in terms of IFR operations or passenger enplanements. The configuration of each Class B airspace area is individually tailored and consists of a surface area and two or more layers (resembling upside-down wedding cakes). The Class B airspace is designed to contain all published instrument procedures once an aircraft enters the airspace. An ATC clearance is required for all aircraft to operate in the area, and all aircraft that are so cleared receive separation services within the airspace.[74]

Class C Airspace: Class C Airspace is, generally, that airspace from the surface to 4,000 feet above the airport elevation (charted in MSL) surrounding those airports that have an operational control tower; are serviced by a radar approach control; and that have a certain number of IFR operations or passenger enplanements. Although the configuration of each Class C airspace area is individually tailored, the airspace usually consists of a

---

[69]United States Department of Transportation, Federal Aviation Regulations Aeronautical Information Manual 617 (Aviation Supplies & Academics, Inc. eds., 2016).

[70]*Id.* at 619.

[71]*Id.* at 628.

[72]*Id.* at 630.

[73]*Id.* at 620.

[74]*Id.* at 620.

5 NM radius core surface area that extends from the surface up to 4,000 feet above the airport elevation, and a 10 NM radius shelf area that extends no lower than 1,200 feet up to 4,000 feet above the airport elevation.[75]

Class D Airspace: Class D Airspace is, generally, that airspace from the surface to 2,500 feet above the airport elevation (chartered in MSL) surrounding those airports that have an operational control tower. The configuration of each Class D airspace area is individually tailored and when instrument procedures are published, the airspace will normally be designed to contain the procedures.[76]

Class E Airspace: Class E Airspace is, generally, that airspace that is not Class A, B, C, or D, and that is controlled airspace.[77]

Class G Airspace: Class G Airspace is uncontrolled. It is that portion of airspace that has not been designated as Class A, B, C, D, or E airspace.[78]

Prohibited Areas: Prohibited areas contain airspace of defined dimensions identified by an area on the surface of the earth within which the flight of aircraft is prohibited. Such areas are established for security or other reasons associated with the national welfare.[79]

Restricted Areas: Restricted areas contain airspace identified by an area on the surface of the earth within which the flight of aircraft, while not wholly prohibited, is subject to restrictions. Activities within these areas must be confined because of their nature or limitations imposed upon aircraft operations that are not a part of those activities or both. Restricted areas denote the existence of unusual, often invisible, hazards to aircraft such as artillery firing, aerial gunnery, or guided missiles. Penetration of restricted areas without authorization from the using or controlling agency may be extremely hazardous to the aircraft and its occupants.[80]

Warning Areas: A warning area is airspace of defined dimensions, extending from 3 NM outward from the coast of the US, which contains activity that may be hazardous to nonparticipating aircraft. The purpose of such warning areas is to warn nonparticipating pilots of the potential danger. A warning area may be located over domestic or international waters or both.[81]

Military Operations Areas (MOAs): Military Operations Areas (MOAs) consist of airspace of defined vertical and lateral limits established for the purpose of separating certain military training activities from IFR traffic. Whenever a MOA is being used, nonparticipating IFR traffic may be cleared through a MOA if IFR separation can be provided by ATC. Otherwise, ATC will reroute or restrict nonparticipating IFR traffic. Examples of

[75] *Id.* at 621.
[76] *Id.* at 625.
[77] *Id.* at 625.
[78] *Id.* at 627.
[79] *Id.* at 628.
[80] *Id.* at 628.
[81] *Id.* at 628.

activities conducted in MOAs include, but are not limited to, air combat tactics, air intercepts, aerobatics, formation training, and low-altitude tactics.[82]

Alert Areas: Alert areas inform nonparticipating pilots of areas that may contain a high volume of pilot training or an unusual type of aerial activity. Pilots should be particularly alert when flying in these areas. All activity within an alert area must be conducted in accordance with the Code of Federal Regulations (CFRs), without waiver, and pilots of participating aircraft as well as pilots transiting the area must be equally responsible for collision avoidance.[83]

Controlled Firing Areas (CFAs): Controlled Firing Areas (CFAs) contain activities that, if not conducted in a controlled environment, could be hazardous to nonparticipating aircraft. The distinguishing feature of the CFA, as compared to other special use airspace, is that its activities are suspended immediately when spotter aircraft, radar, or ground lookout positions indicate an aircraft might be approaching the area.[84]

National Security Areas (NSAs): National Security Areas (NSAs) consist of airspace of defined vertical and lateral dimensions established at locations where there is a requirement for increased security and safety of ground facilities. Pilots are requested to voluntarily avoid flying through the depicted NSA. When it is necessary to provide a greater level of security and safety, flight in NSAs may be temporarily prohibited by regulation.[85]

Airport Advisory/Information Services: There are three advisory type services available at selected airports: (1) Local Airport Advisory (LAA) service; (2) Remote Airport Advisory (RAA); and (3) Remote Airport Information Service (RAIS). LAA service is operated within 10 statute miles (SM) of an airport where a control tower is not operating but where a Flight Service Station (FSS) is located on the airport. At such locations, the FSS provides a complete local airport advisory service to arriving and departing aircraft. RAA service is operated within 10 SM of specified high activity general aviation (GA) airports where a control tower is not operating. RAIS is provided in support of short-term special events like small to medium fly-ins.[86]

Military Training Routes (MTRs): The Military Training Route (MTR) program is a joint venture by the FAA and the Department of Defense (DoD). MTRs are mutually developed for use by the military for the purpose of conducting low-altitude, high-speed training.[87]

Temporary Flight Restrictions (TFRs): The FAA may impose Temporary Flight Restrictions (TFRs) for six different purposes: (1) To protect persons and property in the air or on the surface from an existing or imminent hazard associated with an incident on the surface when the presence of low flying aircraft would magnify, alter, spread, or compound that

---

[82]*Id.* at 628.
[83]*Id.* at 629.
[84]*Id.* at 629.
[85]*Id.* at 629.
[86]*Id.* at 630.
[87]*Id.* at 630.

hazard. (2) To provide a safe environment for the operation of disaster relief aircraft. (3) To prevent an unsafe congestion of sightseeing aircraft above an incident or event, which may generate a high degree of public interest. (4) To protect declared national disasters for humanitarian reasons in the State of Hawaii. (5) To protect the President, Vice President, or other public figures. (6) To provide a safe environment for space agency operations.[88]

Parachute Jump Aircraft Operations: The name is exactly what it implies. Procedures are contained in 14 CFR Part 105.[89]

Published VFR Routes: Published VFR routes for transitioning around, under and through complex airspace such as Class B airspace were developed through a number of FAA and industry initiatives. All of the following terms, VFR Flyway, VFR Corridor, and Class B Airspace VFR Transition Route have been used when referring to the same or different types of routes or airspace.[90]

Terminal Radar Service Area (TRSA): Terminal Radar Service Areas (TRSAs) were originally established as part of the Terminal Radar Program at selected airports. TRSAs were never controlled airspace from a regulatory standpoint because the establishment of TRSAs was never subject to the rulemaking process. The primary airport(s) within the TRSA become(s) Class D airspace. The remaining portion of the TRSA overlies other controlled airspace, which is normally Class E airspace beginning at 700 feet or 1,200 feet and established to transition to/from the en route/terminal environment.[91]

20. Centers of Excellence (COE)

The goal of an FAA Center of Excellence (COE) for UAS is to focus on research, education, and training in areas critical to safe and successful integration of UAS into the nation's airspace. There are several of these COEs selected from among the nation's leading UAS and aviation universities that have a proven commitment to UAS research and development and the necessary resources to provide the matching contribution to the government's investment. They are Drexel University; Embry Riddle Aeronautical University; Kansas State University; Kansas University; Mississippi State University; Montana State University; New Mexico State University; North Carolina State University; Oregon State University; University of Alabama, Huntsville; University of Alaska, Fairbanks; University of North Dakota; and Wichita State University.[92]

The COE research areas are expected to evolve over time, but initially will include: detect and avoid technology; low-altitude operations safety; control and communications; spectrum management; human factors; compatibility with air traffic control operations;

---

[88] *Id.* at 631.
[89] *Id.* at 633.
[90] *Id.* at 633.
[91] *Id.* at 637.
[92] *Centers of Excellence (COE)*, Federal Aviation Administration (FAA), https://www.faa.gov/uas/legislative_programs/coe/.

and training and certification of UAS pilots and other crewmembers, in addition to other areas. Congress appropriated $5 million for the five-year agreement with the COE, which will be matched one-for-one by the team members.[93]

The Alliance for System Safety of UAS through Research Excellence (ASSURE) is comprised of twenty-two of the world's leading research institutions and more than a hundred leading industry/government partners. This alliance features expertise across a broad spectrum of research areas including: Air traffic control interoperability, UAS airport ground operations, control and communications, detect and avoid, human factors, UAS noise reduction, UAS wake signatures, unmanned aircraft pilot training and certification, low-altitude operations safety, spectrum management and UAS traffic management. ASSURE possess the expertise, experience and influence that the FAA Center of Excellence for UAS demands.[94]

21. Air Traffic Management (NASA)

As the numbers of unmanned aircraft continue to grow, there is envisioned a need to find a better air traffic control solution. National Aeronautics and Space Administration (NASA) is researching an Air Traffic Management system that would enable civilian low-altitude airspace and UAS operations to coexist. NASA recognizes the many beneficial civilian applications of UAS, from goods delivery and infrastructure surveillance, to search and rescue, and agricultural monitoring. As UAS operations require interactions with a mix of general aviation aircraft, helicopters, and gliders, there is a strong need to safely accommodate all of these vehicles at lower altitudes. Currently, there is no established infrastructure to enable and safely manage the widespread use of low-altitude airspace and UAS operations, regardless of the type of UAS. A UAS traffic management (UTM) system for low-altitude airspace is needed, much like today's surface vehicles that operate within a system consisting of roads, lanes, stop signs, rules, and lights, regardless of whether the vehicle is automated or driven by a human.[95]

NASA's solution reads as follows. While incorporating lessons learned from the well-established ATM system, which grew from a mid-air collision over the Grand Canyon in the early days of commercial aviation, the UTM system would enable safe and efficient low-altitude airspace operations by providing services such as airspace design, corridors, dynamic geofencing, severe weather and wind avoidance, congestion management, terrain avoidance, route planning and re-routing, separation management, sequencing and spacing, and contingency management. UTM is essential to enable the accelerated development and use of civilian UAS applications.[96]

---

[93] *Id.*
[94] *About Use,* ASSURE, http://www.assureuas.org/about.php.
[95] *UTM,* National Aeronautics and Space Administration (NASA), http://utm.arc.nasa.gov/index.shtml.
[96] *Id.*

One of the attributes of the UTM system is it will not require human operators to monitor every vehicle continuously. The system will provide to human managers the data to make strategic decisions related to initiation, continuation, and termination of airspace operations. This approach would ensure that only authenticated UAS operate in the airspace. In its most mature form, the UTM system will be developed using autonomicity characteristics, which will include self-configuration, self-optimization and self-protection. The self-configuration aspect will determine whether the operations should continue given the current and/or predicted wind/weather conditions.[97]

Two types of UTM systems are envisioned. The first type is a Portable UTM System, which would move between geographical areas and support operations such as precision agriculture and disaster relief. The second type of system is a Persistent UTM System, which would support low-altitude operations and provide continuous coverage for a geographical area. The UTM will require persistent communication, navigation, and surveillance (CNS) coverage to track, ensure, and monitor conformance.[98]

NASA's near-term goal is the development and demonstration of the UTM to safely enable low-altitude airspace and UAS operations within five years. For the longer term (ten to fifteen years in the future), the goal is to safely enable the anticipated dramatic increase in density of all low-altitude airspace operations. Working alongside with many committed government, industry, and academic partners, NASA will lead the research, development, testing, and implementation of the UTM, exploring functional designs, concepts and technology development, and testing of proposed UTM systems utilizing a series of builds, each increasing in capability. The first build, UTM1, will create, analyze and manage trajectories and constraints that enable operations by an interactive system. The focus will be on geo-fencing, altitude "rules of the road," and scheduling of vehicle trajectories. UTM2 will enable increased density and contingency management. Focus areas will include all of UTM1, dynamic adjustments to availability of airspace and contingency management. UTM3 will manage separation by vehicle and/or ground-based capabilities under higher densities. The focus areas will be all of UTM2, active monitoring of trajectory conformance, and a UTM web interface. The final build will be UTM4, which will manage large-scale contingencies. The focus will include all of UTM3 and management of large-scale contingencies such as an "all-land" scenario. During the UTM's development, NASA will collaborate closely with the FAA. After thorough testing, technology transfer of a UTM prototype is expected by 2019. The ultimate goal of this research is to assist all low-altitude operations (e.g., manned and unmanned) in an autonomous manner to accommodate future vehicles and density.[99]

---

[97]*Id.*
[98]*Id.*
[99]*Id.*

22. National Park Service Memo

On June 19, 2014, National Park Service Director Jonathon B. Jarvis signed Policy Memorandum 14–05, *Unmanned Aircraft—Interim Policy.* Its purpose was "to ensure that the use of unmanned aircraft is addressed in a consistent manner by the NPS before a significant level of such use occurs within the National Park System." Each superintendent was directed "to use the authority under 36 CFR 1.5 to close units of the National Park System to launching, landing, or operating unmanned aircraft. . . ." This policy is still in place and the public may not use unmanned aircraft in the national parks.[100]

23. Huerta v. Haughwout

In November 2015, in the privacy of their backyard, a student named Austin Haughwout and his father, Bret Haughwout came to the attention of the FAA because they mounted some type of handgun on their UAS, as well as a flamethrower firing at a turkey roasting on a spit.[101] While the events took place in Connecticut, the YouTube videos were viewed over four million times around the world. The challenge in court, set for July 6, 2016, hinged on whether the FAA has the authority to regulate consumer UAS. The pro bono attorney designated by the judge to represent the Haughwouts is Mario Cerame of the Randazza Legal Group. Mr. Cerame said that at the heart of his challenge to FAA authority is the FAA's definition of "aircraft" which he contends "is crazy" and would cover any contrivance that flies, including "paper airplanes, bullets and flags." At a hearing in March, Judge Jeffrey Meyer ordered the parties to further support their arguments regarding FAA authority over small drones.[102]

In the opinion filed on July 18, 2016, Judge Meyer granted the petition for enforcement of the administrative subpoenas, requiring respondents to comply within thirty days or on such other date as the parties may mutually agree. The Clerk of the Court then closed the case.[103]

---

[100] *Policy Memorandum 14–05,* Unmanned Aircraft—Interim Policy, National Park Service, https://www.nps.gov/fire/aviation/safety/unmanned-aerial-systems.cfm.
[101] *Huerta v. Haughwout,* https://assets.documentcloud.org/documents/2854855/15-Main.pdf.
[102] *Id.*
[103] *Huerta v. Haughwout et al.,* https://www.gpo.gov/fdsys/granule/USCOURTS-ctd-3_16-cv-00358/USCOURTS-ctd-3_16-cv-00358-0.

CHAPTER 3
Commercial UAS Rules

1. Commercial or Nongovernmental Regulations

Earlier in the text, the reader was made aware of the FAA Modernization and Reform Act of 2012. This act led to the distinction between commercial use, public use, and hobby or recreational use, of unmanned aircraft. This section is devoted entirely to the procedure for complying with the rules pertaining to commercial use. Any operation that does not meet the statutory criteria for a public aircraft operation (see Section IV—Government UAS Rules later on in the text for more information) is considered a civil aircraft operation and must be conducted in accordance with all FAA regulations applicable to the operation.

Prior to August 29, 2016, there were two methods of gaining FAA authorization to fly civil (nongovernmental) otherwise known as commercial UAS: (1) Section 333 Exemption and (2) Special Airworthiness Certificate (SAC). A Section 333 Exemption is a grant of exemption in accordance with Section 333 of FMRA of 2012 and a civil Certificate of Waiver or Authorization (COA). This process, which will be explained in more detail in the subsection below, may be used to perform commercial operations in low-risk, controlled environments.[1] With the new Part 107 rule (discussed in the next section) that caters to commercial UAS weighing less than fifty-five pounds a Section 333 Exemption is only required for those UAS weighing more than fifty-five pounds and/or operating outside the parameter of the Part 107 rule.

Special Airworthiness Certificate applicants must be able to describe how their system is designed, constructed, and manufactured, including engineering processes, software development and control, configuration management, and quality assurance procedures used, along with how and where they intend to fly. SAC in the experimental category

[1] *Civil UAS Operations,* FAA, https://www.faa.gov/uas/civil_operations/.

may be used for civil aircraft to perform research and development, crew training, and market surveys; however, carrying persons or property for compensation or hire is prohibited. FAA inspectors use FAA Order 8130.34 (mentioned earlier in Section II—The Federal Legal Environment, Part 11) to issue experimental airworthiness certificates and special flight permits to UAS (for more information contact the FAA Office of Airworthiness Certification Service, AIR-113). If the FAA determines the project does not present an unreasonable safety risk, the local FAA Manufacturing Inspection District Office will issue a SAC in the experimental category with operating limitations applicable to the particular UAS. For a UAS type and airworthiness certificate in the restricted category (14 CFR § 21.25(a)(2) and § 21.185) for a special purpose or a type certificate for production of the UAS under 14 CFR § 21.25(a)(1) or § 21.17 contact the FAA's Los Angeles Aircraft Certification Office.[2]

2. Section 333 Exemptions

Typically, by law, any aircraft operation in the NAS requires a certificated and registered aircraft, a licensed pilot, and operational approval. However, when it comes to unmanned aircraft, Section 333 of the FMRA of 2012 grants the Secretary of Transportation the authority to determine whether an airworthiness certificate is required for a UAS to operate safely in the NAS. Prior to August 29, 2016 this authority was being leveraged to grant case-by-case authorization for certain unmanned aircraft to perform commercial operations prior to the finalization of the small UAS rule that is now the primary method for authorizing small UAS operations (those weighing less than 55 pounds). The Section 333 Exemption process provides operators who wish to pursue safe and legal entry into the NAS a competitive advantage in the UAS marketplace, thus discouraging illegal operations and improving safety. It is anticipated that this activity will result in significant economic benefits, and the FAA Administrator has identified this as a high priority project to address demand for civil operation of UAS for commercial purposes.[3]

The following checklist contains instructions for requesting FAA authorization to operate a UAS for civil (nongovernmental) purposes other than for recreation or hobby.

1. Review the FAA's Guidelines for Submitting a Petition for Exemption. This website details the general exemption process for filing any petition for exemption, and it provides answers to frequently asked questions. A petition is a formal letter requesting an action from the FAA that must be filed in accordance with 14 CFR Part 11 (more specifically 14 CFR 11.81). An exemption is a formal document

[2] *Id.*

[3] Section 333 Exemptions, FAA, https://www.faa.gov/uas/legislative_programs/section_333/.

issued in response to a petition that relieves or denies relief for the petitioner from the requirements of a particular section of 14 CFR.

2. Review the Section 333 Guidance document (available from the FAA's UAS Integration Office or online) that describes what information petitioners should submit specifically to request FAA authorization to operate a UAS in the NAS.
3. Prepare your petition for filing. Before filing your petition, review the FAA's Section 333 Checklist and Guidance document (also available online) to ensure your petition contains all the necessary information.
4. When you have verified that all the necessary information is included, file your petition for exemption on the public docket online. The FAA has a document detailing submission instructions available on their website. If granted, the exemption will typically be good for two years.[4]

After receiving a grant of exemption, petitioners who want to fly outside these blanket parameters will be eligible to apply for a separate COA specific to the airspace required for their operation. Applications must be submitted through the UAS Civil COA Portal online. The COA application must not be submitted to the public docket. The FAA will only accept and process complete COA applications, which must include: (1) an exemption number corresponding to the Federal Register Docket ID for the petition and exemption; and (2) a registration number (N-Number) since all aircraft must be registered with the FAA to be issued a COA.[5] Registration of the commercial small UAS may be accomplished online, starting on April 21, 2016.[6]

The COA application process is separate from the petition for exemption process (hence it is the topic of another subsection detailed below). The COA process makes applicable FAA Air Traffic control facilities aware of proposed UAS operations, and provides the FAA the ability to consider airspace issues unique to UAS operations. Both the COA application and the petition for exemption, however, should be submitted under the same name/company name.[7]

3. 14 CFR Part 107

In Section II—The Federal Legal Environment, Part 2, the FAA Modernization and Reform Act of 2012 was mentioned. This act empowered the FAA to create a rule, 14 CFR Part 107 to govern the commercial use of UAS in the NAS. Mention was also made, in Section II—The Federal Legal Environment, Part 5 of the NPRM that is in the process of creating said Part 107.

---

[4]Section 333 Exemptions, FAA, https://www.faa.gov/uas/legislative_programs/section_333/how_to_file_a_petition/.
[5]*Id.*
[6]*FAA Urges Non-hobby UAS Registration via New System,* FAA, http://www.faa.gov/news/updates/?newsId=85487.
[7]Section 333 Exemptions, FAA, https://www.faa.gov/uas/legislative_programs/section_333/how_to_file_a_petition/.

As anticipated, the FAA released the final rule of 14 CFR Part 107 on June 21, 2016, due to go into effect on August 29, 2016.[8] Again, the FAA reiterated that Part 107 would not apply to model aircraft. Model aircraft operators should continue to satisfy all the criteria specified in Section 336 of Public Law 112-95 (which would now be codified in Part 101), including the stipulation they be operated only for hobby or recreational purposes. In their press release, the FAA also made mention of Privacy Best Practices, which is an effort to build on National Telecommunications and Information Administration guidelines published in May 2016 as the result of a year-long outreach initiative with privacy advocates and industry.[9]

What follows below is a summary of 14 CFR 107 (the actual rule is 624 pages long!).[10] As previously mentioned in the NPRM, there are four categories:

Operational Limitations

- Unmanned aircraft must weigh less than 55 lbs. (25 kg).
- Visual line-of-sight (VLOS) only; the unmanned aircraft must remain within VLOS of the remote pilot in command and the person manipulating the flight controls of the small UAS. Alternatively, the unmanned aircraft must remain within VLOS of the visual observer.
- At all times the small unmanned aircraft must remain close enough to the remote pilot in command and the person manipulating the flight controls of the small UAS for those people to be capable of seeing the aircraft with vision unaided by any device other than corrective lenses.
- Small unmanned aircraft may not operate over any persons not directly participating in the operation, not under a covered structure, and not inside a covered stationary vehicle.
- Daylight-only operations, or civil twilight (30 minutes before official sunrise to 30 minutes after official sunset, local time) with appropriate anticollision lighting.
- Must yield right of way to other aircraft.
- May use visual observer (VO) but not required.
- First-person view camera cannot satisfy "see-and-avoid" requirement but can be used as long as requirement is satisfied in other ways.
- Maximum groundspeed of 100 mph (87 knots).
- Maximum altitude of 400 feet above ground level (AGL) or, if higher than 400 feet AGL, remain within 400 feet of a structure.
- Minimum weather visibility of 3 miles from control station.

---

[8]14 CFR Part 107, FAA Press Release, http://www.faa.gov/uas/media/RIN_2120-AJ60_Clean_Signed.pdf.
[9]Voluntary Best Practices for UAS Privacy, Transparency, and Accountability, FAA, https://www.ntia.doc.gov/files/ntia/publications/voluntary_best_practices_for_uas_privacy_transparency_and_accountability_0.pdf.
[10]Summary of Small Unmanned Aircraft Rule (Part 107), http://www.faa.gov/uas/media/Part_107_Summary.pdf.

- Operations in Class B, C, D, and E airspace are allowed with the required ATC permission.
- Operations in Class G airspace are allowed without ATC permission.
- No person may act as a remote pilot in command or VO for more than one unmanned aircraft operation at one time.
- No operations from a moving aircraft.
- No operations from a moving vehicle unless the operation is over a sparsely populated area.
- No careless or reckless operations.
- No carriage of hazardous materials.
- Requires preflight inspection by the remote pilot in command.
- A person may not operate a small unmanned aircraft if he or she knows or has reason to know of any physical or mental condition that would interfere with the safe operation of a small UAS.
- Foreign-registered small unmanned aircraft are allowed to operate under part 107 if they satisfy the requirements of part 375.
- External load operations are allowed if the object being carried by the unmanned aircraft is securely attached and does not adversely affect the flight characteristics or controllability of the aircraft.
- Transportation of property for compensation or hire allowed provided that:
  - ❑ The aircraft, including its attached systems, payload and cargo weigh less than 55 pounds total;
  - ❑ The flight is conducted within visual line of sight and not from a moving vehicle or aircraft; and
  - ❑ The flight occurs wholly within the bounds of a State and does not involve transport between (1) Hawaii and another place in Hawaii through airspace outside Hawaii; (2) the District of Columbia and another place in the District of Columbia; or (3) a territory or possession of the United States and another place in the same territory or possession.
- Most of the restrictions discussed above are waivable if the applicant demonstrates that his or her operation can safely be conducted under the terms of a certificate of waiver.

Remote Pilot in Command Certification and Responsibilities

- Establishes a remote pilot in command position.
- A person operating a small UAS must either hold a remote pilot airman certificate with a small UAS rating or be under the direct supervision of a person who does hold a remote pilot certificate (remote pilot in command).
- To qualify for a remote pilot certificate, a person must:

- ❑ Demonstrate aeronautical knowledge by either: (1) Passing an initial aeronautical knowledge test at an FAA-approved knowledge testing center; or (2) Hold a part 61 pilot certificate other than student pilot, complete a flight review within the previous 24 months, and complete a small UAS online training course provided by the FAA.
- ❑ Be vetted by the Transportation Security Administration.
- ❑ Be at least 16 years old.

– Part 61 pilot certificate holders may obtain a temporary remote pilot certificate immediately upon submission of their application for a permanent certificate. Other applicants will obtain a temporary remote pilot certificate upon successful completion of TSA security vetting. The FAA anticipates that it will be able to issue a temporary remote pilot certificate within 10 business days after receiving a completed remote pilot certificate application.

– Until international standards are developed, foreign-certificated UAS pilots will be required to obtain an FAA-issued remote pilot certificate with a small UAS rating.

– A remote pilot in command must:
    - ❑ Make available to the FAA, upon request, the small UAS for inspection or testing, and any associated documents/records required to be kept under the rule.
    - ❑ Report to the FAA within 10 days of any operation that results in at least serious injury, loss of consciousness, or property damage of at least $500.
    - ❑ Conduct a preflight inspection, to include specific aircraft and control station systems checks, to ensure the small UAS is in a condition for safe operation.
    - ❑ Ensure that the small unmanned aircraft complies with the existing registration requirements specified in § 91.203(a)(2).

– A remote pilot in command may deviate from the requirements of this rule in response to an in-flight emergency.

Aircraft Requirements

– FAA airworthiness certification is not required. However, the remote pilot in command must conduct a preflight check of the small UAS to ensure that it is in a condition for safe operation.

Model Aircraft

– Part 107 does not apply to model aircraft that satisfy all of the criteria specified in section 336 of Public Law 112-95.

– The rule codifies the FAA's enforcement authority in part 101 by prohibiting model aircraft operators from endangering the safety of the NAS.[11]

[11]Summary of Small Unmanned Aircraft Rule (Part 107), http://www.faa.gov/uas/media/Part_107_Summary.pdf.

In addition to this rule, on the same day the FAA also published their AC 107-2. This fifty-two-page advisory circular provides guidance to remote pilots to assist in meeting the requirements of 14 CFR Part 107.[12]

4. Certificate of Waiver or Authorization (COA)

In the section above dealing with Section 333 Exemptions, mention was made of a Certificate of Waiver or Authorization (COA). In Section II—The Federal Legal Environment—Part 18, the reader was introduced to FAA JO7210-891—UA Operations in the NAS. We shall return to this document to start explaining the COA processing, starting with the following excerpt:

The section on processing prescribes the policies, guidance, and procedures about COA applications for UAS operations.

Application Process: The Air Traffic organization (ATO) issues a COA to a UAS proponent for a specific UA's activity. After a complete application is submitted, the FAA conducts a comprehensive operational and technical review. If necessary, provisions or limitations may be imposed as part of the approval to ensure the UA operates safely.

Application Submission: Electronic applications should be submitted at least 60 business days before the proposed start of UAS operations requiring a COA. The proponent must submit an application for a COA using the online application system at:

Civil Operations: https://oeaaa.faa.gov/oeaaa/external/uas/portal.jsp

Public Operations: https://ioeaaa.faa.gov/oeaaa/

Application Information: The FAA must obtain enough information to assess the proposed operations following current standards and procedures. Because of the dynamic changes in the development of UA technologies, the applicant is responsible for adequately describing the proposed operations so the FAA can conduct an appropriate safety assessment. For this purpose, the following information may be required in a COA application.

1. Organizational and operational points of contact.
2. Operational description (for example, method of navigation, see-and-avoid).
3. Systems description (for example, airframe, control station, communications).
4. Airframe performance characteristics.
5. Airworthiness.
6. Contingency procedures (for example, lost command/control link, lost communications, and emergency).
7. Avionics equipment.

[12]AC 107-2, http://www.faa.gov/documentlibrary/media/advisory_circular/ac_107-2.pdf

8. Lighting.
9. Frequency spectrum analysis.
10. Method of air traffic control (ATC) communications.
11. Surveillance capability (for example, electronic and visual).
12. System monitoring/recording capability.
13. Flight crew qualifications.
14. Flight operations description (flight plan).
15. Special circumstances.
16. Reports of past incidents or accidents (for those applicants who have previously held a COA).

COA Application Processing Procedures:

1. After their application is submitted, FAA Headquarters, FAA Service Center and Air Traffic Facilities (if applicable) will review the proponent's request.
2. COAs must have a termination date not more than 2 years from the effective date unless renewed or extended. The COA expires on the stated termination date unless surrendered sooner by the proponent or revoked by the issuing agency.
3. UAS may be equipped with standard aircraft anti-collision or navigation lights following criteria in 14 CFR, section 23.1401. If installed, these lights must be operating during all phases of flight to enhance flight safety.
4. UAS required to be equipped with an altitude encoding transponder must meet the specifications of 14 CFR, section 91.215. Although paragraph 91.215(b) (2) requires a transponder within 30 NM of an airport listed in the cited appendix (the Mode C Veil), an exception is made for "... any aircraft which was not originally certificated with an engine-driven electrical system or which has not subsequently been certified with such a system installed ... " Many small UAS are battery operated and therefore do not have an engine-driven electrical system and would fall under this exception.
5. The proponent and/or its representatives must be responsible at all times for collision avoidance maneuvers with nonparticipating aviation activities and the safety of persons or property on the surface.
6. The proponent and/or its representatives are responsible for strict compliance with the Incident/Accident and Normal Reporting Provisions contained in the Special Provisions section of each COA. Further guidance can be found in FAA Order JO 7210.3, Facility Management, Chapter 18, Waivers, Authorizations,

and Exemptions. The Certificate of Waiver or Authorization, FAA Form 7711-1, provides additional clarity regarding the strict observance of these terms and conditions.

[NOTES—1. *Approvals for UAS operations require the proponent to provide the UAS with a method that provides an equivalent level of safety comparable to see-and-avoid requirements for manned aircraft. Methods to consider include, but are not limited to, radar observations, forward- or side-looking cameras, electronic detection systems, visual observation from one or more ground sites, monitoring by patrol or chase aircraft, or a combination thereof.* 2. *Risk mitigations that would depend on the establishment of new types and categories of airspace are not considered acceptable. The NAS is established and configured through a rigorous regulatory process. Risk mitigations that result in the prohibition of the public's right to transit airspace will not be considered.*]

COA application coordination between Service Centers and ATC Facilities:

1. Each class of airspace has its own regulatory requirements for entry as established by the CFRs. The COA process provides for an alternate means to comply with certain CFR requirements. During the COA review process, the appropriate Service Center Operations Support Group representative must coordinate with any impacted air traffic facility.
2. Facilities may be contacted for review of COA's even if there is no impact on ATC operations. Facilities must evaluate the COA for impact and if none is found, and no interaction with the UA proponent is necessary, indicate that no contact is necessary for the operations.
3. Air Traffic Facility Managers must ensure any operational requirements, which are necessary for the safe operation of the UA in the facility's airspace, are specifically written into the COA and/or LOA if required. Items to consider when reviewing the proposed COA are: (a) Communication process between the facility and a proponent; (b) Times of operation, notification of beginning and ending of operation; (c) Clearly defined route/altitude information, understandable by the PIC and ATC; (d) Clearly defined route/altitude information and/or graphic illustrations covering lost link procedures /maneuvers for each flight of complex UAS for use by ATC personnel at impacted sectors or positions; (e) Clearly defined notification procedures in the event of an anomaly, if it impacts the NAS; (f) Geographic references for proposed terminal area UA operations should be based off the Airport Reference Point (ARP) using a fix/radial/distance format, additional common visual references can be used only to supplement the location of operations; (g) Some UAS may have multiple PICs for a given flight. Actions to transfer the PIC responsibilities for the UAS should be transparent to ATC.

A method of contacting the appropriate PIC is critical; (h) For IFR operations a clear and concise depiction or description of IFR routes and contingency actions, which may be taken or be executed, must be available to ATC; (i) UAS operations involving towered airports may not be visible from the tower. Consideration of traffic patterns, types of traffic, activity near common reporting points, model/hobbyist club activity, communication methods and weather are critical when reviewing proposed UA operations; and (j) For individual classes of airspace consider the following:

i. Review of UAS COAs in Class A: Procedures by which the UA will depart and transition into Class A airspace, usually via a TFR, MOA, Restricted Area, or Warning Area.
ii. Review of UAS COAs in Class B or C: Unless specified in the COA, the UAS must comply with all Class B or C equipage and communications requirements.
iii. Review of UAS COAs in Class E: Some UAS will enter the Class E airspace above Class A airspace and proceed to their destination. COAs will also contain the procedures and routes on how the UA will depart a terminal environment and transition to Class A airspace or enter SUA.[13]

As of March 23, 2015, the FAA automatically granted a 'blanket' COA for flights at or below 200 feet to any UAS operator with a Section 333 Exemption, provided the aircraft weighs less than 55 pounds, operations are conducted during daytime Visual Flight Rules (VFR) conditions within visual-line-of-sight (VLOS) of the pilot, and certain distances away from airports or heliports (more on this topic in a later section).[14]

And, as of March 29, 2016, after a comprehensive risk analysis, the FAA doubled this aforementioned 'blanket' altitude for Section 333 exemption holders and government aircraft operators to 400 feet. This would allow small UAS to fly up to 400 feet anywhere in the country except restricted airspace and other areas, such as major cities, where the agency prohibits UAS operations.[15]

Under this blanket COA, the FAA would permit flights at or below 400 feet for UAS operators with a Section 333 exemption for aircraft weighing less than 55 pounds and for government UAS operations. Operators must fly under daytime VFR, keep the UAS within VLOS of the pilot and stay certain distances away from airports or heliports as follows: 5 nautical miles (NM) from an airport having an operational control tower; or 3 NM from an airport with a published instrument flight procedure, but not an operational tower; or

[13]JO7210.891, FAA, http://www.faa.gov/documentlibrary/media/notice/notice_uas_7210.891.pdf.
[14]Section 333 Exemptions, FAA, https://www.faa.gov/uas/legislative_programs/section_333/how_to_file_a_petition/.
[15]FAA Doubles 'Blanket' Altitude for Many UAS Flights, http://www.faa.gov/news/updates/?newsId=85264.

2 NM from an airport without a published instrument flight procedure or an operational tower; or 2 NM from a heliport with a published instrument flight procedure.[16]

5. NOTAM for TFR

A Notice to Airmen (NOTAM) is a notice containing information (not known sufficiently in advance to publicize by other means) concerning the establishment, condition, or change in any component (facility, service, or procedure of, or hazards in the National Airspace System) the timely knowledge of which is essential to personnel concerned with flight operations.[17]

A Temporary Flight Restriction (TFR) is a type of NOTAM that defines an area restricted to air travel due to hazardous condition, a special event, or a general warning for the entire FAA airspace.[18]

This NOTAM 4/3621 for TFR was issued in 2014, and the entire text is as follows:

FDC 4/3621 FDC PART 1 OF 3 SPECIAL SECURITY NOTICE SPORTING EVENTS. THIS NOTAM REPLACES FDC NOTAM 9/5151 TO REFLECT A TRANSPORTATION SECURITY ADMINISTRATION (TSA) WEBSITE UPDATE AND ADDITIONAL INFORMATION CONCERNING AIRSPACE WAIVERS. FLIGHT RESTRICTIONS IN THIS NOTAM COMPLY WITH STATUTORY MANDATES DETAILED IN SECTION 352 OF PUBLIC LAW 108-7 AS AMENDED BY SECTION 521 OF PUBLIC LAW 108-199. PURSUANT TO 49 USC 40103(B), THE FEDERAL AVIATION ADMINISTRATION (FAA) CLASSIFIES THE AIRSPACE DEFINED IN THIS NOTAM AS 'NATIONAL DEFENSE AIRSPACE'. ANY PERSON WHO KNOWINGLY OR WILLFULLY VIOLATES THE RULES PERTAINING TO OPERATIONS IN THIS AIRSPACE MAY BE SUBJECT TO CERTAIN CRIMINAL PENALTIES UNDER 49 USC 46307. PILOTS WHO DO NOT ADHERE TO THE FOLLOWING PROCEDURES MAY BE INTERCEPTED, DETAINED AND INTERVIEWED BY LAW ENFORCEMENT/SECURITY PERSONNEL. PURSUANT TO 14 CFR SECTION 99.7, SPECIAL SECURITY INSTRUCTIONS, COMMENCING ONE HOUR BEFORE THE SCHEDULED TIME OF THE EVENT UNTIL ONE HOUR AFTER THE END OF THE EVENT. ALL AIRCRAFT OPERATIONS; INCLUDING PARACHUTE JUMPING, UNMANNED AIRCRAFT AND REMOTE CONTROLLED AIRCRAFT, ARE PROHIBITED WITHIN A 1410271420-PERM END PART 1 OF 3 FDC 4/3621 FDC PART 2 OF 3 SPECIAL 3NMR UP TO AND INCLUDING 3000FT AGL OF ANY STADIUM HAVING A SEATING CAPACITY OF 30,000 OR MORE PEOPLE WHERE EITHER A REGULAR OR POST SEASON MAJOR LEAGUE BASEBALL, NATIONAL FOOTBALL LEAGUE, OR NCAA DIVISION ONE FOOTBALL GAME IS OCCURRING. THIS NOTAM ALSO APPLIES TO NASCAR

[16] *Id.*
[17] NOTAMs, FAA, https://www.faa.gov/air_traffic/publications/notices/.
[18] TFR, FAA, http://tfr.faa.gov/tfr2/about.jsp.

SPRINT CUP, INDY CAR, AND CHAMP SERIES RACES EXCLUDING QUALIFYING AND PRE-RACE EVENTS. FLIGHTS CONDUCTED FOR OPERATIONAL PURPOSES OF ANY EVENT, STADIUM OR VENUE AND BROADCAST COVERAGE FOR THE BROADCAST RIGHTS HOLDER ARE AUTHORIZED WITH AN APPROVED AIRSPACE WAIVER. AN FAA AIRSPACE WAIVER DOES NOT RELIEVE OPERATORS FROM OBTAINING ALL OTHER NECESSARY AUTHORIZATIONS AND COMPLYING WITH ALL APPLICABLE FEDERAL AVIATION REGULATIONS. THE RESTRICTIONS DESCRIBED ABOVE DO NOT APPLY TO THOSE AIRCRAFT AUTHORIZED BY AND IN CONTACT WITH ATC FOR OPERATIONAL OR SAFETY OF FLIGHT PURPOSES, DEPARTMENT OF DEFENSE, LAW ENFORCEMENT, AND AIR AMBULANCE FLIGHT OPERATIONS. ALL PREVIOUSLY ISSUED WAIVERS TO FDC NOTAM 9/5151 REMAIN VALID UNTIL THE SPECIFIED END DATE BUT NOT TO EXCEED 90 DAYS FOLLOWING THE EFFECTIVE DATE OF THIS NOTAM. INFORMATION ABOUT AIRSPACE WAIVER APPLICATIONS 1410271420-PERM END PART 2 OF 3 FDC 4/3621 FDC PART 3 OF 3 SPECIAL AND TSA SECURITY AUTHORIZATIONS CAN BE FOUND AT HTTP://WWW.TSA.GOV/STAKEHOLDERS/AIRSPACE-WAIVERS-0 OR BY CALLING TSA AT 571-227-2071. SUBMIT REQUESTS FOR FAA AIRSPACE WAIVERS AT HTTPS://WAIVERS.FAA.GOV.1410271420-PERM END PART 3 OF 3.[19]

The FAA is leading a public outreach campaign to promote safe and responsible use of UAS. They are accomplishing this by using a digital toolkit entitled "No Drone Zone" with outreach materials to federal, state, and other partners to educate UAS operators that flying in certain areas is prohibited.[20]

6. Washington, DC Flight Restricted Zone (FRZ)

The FAA action effective February 17, 2009, codified special flight rules and airspace and flight restrictions for certain aircraft operations in the Washington, DC Metropolitan Area. The FAA took this action in the interest of national security. This action was necessary to enable the Department of Homeland Security (DHS) and the Department of Defense (DOD) to effectively execute their respective constitutional and Congressionally mandated duties to secure, protect, and defend the United States. It was applicable to 14 CFR Parts 1 and 93, per Docket No. FAA-2004-17005; Amdt. Nos. 1-63 and 93-90, RIN 2120-AI17, and established the Washington, DC Metropolitan Area Special Flight Rules Area.[21]

The FAA Administrator has broad authority to regulate the safe and efficient use of the navigable airspace (Title 49 United States Code (U.S.C.) 40103). The Administrator is also obligated to issue air traffic rules and regulations to govern the flight of aircraft, the navigation, protection, and identification of aircraft for the protection of persons and

[19]NOTAM for TFR, http://tfr.faa.gov/save_pages/detail_4_3621.html.
[20]*No Drone Zone,* Federal Aviation Administration (FAA), https://www.faa.gov/uas/no_drone_zone/.
[21]Washington, DC, Flight Restricted Zone (FRZ), http://www.gpo.gov/fdsys/pkg/FR-2008-12-16/html/E8-29711.htm.

property on the ground and for the efficient use of the navigable airspace. The Administrator is likewise authorized and obligated to issue regulations or orders assigning the use of the airspace to ensure the safety of aircraft as well as the efficient use of the airspace. Additionally, the Administrator is authorized and obligated to prescribe air traffic regulations for the flight of aircraft, to include mandating safe altitudes, for navigating, protecting, and identifying aircraft; protecting individuals and property on the ground; using the navigable airspace efficiently; and preventing collision of aircraft with other airborne objects, land or water vehicles, or other aircraft.[22]

The Administrator is authorized and obligated to establish security provisions governing use of and access to the navigable airspace by civil aircraft, balancing the needs of national security and national defense with the mandate to allow and encourage maximum use of the navigable airspace by civil aircraft. Pursuant to 49 U.S.C. 40103(b)(3)(A), the Administrator is authorized as well as obligated to establish areas in the airspace if the Administrator, after consulting with the Secretary of Defense, determines doing so is necessary in the interest of national security. Since the Department of Homeland Security was established in 2002 after the enactment of the statute referred to above, the Administrator's need and responsibility to consult with the Secretary of Homeland Security in addition to the Secretary of Defense is consistent with the intent and purpose of the statute.[23]

This final rule establishes and defines the DC SFRA, which includes the DC FRZ. It also defines dimensions, procedures and required equipment for operating in the DC SFRA. These procedures include establishing two-way radio communication, filing flight plans, and using discrete transponder codes. In addition, the rule provides for traffic pattern operations at towered and nontowered airports within the DC SFRA, and provides relief from certain procedures for airports located near the boundary of the DC SFRA.[24]

On February 9, 2016, the FAA issued a press release updating its conditions for flying unmanned aircraft in the area between 15 and 30 miles from Washington, D.C. to clarify differences for various types of unmanned aircraft. As of 12:01 a.m. EST, Wednesday, February 10, the operating procedures will allow model aircraft, commercial and public users, to operate in the outer ring of the Special Flight Rules Area (SFRA) under specific conditions.[25]

The SFRA rule still prohibits unmanned aircraft operations within 15 miles of Washington, D.C. in the Flight Restricted Zone without specific FAA authorization. Under the new procedures, hobbyists and recreational unmanned aircraft operators can fly aircraft that weigh less than 55 pounds (including any attachments such as a camera) in the area between 15 and 30 miles of Washington, D.C. if the aircraft are registered and marked,

---

[22] *Id.*

[23] *Id.*

[24] *Id.*

[25] Washington, DC, Flight Restricted Zone (FRZ), News and Updates, https://www.faa.gov/news/updates/?newsId=84787.

and they follow specific operating conditions. The operating conditions require them to fly 400 feet or lower above the ground, stay in the operator's line of sight, only fly in clear conditions, and avoid other aircraft. If hobbyists intend to operate within five miles of an airport or heliport, the new procedures also require them to notify the airport, heliport and air traffic control tower, if there is one, before operating. Commercial and other nonmodel aircraft operators must register and mark their unmanned aircraft, must have an exemption and comply with it, and must notify the FAA an hour before operating to provide specific flight information. Public operators, such as federal, state, or local governments, must also register and mark their aircraft, must have the appropriate FAA authorization to operate, and must complete the same one-hour notification before operating.[26] The NOTAM in its entirety follows below.

FDC 6/2069 ZDC PART 1 OF 8 SECURITY . . . SPECIAL SECURITY INSTRUCTIONS, WASHINGTON, DC, EFFECTIVE 1602100501 UTC UNTIL FURTHER NOTICE. SPECIAL SECURITY INSTRUCTIONS FOR UNMANNED AIRCRAFT OPERATIONS (UAS) IN THE DC SPECIAL FLIGHT RULES AREA (SFRA), INCLUDING THE DC FLIGHT RESTRICTED ZONE (FRZ), ARE IN EFFECT PURSUANT TO 14 CODE OF FEDERAL REGULATIONS (CFR) SECTIONS 93.335, 93.337, 93.339, 93.341, AND 99.7, AND 49 UNITED STATES CODE (USC) SECTION 40103(B)(3). THIS NOTAM CLARIFIES AND SUPPLEMENTS THE OPERATING REQUIREMENTS IN FDC 6/2062 AND 6/2060, WHICH DEFINES THE OVERALL OPERATING REQUIREMENTS FOR THE DC SFRA AND THE DC FRZ, AND THOSE PRESCRIBED BY 14 CFR SECTIONS 93.335, 93.337, 93.339, 93.341, 93.343, 93.345, AND 99.7. SECTION I. SPECIAL NOTES ON UAS OPERATIONS IN THE DC SFRA: A. THIS NOTAM SUPPLEMENTS FDC 6/2062 WITH SPECIAL SECURITY INSTRUCTIONS, INCLUDING OPERATING REQUIREMENTS, THAT ARE SPECIFIC TO UAS OPERATIONS, INCLUDING OPERATIONS BY MODEL AIRCRAFT (HOBBYIST OR RECREATIONAL USE), CIVIL (INCLUDING COMMERCIAL), AND PUBLIC OPERATORS, IN THE DC SFRA. B. THE FAA HAS ESTABLISHED THE DC SFRA, INCLUDING THE DC FRZ, AS "NATIONAL DEFENSE AIRSPACE" PURSUANT TO 49 USC 40103(B)(3). PERSONS 1602100501-PERM END PART 1 OF 8 FDC 6/2069 ZDC PART 2 OF 8 SECURITY . . . SPECIAL SECURITY INSTRUCTIONS, OPERATING UAS IN THE DC SFRA WHO DO NOT ADHERE TO THE PROCEDURES PRESCRIBED BY 14 CFR SECTIONS 93.335, 93.337, 93.339, 93.341, AND 99.7, AND THE FOLLOWING UAS-SPECIFIC SUPPLEMENTAL SPECIAL SECURITY INSTRUCTIONS MAY FACE RESPONSE AND ENFORCEMENT ACTIONS DESCRIBED BY FDC 6/2062 AND 6/2060. C. UAS OPERATIONS ARE PROHIBITED IN THE DC FRZ. SECTION II. OPERATING REQUIREMENTS FOR DC SFRA (MODEL AIRCRAFT UAS OPERATIONS): ALL MODEL AIRCRAFT (FOR

[26]*Id.*

HOBBYIST OR RECREATIONAL USE ONLY) UAS OPERATIONS ARE PROHIBITED WITHIN DC SFRA UNLESS IN COMPLIANCE WITH ALL OF THE REQUIREMENTS LISTED BELOW. REFER TO SECTION VI OF THIS NOTAM FOR THE APPLIED DEFINITION OF MODEL AIRCRAFT OPERATIONS. A. THE UAS IS REGISTERED AND MARKED AS REQUIRED BY THE FAA. REFER TO INSTRUCTIONS PROVIDED AT WWW.FAA.GOV/UAS/REGISTRATION. B. THE UAS WEIGHS LESS THAN 55 LBS., INCLUDING ALL ADDED EQUIPMENT (SUCH AS CAMERAS) ATTACHED TO THE AIRCRAFT. C. OPERATIONS MUST REMAIN AT OR BELOW 400 FEET ABOVE GROUND LEVEL (AGL). D. OPERATIONS MUST REMAIN WITHIN VISUAL LINE OF SIGHT OF THE1602100501-PERM END PART 2 OF 8 FDC 6/2069 ZDC PART 3 OF 8 SECURITY . . . SPECIAL SECURITY INSTRUCTIONS, OPERATOR AND IN COMPLIANCE WITH THE FOLLOWING: 1. OPERATIONS MUST BE CONDUCTED UNDER VISUAL METEOROLOGICAL CONDITIONS (VMC). 2. OPERATIONS MUST NOT BE CONDUCTED DURING NIGHT AS DEFINED IN 14 CFR SECTION 1.1. 3. FLIGHTS UNDER SPECIAL VISUAL FLIGHT RULES (SVFR) ARE NOT AUTHORIZED. E. PRIOR TO OPERATING WITHIN 5 MILES OF AN AIRPORT, THE UAS OPERATOR MUST NOTIFY THE AIRPORT OPERATOR AND, WHEN PRESENT, THE AIR TRAFFIC CONTROL TOWER. F. OPERATIONS MUST NOT INTERFERE WITH AND MUST GIVE WAY TO MANNED AIRCRAFT. G. MODEL AIRCRAFT UAS OPERATIONS ARE COMPLETELY PROHIBITED IN THE DC FRZ.

SECTION III. OPERATING REQUIREMENTS FOR DC SFRA (CIVIL, INCLUDING COMMERCIAL, UAS OPERATIONS): ALL CIVIL, INCLUDING COMMERCIAL, UAS OPERATIONS ARE PROHIBITED WITHIN DC SFRA UNLESS IN COMPLIANCE WITH ALL OF THE REQUIREMENTS LISTED BELOW. REFER TO SECTION VI OF THIS NOTAM FOR THE APPLIED 1602100501-PERM END PART 3 OF 8 FDC 6/2069 ZDC PART 4 OF 8 SECURITY . . . SPECIAL SECURITY INSTRUCTIONS, DEFINITION OF CIVIL, INCLUDING COMMERCIAL, UAS OPERATIONS. A. THE UAS IS REGISTERED AND MARKED AS REQUIRED BY THE FAA. REFER TO INSTRUCTIONS PROVIDED AT WWW.FAA.GOV/UAS/REGISTRATION. B. OPERATIONS MUST COMPLY WITH THE OPERATOR'S APPLICABLE FAA GRANT OF EXEMPTION PURSUANT TO PUBLIC LAW 112-95, SECTION 333, AND FAA CERTIFICATE OF AUTHORIZATION OR WAIVER (COA). C. IN ADDITION TO THE REQUIREMENTS PRESCRIBED BY SUBSECTIONS A AND B OF THIS SECTION III, CIVIL, INCLUDING COMMERCIAL, UAS OPERATORS MUST, AT LEAST ONE HOUR IN ADVANCE OF OPERATIONS, CONTACT THE FAA AT THE NATIONAL CAPITAL REGION COORDINATION CENTER (NCRCC) AT (866) 598-9522 AND PROVIDE LOCATION, START/STOP TIMES, OPERATOR CONTACT, AND COA INFORMATION, AND ANY OTHER INFORMATION REQUESTED BY THE FAA. SECTION IV. OPERATING REQUIREMENTS FOR DC SFRA (PUBLIC UAS OPERATIONS):

ALL PUBLIC UAS OPERATIONS ARE PROHIBITED WITHIN DC SFRA UNLESS IN COMPLIANCE WITH ALL OF THE REQUIREMENTS LISTED BELOW. REFER TO SECTION VII OF THIS NOTAM FOR THE APPLIED DEFINITION OF PUBLIC UAS OPERATIONS. A. THE UAS IS REGISTERED AND MARKED AS REQUIRED BY THE FAA. REFER TO 1602100501-PERM END PART 4 OF 8 FDC 6/2069 ZDC PART 5 OF 8 SECURITY . . . SPECIAL SECURITY INSTRUCTIONS, INSTRUCTIONS PROVIDED AT WWW.FAA.GOV/UAS/REGISTRATION. B. OPERATIONS MUST COMPLY WITH THE OPERATOR'S APPLICABLE CERTIFICATE OF AUTHORIZATION OR WAIVER (COA). C. IN ADDITION TO THE REQUIREMENTS PRESCRIBED BY SUBSECTIONS A AND B OF THIS SECTION IV, PUBLIC UAS OPERATORS MUST, AT LEAST ONE HOUR IN ADVANCE OF OPERATIONS, CONTACT THE FAA AT THE NCRCCAT (866) 598-9522 AND PROVIDE LOCATION, START/STOP TIMES, OPERATOR CONTACT, AND COA INFORMATION, AND ANY OTHER INFORMATION REQUESTED BY THE FAA. SECTION V. ADDITIONAL GENERAL OPERATING REQUIREMENTS AND GUIDANCE FOR DC SFRA (ALL TYPES OF UAS OPERATIONS): A. UAS OPERATIONS IN THE DC SFRA MUST NOT PENETRATE RESTRICTED AREAS, PROHIBITED AREAS, OR TEMPORARY FLIGHT RESTRICTIONS (TFR). RESTRICTED AND PROHIBITED AREAS ARE DEPICTED ON CHARTS AVAILABLE THROUGH THE FAA—REFER TO WWW.FAA.GOV/AIR_TRAFFIC/FLIGHT_INFO/AERONAV. INFORMATION ON CURRENT TFR'S CAN BE OBTAINED AT WWW.TFR.FAA.GOV B. UAS OPERATIONS IN THE DC SFRA MUST NOT PENETRATE THE DC FRZ. UAS OPERATIONS ARE PROHIBITED IN THE DC FRZ. C. UAS OPERATORS SHOULD BE AWARE OF OTHER NOTAMS, WHICH ADDRESS 1602100501-PERM END PART 5 OF 8 FDC 6/2069 ZDC PART 6 OF 8 SECURITY . . . SPECIAL SECURITY INSTRUCTIONS, SECURITY SENSITIVE INCIDENTS, EVENTS, OPERATIONS, AND/OR LOCATIONS SUCH AS MILITARY OR OTHER FEDERAL FACILITIES, CERTAIN STADIUMS, POWER PLANTS, ELECTRIC SUBSTATIONS, DAMS, OIL REFINERIES, NATIONAL PARKS, EMERGENCY, SERVICES AND OTHER INDUSTRIAL COMPLEXES. IN ADDITION TO THE PREVIOUSLY MENTIONED LINK, INFORMATION REGARDING PUBLISHED NOTAMS CAN BE FOUND AT: WWW.FAA.GOV/AIR_TRAFFIC /PUBLICATIONS/NOTICES/ D. UAS OPERATIONS MUST NOT INTERFERE WITH AND MUST GIVE WAY TO MANNED AIRCRAFT. SECTION VI. DEFINITIONS: A. MODEL AIRCRAFT UAS: UAS MUST MEET THE REQUIREMENTS PRESCRIBED BY PUBLIC LAW 112-95, SECTION 336, TO QUALIFY AS A MODEL AIRCRAFT, WHICH IS USED EXCLUSIVELY FOR HOBBYIST OR RECREATIONAL PURPOSES. UAS USED FOR COMMERCIAL ACTIVITY DO NOT QUALIFY AS MODEL AIRCRAFT. B. CIVIL, INCLUDING COMMERCIAL, UAS: CIVIL UAS OPERATIONS GENERALLY COMPRISE FLIGHTS CONDUCTED BY PRIVATE SECTOR ENTITIES FOR COMMERCIAL PURPOSES. REFER TO PUBLIC LAW 112-95,

SECTION 333, FOR ADDITIONAL DETAILS ON UAS OPERATIONS THAT ARE ADDRESSED BY SECTION III OF THIS NOTAM. 1602100501-PERM END PART 6 OF 8 FDC 6/2069 ZDC PART 7 OF 8 SECURITY . . . SPECIAL SECURITY INSTRUCTIONS, C. PUBLIC UAS: PUBLIC UAS OPERATIONS GENERALLY INCLUDE GOVERNMENTAL OPERATIONS, INCLUDING DEPARTMENT OF DEFENSE (DOD) AND NATIONAL GUARD (NG) FLIGHTS. REFER TO 49 USC SECTION 40102(A)(41), WHICH PROVIDES THE DEFINITION OF "PUBLIC AIRCRAFT" AND 49 USC SECTION 40125 PROVIDES THE QUALIFICATIONS FOR PUBLIC AIRCRAFT STATUS. SECTION VII. RESOURCES: A. THE CODE OF FEDERAL REGULATIONS CAN BE FOUND ON THE GOVERNMENT PRINTING OFFICE WEBSITE AT WWW.GPO.GOV/FDSYSACCESS.GOV/CFR/INDEX.HTML, OR WWW.ECFR.GOV. B. ANY UAS OPERATOR QUESTIONS REGARDING DC SFRA OR FRZ PROCEDURES SHOULD BE DIRECTED TO THE FAA SYSTEM OPERATIONS SECURITY REPRESENTATIVE AT THE NATIONAL CAPITAL REGION COORDINATION CENTER (NCRCC) AT 9-AWA-ATS-NCRCC@FAA.GOV OR (866) 598-9522. C. INFORMATION ABOUT FAA/TSA AIRSPACE WAIVER APPLICATIONS AND TSA SECURITY AUTHORIZATIONS CAN BE FOUND AT WWW.TSA.GOV/FOR-INDUSTRY/GENERAL-AVIATION OR BY CONTACTING TSA AT (571) 227-2071. D. FAA INFORMATION TO HELP UAS OPERATORS UNDERSTAND REQUIREMENTS AND RESTRICTIONS, WHICH COULD BE IN EFFECT AT THEIR INTENDED OPERATING 1602100501-PERM END PART 7 OF 8 FDC 6/2069 ZDC PART 8 OF 8 SECURITY . . . SPECIAL SECURITY INSTRUCTIONS, LOCATION, IS AVAILABLE THROUGH THE FAA'S B4UFLY SMARTPHONE APPLICATION—SEE WWW.FAA.GOV/UAS/B4UFLY E. INDIVIDUALS MAY SUBMIT A REQUEST FOR A FAA WAIVER AT WAIVERS.FAA.GOV. AFTER NORMAL BUSINESS HOURS, FOR EMERGENCY OR SHORT NOTICE REQUESTS, CONTACT TSA AT THE NCRCC AT (866) 598-9520. 1602100501-PERM END PART 8 OF 8.[27]

7. 2009 Special Notice

In 2009 Special Notice 4/0811 was issued and reads as follows:

FDC 4/0811 SPECIAL NOTICE

THIS IS A RESTATEMENT OF A PREVIOUSLY ISSUED ADVISORY NOTICE.

IN THE INTEREST OF NATIONAL SECURITY AND TO THE EXTENT PRACTICABLE, PILOTS ARE STRONGLY ADVISED TO AVOID THE AIRSPACE ABOVE, OR IN PROXIMITY TO SUCH SITES AS POWER PLANTS (NUCLEAR, HYDRO-ELECTRIC, OR COAL), DAMS, REFINERIES, INDUSTRIAL COMPLEXES, MILITARY FACILITIES

[27]Washington, DC, Flight Restricted Zone (FRZ), NOTAM FDC 6/2069, http://tfr.faa.gov/save_pages/detail_6_2069.html.

AND OTHER SIMILAR FACILITIES. PILOTS SHOULD NOT CIRCLE AS TO LOITER IN THE VICINITY OVER THESE TYPES OF FACILITIES.[28]

8. State Laws and City Ordinances

As every state in the United States waited impatiently for the FAA to produce its regulation covering UAS operations, more than half of them decided to take matters into their own hands and enact their own legislation. What follows, alphabetically by state, is the current state legislation. City ordinances have been enacted as well, and these were included too. In some cases litigation ensued, challenging those ordinances (for several reasons mentioned above in Section II—The Federal Legal Environment—Part 8 it is the author's opinion that some of these shall be the subject of litigation in the future).

ALABAMA

On April 30, 2013, **House Resolution (HR) 381** was introduced to recognize the benefits of a thriving UAS industry. Its intent was to encourage a partnership between the State of Alabama and the FAA to select Huntsville as a UAS test site.[29]

ALASKA

Alaska has passed House Concurrent Resolution (HCR) 6,[30] HCR 15,[31] House Bill (HB) 255,[32] HB 256,[33] and House Joint Resolution (HJR) 5.[34] Each will be detailed in turn.

House Concurrent Resolution (HCR) 6 reads as follows: Recognizing the Alaska Center for Unmanned Aircraft Systems Integration at the University of Alaska Fairbanks as a national leader in unmanned aircraft research and development; and relating to a Task Force on Unmanned Aircraft Systems.

Be it resolved by the Legislature of the State of Alaska:

WHEREAS the Unmanned Aircraft Systems Program at the University of Alaska Fairbanks has, since 2001, been designing payloads and upgrades for unmanned aircraft systems, testing and evaluating their safety and effectiveness, developing and demonstrating new uses for various potential users, working with the Federal Aviation Administration to make the use of unmanned aircraft systems safe, and developing technology to improve

---

[28]2009 Special Notice 4/0811, https://www.faa.gov/air_traffic/publications/us_restrictions/fdc_notams/pdf/FDC%20NOTAMS%204-4386%20and%204-0811.pdf.

[29]Alabama House Resolution (HR) 381, http://openstates.org/al/bills/2013rs/HR381/#billtext.

[30]House Concurrent Resolution (HCR) 6, The Alaska State Legislature, http://www.legis.state.ak.us/basis/get_bill_text.asp?hsid=HCR006B&session=28.

[31]*House Extends Unmanned Aircraft Systems Task Force*, The Alaska State House of Representatives, http://www.housemajority.org/2014/04/04/house-extends-unmanned-aircraft-systems-task-force/

[32]Section 14.40.082, The Alaska State Legislature, http://www.legis.state.ak.us/basis/statutes.asp#14.40.082.

[33]House Bill (HB) 256, The Alaska State Legislature, http://www.legis.state.ak.us/PDF/29/Bills/HB0256Z.PDF.

[34]House Joint Resolution (HJR) 5, The Alaska State Legislature, http://www.legis.state.ak.us/basis/get_bill.asp?bill=HJR%20%205.

safety and is known across the nation for its leadership in advancing the state of the art and the safe use of unmanned aircraft systems; and WHEREAS the importance of the program's work to the state was recently recognized by the Board of Regents of the University of Alaska when the regents established the Center for Unmanned Aircraft Systems Integration at the University of Alaska Fairbanks to continue research and development of unmanned aircraft systems for the benefit of the state; and

WHEREAS the legislature recognizes the Alaska Center for Unmanned Aircraft Systems Integration at University of Alaska Fairbanks as the state's leader in unmanned aircraft systems research, meeting the goals of industry, the Federal Aviation Administration, and government operations; and

WHEREAS the legislature has provided significant financial support to the Alaska Center for Unmanned Aircraft Systems Integration at University of Alaska Fairbanks to further the center's research initiatives, and to increase its capabilities for use as a possible Federal Aviation Administration test site; and

WHEREAS the State of Alaska, the University of Alaska, the Alaska Aerospace Corporation, the Alaska Aviation Safety Project, and other public and private institutions have developed cooperation with the National Aeronautics and Space Administration, the Federal Aviation Administration, the United States Department of the Interior, the United States Department of Defense, and others to further advance the research, development, and use of technology to benefit and support Alaskans and Americans with the safe use of unmanned aircraft systems; and

WHEREAS Alaska state government and the state's business and research community continue to work with the Federal Aviation Administration to promote the establishment of safe unmanned aircraft system ranges onshore and offshore Alaska, which will help establish procedures for the safe operation of unmanned aircraft systems in the National Airspace System; and

WHEREAS the Alaska Center for Unmanned Aircraft Systems Integration, by virtue of being located in Alaska, has access to an area of geographic and climatic diversity where the air and surface population density is desirable for conducting research and development of unmanned aircraft systems; and

WHEREAS the state has the highest per capita use of general aviation activity in the nation, and Alaska aviation organizations have helped the Federal Aviation Administration develop the next generation of air traffic control for the nation and the world, while making major gains for the state's aviation safety in the process; and

WHEREAS unmanned aircraft systems may present a risk to privacy, but neither the Federal Aviation Administration nor any other state or federal agency currently has specific statutory authority to regulate privacy matters relating to unmanned aircraft systems;

BE IT RESOLVED that the Alaska State Legislature recognizes the Alaska Center for Unmanned Aircraft at the University of Alaska Fairbanks as a leader in aviation pioneering

and further recognizes the significant benefit to the state and nation the center provides in research and development of unmanned aircraft systems; and be it
FURTHER RESOLVED that the Alaska State Legislature fully supports the efforts of the Alaska Center for Unmanned Aircraft Systems Integration at the University of Alaska Fairbanks to become one of the nation's six test sites for unmanned aircraft systems testing and research; and be it
FURTHER RESOLVED by the Alaska State Legislature that the Task Force on Unmanned Aircraft Systems is created in the legislative branch and shall consist of 15 members, who, except for the member from the senate and the member from the house of representatives, shall be appointed jointly by the president of the senate and the speaker of the house of representatives, as follows:

1. one senator selected by the president of the senate, who shall serve as co-chair; one representative selected by the speaker of the house of representatives, who shall serve as co-chair;
2. the commissioner of transportation and public facilities or the commissioner's designee;
3. the commissioner of public safety or the commissioner's designee;
4. the commissioner of natural resources or the commissioner's designee;
5. one member representing municipal law enforcement;
6. the attorney general or the attorney general's designee;
7. the commissioner of fish and game or the commissioner's designee;
8. the adjutant general of the Department of Military and Veterans' Affairs or the adjutant general's designee;
9. a member representing the Alaska Center for Unmanned Aircraft Systems Integration at University of Alaska Fairbanks;
10. a member of the Academy of Model Aeronautics;
11. a member from the Aviation Advisory Board;
12. a member from the Medallion Foundation;
13. one member of the Aerospace States Association;
14. one member of the Alaska UAS Interest Group; and be it

FURTHER RESOLVED that a vacancy on the task force shall be filled in the manner of the original appointment; and be it
FURTHER RESOLVED that the duties of the task force shall include

1. reviewing regulations and guidance from the Federal Aviation Administration regarding unmanned aircraft systems;
2. providing written recommendations, together with suggested legislation, for a comprehensive state policy for unmanned aircraft that protects privacy and

allows the use of unmanned aircraft systems for public and private applications; and

3. submitting, not later than January 15, 2014, an initial report to the legislature and, not later than January 15, 2015, submitting a final report to the legislature; and be it

FURTHER RESOLVED that the task force may meet as frequently as necessary to carry out its responsibilities; and be it

FURTHER RESOLVED that the task force may meet during and between legislative sessions and may request administrative and technical support from the University of Alaska Fairbanks; and be it

FURTHER RESOLVED that the public members of the task force serve without compensation but are entitled to per diem and travel expenses authorized for boards and commissions under AS 39.20.180; and be it

FURTHER RESOLVED that the task force terminates on January 15, 2015.[35]

House Concurrent Resolution (HCR) 15: On April 4, 2014, the Alaska House of Representatives unanimously passed a resolution extending the termination date of the Legislative Task Force on Unmanned Aircraft Systems (UAS), which was created with the passage of HCR6. HCR15 extends the Task Force from June 30, 2014 to June 30, 2017, coinciding with the FAA's test site period for UAS in Alaska. Additionally, the resolution adds five new members to the existing seven on the original Task Force: one public member, two UAS industry members, one representing the Department of Commerce, Community & Economic Development and one representing the Dept. of Transportation & Public Facilities.[36] Alaska House Concurrent Resolution (HCR) No. 15 reads as follows:

Relating to the Task Force on Unmanned Aircraft Systems.

Be it resolved by the Legislature of the State of Alaska:

WHEREAS the University of Alaska has been selected by the Federal Aviation Administration to become a test site for unmanned aircraft systems; and

WHEREAS the legislature has provided significant financial support to the Alaska Center for Unmanned Aircraft Systems Integration at the University of Alaska Fairbanks to further the center's research initiatives and to increase its capabilities for use as a Federal Aviation Administration test site; and

WHEREAS the State of Alaska, the University of Alaska, the Alaska Aerospace Corporation, the Alaska Aviation Safety Project, and other public and private institutions have developed cooperation with the National Aeronautics and Space Administration, the Federal Aviation Administration, the United States Department of the Interior, the United

[35]House Concurrent Resolution (HCR) 6, THE ALASKA STATE LEGISLATURE, http://www.legis.state.ak.us/basis/get_bill_text.asp?hsid=HCR006B&session=28.

[36]House Concurrent Resolution (HCR) No. 15, THE ALASKA STATE HOUSE OF REPRESENTATIVES, http://www.housemajority.org/2014/04/04/house-extends-unmanned-aircraft-systems-task-force/.

States Department of Defense, and others to further advance the research, development, and use of technology to benefit and support Alaskans and Americans with the safe use of unmanned aircraft systems; and

WHEREAS Alaska state government and the state's business and research community continue to work with the Federal Aviation Administration to promote the establishment of safe unmanned aircraft system ranges onshore and offshore Alaska, which will help establish procedures for the safe operation of unmanned aircraft systems in the National Airspace System; and

WHEREAS the Alaska Center for Unmanned Aircraft Systems Integration, by virtue of being located in Alaska, has access to an area of geographic and climatic diversity where the air and surface population density is desirable for conducting research and development of unmanned aircraft systems; and

WHEREAS the state has the highest per capita use of general aviation activity in the nation, and Alaska aviation organizations have helped the Federal Aviation Administration develop the next generation of air traffic control for the nation and the world, while making major gains for the state's aviation safety in the process; and

WHEREAS the Task Force on Unmanned Aircraft Systems created by Legislative Resolve 17, Twenty-Eighth Alaska State Legislature, has performed the tasks required of it but finds that further hearings and research are required to carry out its mission;

BE IT RESOLVED by the Alaska State Legislature that the Task Force on Unmanned Aircraft Systems created in the legislative branch in 2013 by Legislative Resolve 17, Twenty-Eighth Alaska State Legislature, shall be continued; and be it

FURTHER RESOLVED that the task force shall consist of the original seven members and

1. the commissioner of transportation and public facilities or the commissioner's designee;
2. the commissioner of commerce, community, and economic development or the commissioner's designee;
3. one public member who shall be appointed jointly by the president of the senate and the speaker of the house of representatives;
4. two industry representatives who are participants in the unmanned aircraft system industry with experience operating unmanned aircraft systems who shall be appointed jointly by the president of the senate and the speaker of the house of representatives; and be it

FURTHER RESOLVED that a vacancy on the task force shall be filled in the manner of the original appointment; and be it

FURTHER RESOLVED that the duties of the task force shall include

1. reviewing regulations and guidance from the Federal Aviation Administration regarding unmanned aircraft systems;
2. providing written recommendations, together with suggested legislation, for a comprehensive state policy for unmanned aircraft that protects privacy and allows the use of unmanned aircraft systems for public and private applications;
3. evaluating complaints and concerns expressed to the task force;
4. identifying potential privacy and public safety concerns associated with unmanned aircraft systems and determining whether legislation is necessary to address them;
5. considering recommendations for public education related to unmanned aircraft systems;
6. studying the Federal Aviation Administration's "Integration of Civil Unmanned Aircraft Systems (UAS) in the National Airspace System Roadmap," issued November 7, 2013 (1st ed. 2013) and its application to the development of unmanned aircraft systems in the state;
7. conducting a public hearing concerning privacy and the capture of data by unmanned aircraft systems at the University of Alaska's test site;
8. further studying the nonpublic use of unmanned aircraft systems to encourage development of the private sector unmanned aircraft system industry; and
9. further studying and making recommendations with respect to ensuring unmanned aircraft users comply with applicable laws; and be it

FURTHER RESOLVED that the task force may meet as frequently as necessary to carry out its responsibilities; and be it

FURTHER RESOLVED that the task force may meet during and between legislative sessions and may request administrative and technical support from the University of Alaska Fairbanks; and be it

FURTHER RESOLVED that the public and industry members of the task force serve without compensation but are entitled to per diem and travel expenses authorized for boards and commissions under AS 39.20.180; and be it

FURTHER RESOLVED that the task force terminates on June 30, 2017.[37]

House Bill (HB) 255: Alaska now has a number of statutes on its books that are listed below. *Section 14.40.082*, entitled *Unmanned Aircraft System Training*, reads as follows:

[37]House Concurrent Resolution (HCR) No. 15, THE ALASKA STATE LEGISLATURE, http://www.legis.state.ak.us/basis/get_bill_text.asp?hsid=HCR015C&session=28.

The University of Alaska may establish a training program in the operation of unmanned aircraft systems.[38] Alaska's Article 13 entitled *Law Enforcement Use of Unmanned Aircraft Systems*, in *Section 18.65.900*, reads as follows: Except as provided in AS 18.65.900—18.65.909, a law enforcement agency may not use an unmanned aircraft system.[39]

Section 18.605.901, entitled *Operational Requirements for Unmanned Aircraft Systems*, reads as follows: (a) A law enforcement agency shall adopt procedures for the use of unmanned aircraft systems. The procedures adopted under this section must require, at a minimum, that the law enforcement agency

1. obtain any authorization, permit, or certificate required by the Federal Aviation Administration to operate the unmanned aircraft system;
2. allow the unmanned aircraft system to be operated only by unmanned aircraft system pilots and crew members who have been trained and certified in the operation of the unmanned aircraft system and only under the supervision of officials trained in the policies and procedures governing the use of the unmanned aircraft system;
3. provide that the flight of an unmanned aircraft system be approved by the commissioner or deputy commissioner of public safety or the chief administrative officer of the law enforcement agency or the officer's designee;
4. ensure that the flight of an unmanned aircraft system be for a public purpose;
5. maintain a record of each flight, including the time, date, and purpose of the flight, and the identity of the authorizing official;
6. establish an auditable flight record system, including the documentation of a change in a flight time record;
7. establish a method for notifying the public of the operation of an unmanned aircraft system, unless notifying the public would endanger the safety of a person;
8. provide for community involvement in the development of the policies required in this section, including the consideration of public comment.

(b) In this section, "chief administrative officer" has the meaning given in AS 18.65.290.[40]

Section 18.65.902, entitled *Use of an Unmanned Aircraft System by a Law Enforcement Agency*, reads as follows: A law enforcement agency may use an unmanned aircraft system (1) to gather evidence in a criminal investigation

A. under the express terms of a search warrant issued by a court; or

[38]Section 14.40.082, The Alaska State Legislature, http://www.legis.state.ak.us/basis/statutes.asp#14.40.082.
[39]Section 18.65.900, The Alaska State Legislature, http://www.legis.state.ak.us/basis/statutes.asp#18.65.900.
[40]Section 18.605.901, The Alaska State Legislature, http://www.legis.state.ak.us/basis/statutes.asp#18.65.901.

B. in accordance with a judicially recognized exception to the warrant requirement; or

3. in situations and for uses not involving a criminal investigation and not intended to lead to the production of evidence for use in a criminal investigation, if the use does not constitute an unwarranted invasion of personal privacy and is consistent with the procedures in AS 18.65.901.[41]

Section 18.65.903, entitled *Retention of Images*, reads as follows: (a) A law enforcement agency may not retain images captured by an unmanned aircraft system unless retention of the image is required

1. as part of an investigation or prosecution;
2. for training purposes; or
3. by federal or state law or by municipal ordinance.

d. (b) Images that may not be retained under (a) of this section are confidential and are not public records under AS 40.25.100—40.25.295.[42]

Section 18.65.909, entitled *Definitions*, reads as follows: In AS 18.65.900—18.65.909,

1. "law enforcement agency" has the meaning given in AS 12.36.090;
2. "unmanned aircraft system" means an unmanned aircraft that is operated without direct human intervention from inside or on the aircraft and includes the associated support equipment, control station, data links, telemetry, communications, and navigation equipment necessary to operate the unmanned aircraft;
3. "unmanned aircraft system crew member" means a person other than an unmanned aircraft system pilot who is assigned to duties related to an unmanned aircraft system during flight;
4. "unmanned aircraft system pilot" means a person exercising control over an unmanned aircraft system during flight.[43]

Section 29.35.146, entitled *Regulation of Unmanned Aircraft Systems*, reads as follows: (a) A municipality may not adopt an ordinance that permits the release of images captured by an unmanned aircraft system in a manner inconsistent with AS 18.65.903.

e. (b) In this section, "unmanned aircraft system" has the meaning given in AS 18.65.909.[44]

---

[41]Section 18.65.902, The Alaska State Legislature, http://www.legis.state.ak.us/basis/statutes.asp#18.65.902.
[42]Section 18.65.903, The Alaska State Legislature, http://www.legis.state.ak.us/basis/statutes.asp#18.65.903.
[43]Section 18.65.909, The Alaska State Legislature, http://www.legis.state.ak.us/basis/statutes.asp#18.65.909.
[44]Section 29.35.146, The Alaska State Legislature, http://www.legis.state.ak.us/basis/statutes.asp#29.35.146.

House Bill (HB) 256: On July 1, 2016, this law went into effect requesting the Department of Fish and Game to evaluate the use of UAS for aerial survey work and report findings related to safety and cost-savings compared to manned aircraft.[45]

House Joint Resolution (HJR) 5: On October 24, 2015, Alaska recognized the Academy of Model Aeronautics (AMA) for establishing safety guidelines and best practices, and it reads as follows:

Be it Resolved by the Legislature of the State of Alaska:

WHEREAS the Academy of Model Aeronautics is a community-based organization that has a long history of overseeing the nation's expansive aeromodeling community and that has achieved an exceptional safety record; and

WHEREAS the Academy of Model Aeronautics has had an active presence in the state since 1957 through flying sites and organized flying clubs; and

WHEREAS the Task Force on Unmanned Aircraft Systems, created by the Twenty-Eighth Alaska State Legislature, has been working with the Federal Aviation Administration, the University of Alaska, the Association for Unmanned Vehicles Systems International, the Academy of Model Aeronautics, and other stakeholders to address a broad range of concerns related to operations, safety, and privacy rights associated with unmanned aircraft systems; and

WHEREAS the state intends to continue to work with those organizations and other stakeholders to develop meaningful regulation of unmanned aircraft systems in the state; and

WHEREAS the Academy of Model Aeronautics has, since 1936, published, maintained, and established safety standards and best practices that follow national guidelines and regulations of the Federal Aviation Administration, the Federal Communications Commission, and other agencies for the benefit of Academy of Model Aeronautics members and the greater aeromodeling community; and

WHEREAS the Academy of Model Aeronautics has provided insurance coverage for operators and flying field locations that is consistent with national guidelines and regulations; and

WHEREAS the Academy of Model Aeronautics is recognized nationally by the

National Aeronautic Association and internationally by the Federation Aeronautique Internationale as the nationwide, community-based organization for recreational and sport aeromodeling activity in the United States;

BE IT RESOLVED by the Alaska State Legislature that the Legislative Task Force on Unmanned Aircraft Systems recognizes the Academy of Model Aeronautics as a community-based organization that promotes industry and community safety guidelines and best practices in the state for the recreational, hobby-related, and sport-related use of model aviation and unmanned aircraft systems; and be it

---

[45]House Bill (HB 256), The Alaska State Legislature, http://www.legis.state.ak.us/PDF/29/Bills/HB0256Z.PDF.

FURTHER RESOLVED that the Alaska State Legislature urges the state to work with federal agencies to promote best practices and communications that result in meaningful guidelines and regulations for community-based aeromodeling organizations that protect personal privacy and safety and that can be understood by the public.[46]

ARIZONA

The state of Arizona has had State Bill (SB) 1449 pass into law on May 12, 2016.[47] Prior to that date, a few bills had been introduced, like House Bill (HB) 2574 in 2013;[48] HB 2538 in 2014;[49] and HB 2073 in 2016.[50]

Senate Bill (SB) 1449: This bill was turned into three different codes, one criminal and two transportation ones. Arizona Revised Statute (ARS) 13-3729 reads as follows:[51]

13-3729. Unlawful operation of model or unmanned aircraft; state preemption; classification; definitions

A. It is unlawful for a person to operate a model aircraft or a civil unmanned aircraft if the operation:
   1. Is prohibited by a federal law or regulation that governs aeronautics, including federal aviation administration regulations.
   2. Interferes with a law enforcement, firefighter or emergency services operation.

B. It is unlawful for a person to operate or use an unmanned aircraft or unmanned aircraft system to intentionally photograph or loiter over or near a critical facility in the furtherance of any criminal offense.

C. Except as authorized by law, a city, town or county may not enact or adopt any ordinance, policy or rule that relates to the ownership or operation of an unmanned aircraft or unmanned aircraft system or otherwise engage in the regulation of the ownership or operation of an unmanned aircraft or an unmanned aircraft system. Any ordinance, policy or rule that violates this subsection, whether enacted or adopted by the city, town or county before or after August 6, 2016, is void.

D. This section does not:
   1. Apply to a person or entity that is authorized or allowed by the federal aviation administration to operate or use an unmanned aircraft system if

---

[46]House Joint Resolution (HJR) 5, The Alaska State Legislature, http://www.legis.state.ak.us/basis/get_bill.asp?bill=HJR%20%205.

[47]Arizona State Bill (SB) 1449, http://www.azleg.gov//FormatDocument.asp?inDoc=/legtext/52leg/2r/bills/sb1449p.htm&Session_ID=115.

[48]Arizona House Bill (HB) 2574, https://legiscan.com/AZ/bill/HB2574/2013.

[49]Arizona House Bill (HB) 2538, https://legiscan.com/AZ/text/HB2538/id/942355.

[50]Arizona House Bill (HB) 2073, http://www.azleg.gov/legtext/52leg/2r/bills/hb2073p.pdf.

[51]Arizona Revised Statute (ARS) 13-3729, http://www.azleg.gov/search/oop/qfullhit.asp?CiWebHitsFile=/ars/13/03729.htm&CiRestriction=unmanned.

the person's or entity's operation or use complies with the authorization granted to the person or entity or with federal aviation administration rules.

2. Prohibit a city, town or county from enacting or adopting ordinances or rules on the operation or use of a public unmanned aircraft that is owned by the city, town or county.
3. Prohibit a city, town or county from enacting or adopting ordinances or rules that regulate the takeoff or landing of a model aircraft in a park or preserve owned by the city, town or county if:
   a. There are other parks or preserves that are within the city, town or county and that are available for model aircraft operation.
   b. The city, town or county only has one park or preserve that is within the city, town or county.
4. Apply to the operation of an unmanned aircraft, including a public unmanned aircraft, by a first responder as defined in section 36-661 while acting in the first responder's official capacity or an emergency worker while engaged in or supporting authorized emergency management activities or performing emergency functions pursuant to title 26, chapter 2.

E. A violation of subsection B of this section is a class 6 felony, except that a second or subsequent violation is a class 5 felony. A violation of subsection A of this section is a class 1 misdemeanor.

F. For the purposes of this section:

1. "Civil unmanned aircraft" means an unmanned aircraft or unmanned aircraft system that is operated by a person for any purpose other than strictly for hobby or recreational purposes, including commercial purposes, or in furtherance of or incidental to any business or media service or agency.
2. "Commercial purposes" means the use of an unmanned aircraft in return for financial compensation and includes aerial photography, aerial mapping or geospatial imaging.
3. "Critical facility" means any of the following:
   a. A petroleum or alumina refinery.
   b. A petroleum, chemical or rubber production, transportation, storage or processing facility.
   c. A chemical manufacturing facility.
   d. A water or wastewater treatment facility and water development, distribution or conveyance system, including a dam.
   e. An electric generation facility, as defined in section 42-14156, and any associated substation or switchyard.
   f. An electrical transmission or distribution substation.
   g. An electrical transmission line of at least sixty-nine thousand volts.

h. An electronic communication station or tower.
i. An energy control center.
j. A distribution operating center.
k. A facility that transfers or distributes natural gas, including a compressor station, regulator station, city gate station or pressure limiting station or a liquefied natural gas facility or supplier tap facility.
l. Any railroad infrastructure or facility.
m. A federal, state, county or municipal court.
n. A public safety or emergency operation facility.
o. A federal, state, county or municipal jail or prison or other facility in which persons are incarcerated.
p. A federal or state military installation or facility.
q. A hospital that receives air ambulance services.

4. "Model aircraft" has the same meaning prescribed in section 336 of the FAA modernization and reform act of 2012 (P.L. 112-95), as amended.
5. "Person" means a corporation, firm, partnership, association, individual or organization or any other group acting as a unit.
6. "Public unmanned aircraft" means an unmanned aircraft or unmanned aircraft system that is operated by a public agency for a government-related purpose.
7. "Unmanned aircraft" means an aircraft, including an aircraft commonly known as a drone, that is operated without the possibility of direct human intervention from within or on the aircraft.
8. "Unmanned aircraft system" means an unmanned aircraft and associated elements, including any communication links and components that control the unmanned aircraft.[52]

ARS 28-8280 reads as follows:[53]

28-8280. Careless or reckless aircraft operation; violation; classification; definitions

A. A person who operates an aircraft in the air, on the ground or on the water in a careless or reckless manner that endangers the life or property of another is guilty of a class 1 misdemeanor. In determining whether the operation was careless or reckless, the court shall consider the standards for safe operation of aircraft prescribed by federal statutes or regulations governing aeronautics.
B. For the purposes of this section:
    1. "Aircraft" includes a model aircraft and civil unmanned aircraft.

---

[52]*Id.*

[53]Arizona Revised Statute (ARS)28-8280, http://www.azleg.gov/search/oop/qfullhit.asp?CiWebHitsFile=/ars/28/08280.htm&CiRestriction=unmanned.

2. "Civil unmanned aircraft" has the same meaning prescribed in section 13-3729.
3. "Model aircraft" has the same meaning prescribed in section 13-3729.[54]

ARS 28-8242 in part reads as follows:[55]

C. The director shall:

1. Provide on the department's website pictures that show examples of critical facilities, as defined in section 13-3729, to provide unmanned aircraft operators with information on what is considered a critical facility. A picture or any written description on the website may not identify the owner or operator of the critical facility or the location of the critical facility.[56]

In addition to state law, the city of Phoenix has a UAS ordinance;[57] the city of Mesa has a Park Ordinance;[58] and the town of Scottsdale just passed an ordinance on September 20, 2016.[59] The Town of Paradise Valley has Ordinance 691 that reads as follows:[60]
Town of Paradise Valley Ordinance 691: An ordinance of the Town of Paradise Valley, Arizona amending the Paradise Valley Town Code, Chapter 10, by adding Article 10-12, regulations related to the operation of unmanned aerial vehicles; and declaring an emergency. Be it ordained by the Mayor and Town Council of the Town of Paradise Valley, Arizona:
Section 1. Chapter 10, offenses, is hereby amended (with Article 10-12 added):
Article 10-12 Restrictions and exceptions to unmanned aerial vehicle operations (UAVs).

Section 10-12-1 Purpose; Harmony with other governmental regulations.

The Town Council hereby recognizes that UAVs, also known as drones, can pose unique safety, nuisance, and privacy invasion risks; thus regulating the operation of UAVs within the Town is needed to promote the public safety and welfare of the Town and its residents. These regulations are to be read in harmony with all other regulations regarding the use of UAVs, specifically including any rules promulgated by the FAA. Further, compliance with these regulations should not be interpreted as express, implied or tacit approval to operate a UAV in violation of any other governmental regulations or in a manner that jeopardizes the health, safety, or welfare of the general public. Nothing in this article shall be construed to authorize the operation of any UAV in Town airspace in violation of any Federal statute or rules promulgated thereunder. Operators of UAVs should familiarize

---

[54]*Id.*
[55]ARS 28-8242, http://www.azleg.gov/search/oop/qfullhit.asp?CiWebHitsFile=/ars/28/08242.htm&CiRestriction=unmanned.
[56]*Id.*
[57]City of Phoenix UAS Ordinance, http://www.codepublishing.com/AZ/Phoenix/?Phoenix24/Phoenix2449.html.
[58]City of Mesa Park Ordinance, http://www.mesaaz.gov/things-to-do/parks-recreation-commercial-facilities/parks/park-rules-regulations.
[59]Town of Scottsdale Model Aircraft (Drone) Operation in City Parks, http://www.scottsdaleaz.gov/Asset67324.aspx.
[60]ARIZONA, TOWN OF PARADISE VALLEY, Ordinance 691, http://www.paradisevalleyaz.gov/DocumentCenter/View/4185.

themselves with all applicable regulations relating to the use of a UAV, including the FAA requirements regarding notification of an airport operator and control tower, where applicable, prior to operating a UAV within five miles of an airport.

Section 10-12-2 Definitions.

The following words, terms and phrases, when used in this article, have the meanings ascribed to them in this section, except where the context clearly indicates a different meaning:

1. "Commercial use" means the use of a UAV within the Town for a commercial purpose, including aerial photography, aerial mapping, or geospatial imaging.
2. "Commercial user" means any company, entity or person who is in the business of flying a UAV for a commercial use.
3. "FAA" means the Federal Aviation Administration.
4. "Law Enforcement Agency" means a lawfully established federal, state, or local public law enforcement agency that is responsible for the prevention and detection of crime, local government code enforcement, and the enforcement of penal, traffic, regulatory, game, or controlled substance laws.
5. "Private Property" means all parcels of land within the Town of Paradise Valley limits that are not Public Property, including, but not limited to, residences, schools, churches, resorts, utility substations, golf course, or canals.
6. "Public Property" means streets, rights of way, parks, mountain preserves, and other parcels of land owned by the Town of Paradise Valley or the Mummy Mountain Preserve Trust.
7. "PVPD" means the Town of Paradise Valley Police Department.
8. "Unmanned Aerial Vehicle" or "UAV" (aka "Drone") means an aircraft that may be flown without a pilot or operator in or touching the aircraft.

Section 10-12-3 Restrictions; Exceptions.

A. Private Property. It is unlawful for a person to use a UAV over Private Property at a level between zero feet and five hundred feet above the ground level of the Private Property without the express permission of the owner of the Private Property over which the UAV is flying. An offence under this section will be considered a criminal trespass and subject to civil and criminal penalties pursuant to subsection 10-12-4 below and Article 1-9 of the Town Code.

B. Public Property. The use of a UAV over Public Property within the Town is unlawful unless the use has been authorized by the issuance of a Special Event Permit pursuant to Section 8-8-3 of the Paradise Valley Town Code.

C. Commercial Use on Private or Public Property. The Commercial Use of a UAV within the Town is unlawful unless the Commercial User has met the requirements provided in subsection A and B above with respect to operation of a UAV on either Private Property or Public Property and the following additional requirements:

1. Registered as a Commercial User with the PVPD, said registration to include proof of having obtained the appropriate certifications or registrations required from the FAA for the type of Commercial Use(s) to be conducted within the Town by the Commercial User;
2. Provided a list of the types and number of UAVs that the Commercial User plans to use within the Town limits, as well as: the brand and model of each UAV; any registration number, license number or other identifying information for each such UAV, specifically including registration numbers supplied by the FAA; and photos of each UAV, at least one such photo to show the particular registration or license number for each UAV; and
3. At least four hours prior to each Commercial Use of a UAV, the Commercial User will first have provided notification to the PVPD of the planned date, time, location of the Commercial Use, contact information for the Commercial User, and other information required by the Police Department through its online UAV Commercial User web portal.

D. Exceptions. In addition to the uses allowed above, the use of a UAV is allowed within the Town for:

1. Recreational use of a UAV within one's own property so long as such use is at a height of less than five hundred (500) feet and not in violation of the additional violations set forth in subsection E below; or
2. By a Law Enforcement Agency in response to an emergency situation or after obtaining a warrant based upon probable cause that criminal activity is occurring.

E. Additional Violations. In addition to those violations enumerated above, the following are violations of this Article, even under an otherwise authorized use or exception. No person or entity shall operate a UAV within the Town:

1. In a careless or reckless manner that poses an apparent or actual threat of harm, or actual harm to persons or property, or
2. Without the express permission of a person, owner, or authorized representative, to capture, view, record, or transmit any visual image or audio recording of such person or their private real property, located in the Town, under circumstances in which the subject person or owner of the private property has a reasonable expectation of privacy (including, but not limited to, inside a private residence, school room, resort room, office, or inside an enclosed yard);
3. Or in such a manner as to intentionally harass, annoy, or assault a person or persons or to cause a public nuisance.

Section 10-12-4 Penalties/Seizure of Evidence.

For a first violation of this ordinance (except those violations which are accompanied by intentional or reckless acts), the penalty shall be a civil violation with a fine not to exceed $500. For a second violation or any violation accompanied by intentional or reckless acts, the charge or violation shall be as provided for in Article 1-9 of the Town Code, with a maximum penalty to include a criminal misdemeanor charge, which, upon conviction thereof shall be a fine not to exceed two thousand five hundred dollars ($2,500) or imprisonment for a period of not to exceed six months or both fine and imprisonment. Each day that a violation continues shall be a separate offense punishable as described.

If the PVPD police chief or a Town code enforcement officer, or any of their duly authorized enforcement officers or designees, have a reasonable basis to believe that any UAV is or has been operating in violation of this section, said UAV may be seized by such duly authorized enforcement official, followed by an opportunity for an administrative hearing, with notice to the owner within seven calendar days of such seizure, for the purpose of reviewing the appropriateness of the seizure, and shall be held by the Town until such time that the owner of such UAV reimburses the Town for the actual costs incurred in connection with the seizure and storage of the UAV. If criminal charges involving the use, condition or operation of the UAV are pending, the UAV shall be held until disposition of the criminal charges. If it is determined at an administrative hearing, by a preponderance of the evidence, that the UAV was not being operated in violation of this article, such UAV shall be returned to its owner without charge.

Section 2. Due to the uncertainty regarding future regulation of UAVs by other governmental agencies, including the FAA and the State of Arizona, and potential unforeseen issues related to the rapidly evolving technology improvements in UAVs, this ordinance shall be brought back to the Council for review within one year after its adoption for the Council to consider whether any changes or modifications should be made to the provisions provided for in this ordinance.

Section 3. If any section, subsection, sentence, clause, phrase or portion of this ordinance or any part of these amendments to the Town Code adopted herein by reference is for any reason held to be invalid or unconstitutional by the decision of nay court of competent jurisdiction, such decision shall not affect the validity of the remaining portions thereof.

Section 4. Whereas the immediate operation of the provision of this ordinance is necessary for the preservation of the public peace, health and safety, an emergency is hereby declared to exist and this ordinance shall be in full force and effect from and after its passage by the council.

Passed and adopted by the Mayor and Council of the Town of Paradise Valley, Arizona, this 3rd day of December 2015.[61]

---

[61]Ordinance 691, Arizona, Town of Paradise Valley, http://www.paradisevalleyaz.gov/DocumentCenter/View/4185.

ARKANSAS

Arkansas has enacted two house bills in 2015, HB 1349,[62] and HB 1770.[63] They are detailed below:

House Bill (HB) 1349: On February 27, 2015, HB 1349 was approved and passed into Act 293 amending the offenses of voyeurism and video voyeurism to include the use of an unmanned vehicle or aircraft. Act 293 entitled, An Act Concerning the Use of an Unmanned Vehicle or Aircraft to Commit the Offense of Voyeurism or Video Voyeurism; and for other purposes, reads as follows:

Be it enacted by the General Assembly of the State Arkansas:

SECTION 1. Arkansas Code § 5-16-101(b), concerning the offense of video voyeurism, is amended to read as follows:

a. It is unlawful to knowingly use an unmanned vehicle or aircraft, a camcorder, a motion picture camera, a photographic camera of any type, or other equipment that is concealed, flown in a manner to escape detection, or disguised to secretly or surreptitiously videotape, film, photograph, record, or view by electronic means a person:
    1. For the purpose of viewing any portion of the person's body that is covered with clothing and for which the person has a reasonable expectation of privacy;
    2. Without the knowledge or consent of the person being videotaped, filmed, photographed, recorded, or viewed by electronic means; and
    3. Under circumstances in which the person being videotaped, filmed, photographed, recorded, or viewed by electronic means has a reasonable expectation of privacy.

SECTION 2. Arkansas Code § 5-16-102(b), concerning the offense of voyeurism, is amended to read as follows:

b. A person commits the offense of voyeurism if for the purpose of sexual arousal or gratification, he or she knowingly:
    1. Without the consent of each person who is present in the private place, personally or through the use of an unmanned vehicle or aircraft, looks into a private place that is, or is part of, a public accommodation and in which a person may reasonably be expected to be nude or partially nude; or

---

[62] House Bill 1349, The Arkansas State Legislature, http://www.arkleg.state.ar.us/assembly/2015/2015R/Pages/BillInformation.aspx?measureno=hb1349.

[63] House Bill 1770, The Arkansas State Legislature, http://www.arkleg.state.ar.us/assembly/2015/2015R/Acts/Act1019.pdf.

2. Enters personally or through the use of an unmanned vehicle or aircraft another person's private property without the other person's consent and looks into any person's dwelling unit if all of the following apply:
   A. The person looks into the dwelling with the purpose to intrude upon or interfere with a person's privacy;
   B. The person looks into a part of the dwelling in which an individual is present;
   C. The individual present has a reasonable expectation of privacy in that part of the dwelling; and
   D. The individual present does not consent to the person's looking into that part of the dwelling.[64]

House Bill (HB) 1770: On April 2, 2015, HB 1770 was approved and passed into Act 1019. Act 1019 entitled, An Act Concerning the Regulation of Aeronautics; To Restrict the Use of Unmanned Aircraft Systems under Certain Circumstances; To Provide for Criminal Penalties and Civil Liability; and for other purposes, reads as follows:

Be it enacted by the General Assembly of the State of Arkansas:

SECTION 1. Arkansas Code Title 5, Chapter 60, Subchapter 1, is amended to add an additional section to read as follows: 5-60-103. Unlawful use of unmanned aircraft system.

(a) As used in this section:

1. "Unmanned aircraft system" means an unmanned, powered aircraft that:
   i. Does not carry a human operator;
   ii. Can be autonomous or remotely piloted or operated; and
   iii. Can be expendable or recoverable.
   A. "Unmanned aircraft system" does not include:
      i. A satellite orbiting the earth;
      ii. An unmanned aircraft system used by the federal government or a person who is acting pursuant to contract with the federal government to conduct surveillance of specific critical infrastructure;
      iii. An unmanned aircraft system used by the state after consultation with the Governor or a person who is acting under contract with the state to conduct surveillance of specific critical infrastructure;
      iv. (a) An unmanned aircraft system used pursuant to prior written authorization of the Arkansas Department of Emergency Management. (b) The department shall provide standards for authorizations under subdivision (a)(1)(B)(iv)(a) of this section in rules adopted in

[64]House Bill 1349, The Arkansas State Legislature, http://www.arkleg.state.ar.us/assembly/2015/2015R/Pages/BillInformation.aspx?measureno=hb1349.

accordance with the Administrative Procedure Act, § 25-15-201, et seq., after consultation with the Governor;

v. An unmanned aircraft system used under a certificate of authorization issued by the Federal Aviation Administration; or

vi. An unmanned aircraft used by a law enforcement agency, emergency medical service agency, hazardous materials response team, disaster management agency, or other emergency management agency for the purpose of incident command, area reconnaissance, personnel and equipment deployment monitoring, training, or a related purpose; and

2. "Critical infrastructure" means:
   A. An electrical power generation or delivery system;
   B. A petroleum refinery;
   C. A chemical or rubber manufacturing facility; or
   D. A petroleum or chemical storage facility.

(b) A person commits the offense of unlawful use of an unmanned aircraft system if he or she knowingly uses an unmanned aircraft system to conduct surveillance of, gather evidence or collect information about, or photographically or electronically record critical infrastructure without the prior written consent of the owner of the critical infrastructure.
(c) This section does not prohibit:

1. (1)(A) A person from using an unmanned aircraft system to conduct surveillance of, gather evidence or collect information about, or photographically or electronically record his or her own property that is:
   i. Located on his or her own immovable property; or
   ii. Located on immovable property owned by another person under a valid lease, servitude, right-of-way, right of use, permit, license, or other right.
      B. A third person retained by the owner of the property described in subdivision (c)(1)(A) of this section to conduct activities described in subdivision (c)(1)(A) of this section is not prohibited under this section from using an unmanned aircraft system to conduct the activities described in subdivision (c)(1) of this section; or
2. An insurance company or a person acting on behalf of an insurance company from using an unmanned aircraft system for purposes of underwriting an insurance risk or investigating damage to insured property.
   a. (d) Unlawful use of unmanned aircraft system is:
      1. A Class B misdemeanor; or
      2. A Class A misdemeanor for a second or subsequent offense.

SECTION 2. Arkansas Code Title 16, Chapter 118, is amended to add an additional section to read as follows: 16-118-111. Civil actions for against operators of an unmanned aircraft system. A person who violates § 5-60-103 is also liable to the owner of the critical infrastructure that is the subject of the violation as follows:

1. Any actual damages sustained as a result of the violation, or ten thousand dollars ($10,000), whichever is greater;
2. Three (3) times actual damages, or ten thousand dollars ($10,000), whichever is greater, in a case in which the violation resulted in profit or monetary gain; and
3. The costs of an action brought under this section, together with reasonable attorney's fees as determined by the court.[65]

CALIFORNIA

The state of California has passed Assembly Bill (AB) 856 that became effective on December 2, 2015.[66] The governor vetoed one bill, Senate Bill (SB) 142, in September 2015.[67] There was also Assembly Joint Resolution (AJR) 6 in 2013 to consider California a test site;[68] Senate Concurrent Resolution (SCR) 16 in 2013 to recognize the contributions of the aerospace industry by proclaiming March 2013 as California Aerospace Month;[69] AB 14 in 2014 to add Division 16.8 to the Vehicle Code relating to unmanned aircraft;[70] AB 1592 in 2016 to add Section 38755 to the Vehicle Code, relating to autonomous vehicles;[71] AB 1662 in 2016 to add Part 6.5 to Division 9 of the Public Utilities Code, relating to aviation;[72] AB 37 in 2014 to add Section 6254.31 to the Government Code and to add Title 14 to Part 4 of the Penal Code, relating to unmanned aircraft systems;[73] AB 56 in 2015 to add Title 14 to Part 4 of the Penal Code, relating to unmanned aircraft systems;[74]

---

[65]House Bill 1770, The Arkansas State Legislature, http://www.arkleg.state.ar.us/assembly/2015/2015R/Acts/Act1019.pdf.

[66]AB 856, California Legislative Information, http://www.leginfo.ca.gov/pub/15-16/bill/asm/ab_0851-0900/ab_856_cfa_20150813_131420_sen_floor.html.

[67]SB 142, California Legislative Information, https://leginfo.legislature.ca.gov/faces/billHistoryClient.xhtml?bill_id=201520160SB142.

[68]AJR 6, California Legislative Information, https://leginfo.legislature.ca.gov/faces/billHistoryClient.xhtml?bill_id=201320140AJR6.

[69]SCR 16, California Legislative Information, http://leginfo.legislature.ca.gov/faces/billHistoryClient.xhtml?bill_id=201320140SCR16.

[70]AB 14, California Legislative Information, https://legiscan.com/CA/text/AB14/2015.

[71]AB 1592, California Legislative Information, available at http://leginfo.legislature.ca.gov/faces/billNavClient.xhtml?bill_id=201520160AB1592.

[72]AB 1662, California Legislative Information, https://leginfo.legislature.ca.gov/faces/billTextClient.xhtml?bill_id=201520160AB1662.

[73]AB 37, California Legislative Information, http://www.leginfo.ca.gov/pub/15-16/bill/asm/ab_0001-0050/ab_37_bill_20141201_introduced.html.

[74]AB 56, California Legislative Information, https://leginfo.legislature.ca.gov/faces/billNavClient.xhtml?bill_id=201520160AB56.

SB 167 in 2015 relating to interference by unmanned aircraft during forest fires;[75] SB 168 in 2015 to add Section 43.101 to the Civil Code, to add Chapter 4.5 to Part 2 of Division 3.6 of Title 1 of the Government Code, and to add Section 402.5 to the Penal Code, relating to unmanned aircraft, and declaring the urgency thereof, to take effect immediately;[76] SB 170 in 2015 to add Section 4577 to the Penal Code;[77] SB 262 in 2015 to add Chapter 4 to Title 1 of Part 4 of the Penal Code;[78] SB 263 in 2015 to amend Section 273.6 of the Penal Code relating to protective orders and the use of UAS;[79] SB 271 in 2015 to add Section 626.12 to the Penal Code;[80] SB 868 in 2016 also known as the State Remote Piloted Aircraft Act;[81] Senate Joint Resolution (SJR) 18 in 2015 relating to small UAS;[82] SB 807 in 2016 to add Section 43.101 to the Civil Code and to add Chapter 4.5 to Part 2 of Division 3.6 of Title 1 of the Government Code;[83] SB 809 in 2016 to add Section 626.12 to the Penal Code;[84] SB 810 in 2016 to add Section 402.5 to the Penal Code;[85] and SB 811 in 2016 to add Section 4577 to the Penal Code.[86]

Additionally, there are also a number of ordinances, like ordinance No. 942-2015 in Barstow to protect the safety and privacy of Barstow residents regarding the use of unmanned aircraft within the corporate city limits;[87] a No Drone Zone Ordinance in Berkeley in 2015 establishing a two year moratorium on drones;[88] Code 12.36.050 in Daly

---

[75]SB 167, California Legislative Information, http://www.leginfo.ca.gov/pub/15-16/bill/sen/sb_0151-0200/sb_167_bill_20150714_amended_sen_v98.html.

[76]SB 168, California Legislative Information, https://leginfo.legislature.ca.gov/faces/billNavClient.xhtml?bill_id=201520160SB168.

[77]SB 170, California Legislative Information, https://leginfo.legislature.ca.gov/faces/billNavClient.xhtml?bill_id=201520160SB170.

[78]SB 262, California Legislative Information, https://leginfo.legislature.ca.gov/faces/billNavClient.xhtml?bill_id=201520160SB262.

[79]SB 263, California Legislative Information, https://leginfo.legislature.ca.gov/faces/billNavClient.xhtml?bill_id=201520160SB263.

[80]SB 271, California Legislative Information, https://leginfo.legislature.ca.gov/faces/billNavClient.xhtml?bill_id=201520160SB271.

[81]SB 868, California Legislative Information, http://leginfo.legislature.ca.gov/faces/billNavClient.xhtml?bill_id=201520160SB868.

[82]SJR 18, California Legislative Information, https://leginfo.legislature.ca.gov/faces/billNavClient.xhtml?bill_id=201520160SJR18.

[83]SB 807, California Legislative Information, http://www.leginfo.ca.gov/pub/15-16/bill/sen/sb_0801-0850/sb_807_bill_20160104_introduced.html.

[84]SB 809, California Legislative Information, https://leginfo.legislature.ca.gov/faces/billNavClient.xhtml?bill_id=201520160SB809.

[85]SB 810, California Legislative Information, http://www.leginfo.ca.gov/pub/15-16/bill/sen/sb_0801-0850/sb_810_bill_20160104_introduced.html.

[86]SB 811, California Legislative Information, http://www.leginfo.ca.gov/pub/15-16/bill/sen/sb_0801-0850/sb_811_bill_20160104_introduced.html.

[87]City of Barstow Ordinance No. 942-2015, http://uavs.insct.org/wp-content/uploads/2016/04/Ordinance-No-942-2015.pdf.

[88]Berkeley Moratorium on Drones, http://www.mercurynews.com/2015/02/25/berkeley-council-passes-one-year-moratorium-on-police-drones/https://webcache.googleusercontent.com/search?q=cache:Uo0Oiq4ZFkMJ:https://www.cityofberkeley.info/Clerk/City_Council/2015/02_Feb/Documents/2015-02-24_Item_26_Establishing_a_Two.aspx+&cd=2&hl=en&ct=clnk&gl=us.

City prohibiting use of UAS of any size;[89] Ordinance No. 183912 in the city of Los Angeles adding Section 56.31 to Article 6 of Chapter V of the Los Angeles Municipal Code to impose community-based safety requirements on the operation of Model Aircraft and to impose restrictions consistent with certain FAA rules on the operation of both Model Aircraft and Civil UAS;[90] and an ordinance in Rancho Mirage, which will prohibit the flying of drones in residentially zoned areas of the city with some exceptions.[91]

Assembly Bill (AB) 856: An act to amend Section 1708.8 of the Civil Code, relating to privacy. The Governor approved this bill on October 6, 2015, and it reads as follows:

Under existing law, except as specified, a person is liable for physical invasion of privacy when that person knowingly enters onto the land of another person without permission or otherwise trespasses to capture any type of visual image, sound recording, or other physical impression of a person engaging in a private, personal, or familial activity and the invasion occurs in a manner that is offensive to a reasonable person. Under existing law, a person is liable for constructive invasion of privacy for the same activity, as specified, through the use of any device, regardless of whether there is a physical trespass. Existing law subjects a person who commits physical or constructive invasion of privacy to specified damages and civil fines. This bill expanded liability for physical invasion of privacy to additionally include a person knowingly entering into the airspace above the land of another person without permission, as provided.

The people of the State of California do enact as follows:

SECTION 1. Section 1708.8 of the Civil Code is amended to read:

1708.8. (a) A person is liable for physical invasion of privacy when the person knowingly enters onto the land or into the airspace above the land of another person without permission or otherwise commits a trespass in order to capture any type of visual image, sound recording, or other physical impression of the plaintiff engaging in a private, personal, or familial activity and the invasion occurs in a manner that is offensive to a reasonable person.

b. A person is liable for constructive invasion of privacy when the person attempts to capture, in a manner that is offensive to a reasonable person, any type of visual image, sound recording, or other physical impression of the plaintiff engaging in a private, personal, or familial activity, through the use of any device, regardless of whether there is a physical trespass, if this image, sound recording, or other

[89]Daly City Code 12.36.050, http://uavs.insct.org/wp-content/uploads/2016/04/Daly-City-CA-Code-of-Ordinances-12.36.050.pdf.

[90]Los Angeles Municipal Code Section 56.31, http://clkrep.lacity.org/onlinedocs/2015/15-0927_ord_183912_12-02-15.pdf.

[91]City of Rancho Mirage, Drone Prohibition Ordinance, http://uavs.insct.org/wp-content/uploads/2013/06/Rancho-Mirage_CA_Drone-Ordinance.pdf.

physical impression could not have been achieved without a trespass unless the device was used.

c. An assault or false imprisonment committed with the intent to capture any type of visual image, sound recording, or other physical impression of the plaintiff is subject to subdivisions (d), (e), and (h).

d. A person who commits any act described in subdivision (a), (b), or (c) is liable for up to three times the amount of any general and special damages that are proximately caused by the violation of this section. This person may also be liable for punitive damages, subject to proof according to Section 3294. If the plaintiff proves that the invasion of privacy was committed for a commercial purpose, the person shall also be subject to disgorgement to the plaintiff of any proceeds or other consideration obtained as a result of the violation of this section. A person who comes within the description of this subdivision is also subject to a civil fine of not less than five thousand dollars ($5,000) and not more than fifty thousand dollars ($50,000).

e. A person who directs, solicits, actually induces, or actually causes another person, regardless of whether there is an employer-employee relationship, to violate any provision of subdivision (a), (b), or (c) is liable for any general, special, and consequential damages resulting from each said violation. In addition, the person that directs, solicits, actually induces, or actually causes another person, regardless of whether there is an employer-employee relationship, to violate this section shall be liable for punitive damages to the extent that an employer would be subject to punitive damages pursuant to subdivision (b) of Section 3294. A person who comes within the description of this subdivision is also subject to a civil fine of not less than five thousand dollars ($5,000) and not more than fifty thousand dollars ($50,000).

f. (1) The transmission, publication, broadcast, sale, offer for sale, or other use of any visual image, sound recording, or other physical impression that was taken or captured in violation of subdivision (a), (b), or (c) shall not constitute a violation of this section unless the person, in the first transaction following the taking or capture of the visual image, sound recording, or other physical impression, publicly transmitted, published, broadcast, sold, or offered for sale the visual image, sound recording, or other physical impression with actual knowledge that it was taken or captured in violation of subdivision (a), (b), or (c), and provided compensation, consideration, or remuneration, monetary or otherwise, for the rights to the unlawfully obtained visual image, sound recording, or other physical impression.

   2. For the purposes of paragraph (1), "actual knowledge" means actual awareness, understanding, and recognition, obtained prior to the time at which

the person purchased or acquired the visual image, sound recording, or other physical impression, that the visual image, sound recording, or other physical impression was taken or captured in violation of subdivision (a), (b), or (c). The plaintiff shall establish actual knowledge by clear and convincing evidence.

3. Any person that publicly transmits, publishes, broadcasts, sells, or offers for sale, in any form, medium, format, or work, a visual image, sound recording, or other physical impression that was previously publicly transmitted, published, broadcast, sold, or offered for sale by another person, is exempt from liability under this section.
4. If a person's first public transmission, publication, broadcast, or sale or offer for sale of a visual image, sound recording, or other physical impression that was taken or captured in violation of subdivision (a), (b), or (c) does not constitute a violation of this section, that person's subsequent public transmission, publication, broadcast, sale, or offer for sale, in any form, medium, format, or work, of the visual image, sound recording, or other physical impression, does not constitute a violation of this section.
5. This section applies only to a visual image, sound recording, or other physical impression that is captured or taken in California in violation of subdivision (a), (b), or (c) after January 1, 2010, and shall not apply to any visual image, sound recording, or other physical impression taken or captured outside of California.
6. Nothing in this subdivision shall be construed to impair or limit a special motion to strike pursuant to Section 425.16, 425.17, or 425.18 of the Code of Civil Procedure.
7. This section shall not be construed to limit all other rights or remedies of the plaintiff in law or equity, including, but not limited to, the publication of private facts.

g. This section shall not be construed to impair or limit any otherwise lawful activities of law enforcement personnel or employees of governmental agencies or other entities, either public or private, who, in the course and scope of their employment, and supported by an articulable suspicion, attempt to capture any type of visual image, sound recording, or other physical impression of a person during an investigation, surveillance, or monitoring of any conduct to obtain evidence of suspected illegal activity or other misconduct, the suspected violation of any administrative rule or regulation, a suspected fraudulent conduct, or any activity involving a violation of law or business practices or conduct of public officials adversely affecting the public welfare, health, or safety.

h. In any action pursuant to this section, the court may grant equitable relief, including, but not limited to, an injunction and restraining order against further violations of subdivision (a), (b), or (c).

i. The rights and remedies provided in this section are cumulative and in addition to any other rights and remedies provided by law.

j. It is not a defense to a violation of this section that no image, recording, or physical impression was captured or sold.

k. For the purposes of this section, "for a commercial purpose" means any act done with the expectation of a sale, financial gain, or other consideration. A visual image, sound recording, or other physical impression shall not be found to have been, or intended to have been, captured for a commercial purpose unless it is intended to be, or was in fact, sold, published, or transmitted.

l. (1) For the purposes of this section, "private, personal, and familial activity" includes, but is not limited to:

   A. Intimate details of the plaintiff's personal life under circumstances in which the plaintiff has a reasonable expectation of privacy.
   B. Interaction with the plaintiff's family or significant others under circumstances in which the plaintiff has a reasonable expectation of privacy.
   C. If and only after the person has been convicted of violating Section 626.8 of the Penal Code, any activity that occurs when minors are present at any location set forth in subdivision (a) of Section 626.8 of the Penal Code.
   D. Any activity that occurs on a residential property under circumstances in which the plaintiff has a reasonable expectation of privacy.
   E. Other aspects of the plaintiff's private affairs or concerns under circumstances in which the plaintiff has a reasonable expectation of privacy.

(2) "Private, personal, and familial activity" does not include illegal or otherwise criminal activity as delineated in subdivision (g). However, "private, personal, and familial activity" shall include the activities of victims of crime in circumstances under which subdivision (a), (b), or (c) would apply.

m. (1) A proceeding to recover the civil fines specified in subdivision (d) or (e) may be brought in any court of competent jurisdiction by a county counsel or city attorney.

(2) Fines collected pursuant to this subdivision shall be allocated, as follows:

   A. One-half shall be allocated to the prosecuting agency.
   B. One-half shall be deposited in the Arts and Entertainment Fund, which is hereby created in the State Treasury.

(3) Funds in the Arts and Entertainment Fund created pursuant to paragraph may be expended by the California Arts Council, upon appropriation by the Legislature, to issue grants pursuant to the Dixon-Zenovich-Maddy California Arts Act

of 1975 (Chapter 9 (commencing with Section 8750) of Division 1 of Title 2 of the Government Code).

(4) The rights and remedies provided in this subdivision are cumulative and in addition to any other rights and remedies provided by law.

n. The provisions of this section are severable. If any provision of this section or its application is held invalid, that invalidity shall not affect other provisions or applications that can be given effect without the invalid provision or application.[92]

City of Los Angeles Ordinance 183912: The city of Los Angeles Ordinance 183912, Council File No. 15-0927, became effective on December 2, 2015, and reads as follows:[93]

An ordinance adding Section 56.31 to Article 6 of Chapter V of the Los Angeles Municipal Code to impose community-based safety requirements on the operation of Model Aircraft and to impose restrictions consistent with certain Federal Aviation Rules on the operation of both Model Aircraft and Civil Unmanned Aircraft Systems (UASs), commonly known as drones.

WHEREAS, the operation of Unmanned Aircraft such as Model Aircraft and Civil UASs can at times pose a hazard to full-scale aircraft in flight and to persons and property on the ground; and

WHEREAS, imposing community-based safety requirements on the operation of Model Aircraft and imposing restrictions on the operation of both Model Aircraft and Civil UASs consistent with Federal Aviation Rules is necessary to mitigate such risks and to protect the public from the hazards associated with the operation of Unmanned Aircraft.

NOW, THEREFORE, THE PEOPLE OF THE CITY OF LOS ANGELES DO ORDAIN AS FOLLOWS:

Section 1. A new Section 56.31 is added to Article 6 of Chapter V of the Los Angeles Municipal Code to read as follows:

SEC. 56.31. UNMANNED AIRCRAFT SYSTEMS.

a. For purposes of this section:
    1. "Unmanned Aircraft" shall mean an aircraft, including, but not limited to, an aircraft commonly known as a drone, that is operated without the possibility of direct human intervention from within or on the aircraft.
    2. "Unmanned Aircraft System" shall mean an Unmanned Aircraft and associated elements, including, but not limited to, any communication links and components that control the Unmanned Aircraft.

[92]AB 856, California Legislative Information, http://www.leginfo.ca.gov/pub/15-16/bill/asm/ab_0851-0900/ab_856_cfa_20150813_131420_sen_floor.html.

[93]California, City of Los Angeles, Ordinance 183912, http://clkrep.lacity.org/onlinedocs/2015/15-0927_ord_183912_12-02-15.pdf.

3. "Person" shall have the same meaning as set forth in Subsection (a) of Section 11.01 of this Code.
4. "Model Aircraft" shall mean an Unmanned Aircraft or Unmanned Aircraft System operated by any Person strictly for hobby or recreational purposes.
5. "Civil UAS" shall mean an Unmanned Aircraft or Unmanned Aircraft System operated by any Person for any purposes other than strictly hobby or recreational purposes, including, but not limited to, commercial purposes or in furtherance of, or incidental to, any business or media service or agency.
6. "Public UAS" shall mean an Unmanned Aircraft or Unmanned Aircraft System operated by any public agency for government related purposes.

b. The following shall apply to the operation of any Model Aircraft within the City of Los Angeles:
   1. No Person shall operate any Model Aircraft within the City of Los Angeles and within 5 miles of an airport without the prior express authorization of the airport air traffic control tower.
   2. No Person shall operate any Model Aircraft within the City of Los Angeles in a manner that interferes with manned aircraft, and shall always give way to any manned aircraft.
   3. No Person shall operate any Model Aircraft within the City of Los Angeles beyond the visual line of sight of the person operating the Model Aircraft. The operator must use his or her own natural vision (which includes vision corrected by standard eyeglasses or contact lenses) to observe the Model Aircraft. People other than the operator may not be used in lieu of the operator for maintaining visual line of sight. Visual line of sight means that the operator has an unobstructed view of the Model Aircraft. The use of vision-enhancing devices, such as binoculars, night vision goggles, powered vision magnifying devices, and goggles or other devices designed to provide a "first-person view" from the model, do not constitute the visual line of sight of the person operating the Model Aircraft.
   4. No Person shall operate any Model Aircraft within the City of Los Angeles other than during daylight hours defined as between official sunrise and official sunset for local time.
   5. No Person shall operate any Model Aircraft within the City of Los Angeles more than 400 feet above the earth's surface.
   6. Excluding takeoff and landing, no Person shall operate any Model Aircraft within the City of Los Angeles closer than 25 feet to any individual, except the operator or the operator's helper(s).

c. The following shall apply to the operation of any Model Aircraft or Civil UAS within the City of Los Angeles:

1. No Person shall operate any Model Aircraft or Civil UAS within the City of Los Angeles in a manner that is prohibited by any federal statute or regulation governing aeronautics.
2. No Person shall operate any Model Aircraft or Civil UAS within the City of Los Angeles in violation of any temporary flight restriction (TFR) or notice to airmen (NOTAM) issued by the Federal Aviation Administration.
3. No Person shall operate any Model Aircraft or Civil UAS within the City of Los Angeles in a careless or reckless manner so as to endanger the life or property of another. The standard for what constitutes careless and reckless operation under this section shall be the same as the standard set forth in any federal statutes or regulations governing aeronautics including but not limited to Federal Aviation Rule 91.13.

d. It shall be unlawful for any Person to violate or fail to comply with this section. Any Person violating the provisions of this section shall be guilty of a misdemeanor and subject to the provisions of Subsection (m) of Section 11.00 of this Code.

e. This section shall not apply to any Public UAS operated pursuant to, and in compliance with, the terms and conditions of any current and enforceable authorization granted by the Federal Aviation Administration.[94]

According to the Los Angeles Times, in January 2016, two men, Michael Ponce and Arvel Chappel, accused of flying their UAS in the vicinity of a hospital and police heliports became the first to face criminal charges under this ordinance. City prosecutors said that Chappel's UAS forced an LAPD air unit to change its landing path when he flew his device within a quarter-mile of the Hooper Heliport downtown. Days earlier, Ponce was cited for flying a UAS within three miles of several hospital heliports in Griffith Park. Police seized the UAS in both cases.[95]

COLORADO

On February 25, 2015, Colorado failed to pass SB 59, concerning the use of UAVs in limited circumstances.[96] Similarly, on April 21, 2015, HB 1115, an act concerning privacy in regard to emerging technologies, also died in chamber.[97] On January 26, 2016, HB 1020, entitled no drones near airports or jails, failed when the House Committee on Judiciary postponed indefinitely.[98] Colorado does have general provisions in the Colorado Parks and Wildlife Laws, Chapter W-0 making it unlawful to use a drone to look for, scout, or

[94]Ordinance 183912, California, City of Los Angeles, http://clkrep.lacity.org/onlinedocs/2015/15-0927_ord_183912_12-02-15.pdf.

[95]James Queally, *Los Angeles Times,* http://www.latimes.com/local/lanow/la-me-ln-city-attorney-drones-20160120-story.html.

[96]Senate Bill 59, Colorado General Assembly, https://legiscan.com/CO/text/SB059/id/1066405.

[97]House Bill 1115, Colorado General Assembly, https://legiscan.com/CO/bill/HB1115/2015.

[98]House Bill 1020, Colorado General Assembly, https://legiscan.com/CO/bill/HB1020/2016.

detect wildlife as an aid in the hunting or taking of wildlife.[99] There are two ordinances, one in Cherry Hills Village to establish a registration system for hobby operators and regulations pertaining to any UAS operating within the jurisdictional borders of the City,[100] and the other in the Town of Deer Trail to defend the sovereign airspace of the Town of Deer Trail, Colorado, and that of its citizens, their homes, businesses, related properties and interests, from unwanted incursions by small UAVs.[101]

CONNECTICUT

On June 2, 2015, SB 974, an act implementing the recommendations of the Legislative Program Review and Investigations Committee concerning the use of drones, died in chamber.[102] HB 5217, an act establishing the criminal use of an unmanned aircraft, also met a similar fate.[103]

DELAWARE

HB 195, an act to amend Title 11 of the Delaware Code relating to UAS, passed the house on June 23, 2015.[104]

FLORIDA

Florida currently has two bills that were signed into law: SB 92[105] and SB 766.[106] Additionally, there is HB 1301 in 2016, an act providing criminal penalties for convicted sexual predators who use or operate a drone to view or record an image of a minor under certain circumstances;[107] HB 459 providing for liability for damage to person or property in this state for which operation of drone was substantial contributing factor;[108] SB 510 providing criminal penalties for convicted sexual predators who use or operate a drone to view or record an image of a minor under certain circumstances;[109] and SB 642 declaring that a drone is a dangerous instrumentality.[110]

---

[99]Colorado General Provisions of the Parks and Wildlife Laws, http://cpw.state.co.us/Documents/RulesRegs/Regulations/Ch00.pdf.
[100]Cherry Hills Village, Municipal Code, http://uavs.insct.org/wp-content/uploads/2016/04/Cherry-Hills-Village-CO-Municipal-Code.pdf.
[101]Town of Deer Trail, Drone Ordinance, http://www.deertrailcolorado.com/drones.html.
[102]SB 974, Connecticut Legislative Information, https://legiscan.com/CT/text/SB00974/2015.
[103]HB 5217, Connecticut Legislative Information, https://www.cga.ct.gov/asp/cgabillstatus/cgabillstatus.asp?selBillType=Bill&bill_num=5217&which_year=2014.
[104]Delaware House Bill 195, http://openstates.org/de/bills/148/HB195/.
[105]SB 92, The Florida Senate, https://www.flsenate.gov/Session/Bill/2013/0092/BillText/Filed/PDF.
[106]SB 766, The Florida Senate, https://www.flsenate.gov/Session/Bill/2015/0766/BillText/er/PDF.
[107]Florida House Bill 1301, http://flsenate.gov/Session/Bill/2016/1301/BillText/Filed/PDF.
[108]Florida House Bill 459, https://www.flsenate.gov/Session/Bill/2016/0459.
[109]Florida Senate Bill 510, https://www.flsenate.gov/Session/Bill/2016/0510.
[110]Florida Senate Bill 642, https://www.flsenate.gov/Session/Bill/2016/0642.

Furthermore, there are city ordinances in Bonita Springs Section 28-41[111] and Miami Section 37-12.[112]

Senate Bill (SB) 92: On April 25, 2013, Governor Rick Scott signed into law SB 92, an act relating to searches and seizures; creating the "Freedom from Unwarranted Surveillance Act"; defining the terms "drone" and "law enforcement agency"; prohibiting a law enforcement agency from using a drone to gather evidence or other information; providing an exception; authorizing an aggrieved party to initiate a civil action in order to prevent or remedy a violation of the act; prohibiting a law enforcement agency from using in any court of law in this state evidence obtained or collected in violation of the act; providing an effective date.

Be it enacted by the Legislature of the State of Florida:

Section 1. Searches and seizure using a drone.

1. SHORT TITLE. This act may be cited as the "Freedom from Unwarranted Surveillance Act."
2. DEFINITIONS. As used in this act, the term:
    a. "Drone" means a powered, aerial vehicle that:
        1. Does not carry a human operator;
        2. Uses aerodynamic forces to provide vehicle lift;
        3. Can fly autonomously or be piloted remotely;
        4. Can be expendable or recoverable; and
        5. Can carry a lethal or nonlethal payload.
    b. "Law enforcement agency" means a lawfully established state or local public agency that is responsible for the prevention and detection of crime and the enforcement of penal, traffic, regulatory, game, or controlled substance laws.
3. PROHIBITED USE OF DRONES. A law enforcement agency may not use a drone to gather evidence or other information.
4. EXCEPTIONS. This act does not prohibit the use of a drone to counter a high risk of a terrorist attack by a specific individual or organization if the United States Secretary of Homeland Security determines that credible intelligence indicates that there is such a risk.
5. REMEDIES FOR VIOLATION. An aggrieved party may initiate a civil action against a law enforcement agency to obtain all appropriate relief in order to prevent or remedy a violation of this act.

[111]Ordinance Section 28-41, Bonita Springs, http://uavs.insct.org/wp-content/uploads/2016/04/Bonita-Springs-FL-Code-of-Ordinances-28-41.pdf.

[112]Ordinance Section 37-12, City of Miami, https://www2.municode.com/library/fl/miami/codes/code_of_ordinances?nodeId=PTIITHCO_CH37OFIS_S37-12PUSAUNAISYCOKNDR.

6. PROHIBITION ON USE OF EVIDENCE. Evidence obtained or collected in violation of this act is not admissible as evidence in a criminal prosecution in any court of law in this state. Section 2. This act shall take effect July 1, 2013.[113]

Senate Bill (SB) 766: On May 15, 2015, SB 766 was signed into law. This was an act relating to surveillance by a drone; amending s. 934.50, F.S.; defining terms; prohibiting a person, a state agency, or a political subdivision from using a drone to capture an image of privately owned real property or of the owner, tenant, occupant, invitee, or licensee of such property with the intent to conduct surveillance without his or her written consent if a reasonable expectation of privacy exists; specifying when a reasonable expectation of privacy may be presumed; authorizing the use of a drone by a person or entity engaged in a business or profession licensed by the state in certain circumstances; authorizing the use of a drone by an employee or contractor of a property appraiser for the purpose of assessing property for ad valorem taxation; authorizing the use of a drone by or on behalf of certain utilities for specified purposes; authorizing the use of a drone for aerial mapping under certain circumstances; authorizing the use of a drone for delivering cargo under certain circumstances; authorizing the use of a drone to capture certain images under certain circumstances; providing that an owner, tenant, occupant, invitee, or licensee may initiate a civil action for compensatory damages and may seek injunctive relief against a person, a state agency, or a political subdivision that violates the act; providing for construction; providing for the recovery of attorney fees and punitive damages; specifying that remedies provided by the act are cumulative to other remedies; providing an effective date.
Be it enacted by the Legislature of the State of Florida:
Section 1. Section 934.50, Florida Statutes, is amended to read:
934.50 Searches and seizure using a drone.

1. SHORT TITLE. This act may be cited as the "Freedom from Unwarranted Surveillance Act."
2. DEFINITIONS. As used in this act, the term:
    a. "Drone" means a powered, aerial vehicle that:
        1. Does not carry a human operator;
        2. Uses aerodynamic forces to provide vehicle lift;
        3. Can fly autonomously or be piloted remotely;
        4. Can be expendable or recoverable; and
        5. Can carry a lethal or nonlethal payload.
    b. "Image" means a record of thermal, infrared, ultraviolet, visible light, or other electromagnetic waves; sound waves; odors; or other physical phenomena

[113]SB 92, The Florida Senate, https://www.flsenate.gov/Session/Bill/2013/0092/BillText/Filed/PDF.

which captures conditions existing on or about real property or an individual located on that property.

c. "Imaging device" means a mechanical, digital, or electronic viewing device; still camera; camcorder; motion picture camera; or any other instrument, equipment, or format capable of recording, storing, or transmitting an image.

d. "Law enforcement agency" means a lawfully established state or local public agency that is responsible for the prevention and detection of crime, local government code enforcement, and the enforcement of penal, traffic, regulatory, game, or controlled substance laws.

e. "Surveillance" means:

   1. With respect to an owner, tenant, occupant, invitee, or licensee of privately owned real property, the observation of such persons with sufficient visual clarity to be able to obtain information about their identity, habits, conduct, movements, or whereabouts; or

   2. With respect to privately owned real property, the observation of such property's physical improvements with sufficient visual clarity to be able to determine unique identifying features or its occupancy by one or more persons.

3. PROHIBITED USE OF DRONES.

   a. A law enforcement agency may not use a drone to gather evidence or other information. (b) A person, a state agency, or a political subdivision as defined in s. 11.45 may not use a drone equipped with an imaging device to record an image of privately owned real property or of the owner, tenant, occupant, invitee, or licensee of such property with the intent to conduct surveillance on the individual or property captured in the image in violation of such person's reasonable expectation of privacy without his or her written consent. For purposes of this section, a person is presumed to have a reasonable expectation of privacy on his or her privately owned real property if he or she is not observable by persons located at ground level in a place where they have a legal right to be, regardless of whether he or she is observable from the air with the use of a drone.

4. EXCEPTIONS. This section does not prohibit the use of a drone:

   a. To counter a high risk of a terrorist attack by a specific individual or organization if the United States Secretary of Homeland Security determines that credible intelligence indicates that there is such a risk.

   b. If the law enforcement agency first obtains a search warrant signed by a judge authorizing the use of a drone.

   c. If the law enforcement agency possesses reasonable suspicion that, under particular circumstances, swift action is needed to prevent imminent danger to life or serious damage to property, to forestall the imminent escape of a suspect or the destruction of evidence, or to achieve purposes including, but not limited to, facilitating the search for a missing person.

d. By a person or an entity engaged in a business or profession licensed by the state, or by an agent, employee, or contractor thereof, if the drone is used only to perform reasonable tasks within the scope of practice or activities permitted under such person's or entity's license. However, this exception does not apply to a profession in which the licensee's authorized scope of practice includes obtaining information about the identity, habits, conduct, movements, whereabouts, affiliations, associations, transactions, reputation, or character of any society, person, or group of persons.
e. By an employee or a contractor of a property appraiser who uses a drone solely for the purpose of assessing property for ad valorem taxation.
f. To capture images by or for an electric, water, or natural gas utility:
    1. For operations and maintenance of utility facilities, including facilities used in the generation, transmission, or distribution of electricity, gas, or water, for the purpose of maintaining utility system reliability and integrity;
    2. For inspecting utility facilities, including pipelines, to determine construction, repair, maintenance, or replacement needs before, during, and after construction of such facilities;
    3. For assessing vegetation growth for the purpose of maintaining clearances on utility rights-of-way;
    4. For utility routing, siting and permitting for the purpose of constructing utility facilities or providing utility service; or
    5. For conducting environmental monitoring, as provided by federal, state, or local law, rule, or permit.
g. For aerial mapping, if the person or entity using a drone for this purpose is operating in compliance with Federal Aviation Administration regulations.
h. To deliver cargo, if the person or entity using a drone for this purpose is operating in compliance with Federal Aviation Administration regulations.
i. To capture images necessary for the safe operation or navigation of a drone that is being used for a purpose allowed under federal or Florida law.

5. REMEDIES FOR VIOLATION.
    a. An aggrieved party may initiate a civil action against a law enforcement agency to obtain all appropriate relief in order to prevent or remedy a violation of this section.
    b. The owner, tenant, occupant, invitee, or licensee of privately owned real property may initiate a civil action for compensatory damages for violations

of this section and may seek injunctive relief to prevent future violations of this section against a person, state agency, or political subdivision that violates paragraph (3)(b). In such action, the prevailing party is entitled to recover reasonable attorney fees from the nonprevailing party based on the actual and reasonable time expended by his or her attorney billed at an appropriate hourly rate and, in cases in which the payment of such a fee is contingent on the outcome, without a multiplier, unless the action is tried to verdict, in which case a multiplier of up to twice the actual value of the time expended may be awarded in the discretion of the trial court. (c) Punitive damages for a violation of paragraph (3)(b) may be sought against a person subject to other requirements and limitations of law, including, but not limited to, part II of chapter 768 and case law.

c. The remedies provided for a violation of paragraph (3)(b) are cumulative to other existing remedies.

6. PROHIBITION ON USE OF EVIDENCE. Evidence obtained or collected in violation of this act is not admissible as evidence in a criminal prosecution in any court of law in this Section 2. This act shall take effect July 1, 2015.[114]

GEORGIA

Georgia currently has three enacted bills pertaining to UAS, namely, HR 80,[115] HR 81,[116] and SR 172.[117]

Additionally, there is HB 157 relating to lawful flight over lands and waters of this state,[118] HB 44 relating to the regulation of UAS and their prohibition in certain areas,[119] HB 5 relating to invasions of privacy using unmanned aircraft and the prohibition of capturing certain images by unmanned aircraft,[120] HB 779 relating to crimes and offenses and the use of UAS for capturing images,[121] HR 744 creating the House Study Committee on the use of drones,[122] and SB 113 to create a new class of motor vehicles to be known as autonomous vehicles.[123]

Furthermore, there are ordinances in Augusta to place regulations on the launching and operation of UAS[124] and Cherokee County Section 42-55 governing the use of

[114]SB 766, The Florida Senate, https://www.flsenate.gov/Session/Bill/2015/0766/BillText/er/PDF.
[115]HR 80, The Georgia General Assembly, http://www.legis.ga.gov/Legislation/en-US/display/20132014/HR/80.
[116]HR 81, The Georgia General Assembly, http://www.legis.ga.gov/Legislation/20132014/128994.pdf.
[117]SR 172, The Georgia General Assembly, http://www.legis.ga.gov/Legislation/20132014/129816.pdf.
[118]HB 157, Georgia General Assembly, http://www.legis.ga.gov/Legislation/en-US/display/20152016/HB/157.
[119]HB 44, Georgia General Assembly, http://www.legis.ga.gov/Legislation/en-US/display/20152016/HB/44.
[120]HB 5, Georgia General Assembly, http://www.legis.ga.gov/Legislation/en-US/display/20152016/HB/5.
[121]HB 779, Georgia General Assembly, http://www.legis.ga.gov/legislation/en-US/Display/20152016/HB/779.
[122]HR 744, Georgia General Assembly, http://www.legis.ga.gov/Legislation/en-US/display/20152016/HR/744.
[123]SB 113, Georgia General Assembly, http://www.legis.ga.gov/Legislation/en-US/display/20152016/SB/113.
[124]Augusta Ordinance, http://uavs.insct.org/wp-content/uploads/2016/04/Augusta-Georiga.pdf.

engine-powered models, toys, and UAS making it unlawful to start, fly, or use them in certain areas.[125]

House Resolution (HR) 80: On February 5, 2013, Georgia General Assembly adopted HR 80, and on February 14, 2014, passed it on to the Senate. This is a resolution: Recognizing and commending the Georgia Aerospace Industry for its significant beneficial impact to the people and economy of the State of Georgia and invite Donald Mitchell, Co-Chair of Georgia Aerospace Policy Working Group, and Chance McColl, Co-Chair of Georgia Aerospace Policy Working Group, to be recognized by the House of Representatives; and for other purposes.

WHEREAS, Georgia is a world leader aerospace as the state is home to more than 500 companies employing more than 84,000 Georgians; and

WHEREAS, Delta Air Lines is the world's second largest airline with headquarters in Atlanta and half of its 60,000 member workforce located in Georgia; and

WHEREAS, Gulfstream Aerospace employs more than 7,000 highly skilled workers in Savannah and Brunswick, and Lockheed Martin employs approximately 7,000 workers at its Marietta plant; and

WHEREAS, Thrush and Maule operate aircraft manufacturing facilities in Albany and Moultrie, respectively, while companies throughout the state manufacture key components for major aircraft engine programs throughout the world; and

WHEREAS, Georgia's aerospace industry is a world leader in aerospace exports; between 2002 and 2011, the value of Georgia's aerospace exports grew a remarkable 838% to over $5.75 billion in 2011 (increasing over 30% just between 2010 and 2011), ranking Georgia fourth among U.S. states; and

WHEREAS, Georgia is home to the world's busiest airport, Hartsfield-Jackson Atlanta International Airport, where aviation activities and transportation hubs contribute to the employment of 471,175 people and $62.6 billion in statewide total economic impact; and

WHEREAS, Georgia aerospace is a leader in the maintenance, repair, and overhaul (MRO), of military and civil aircraft engines and equipment and is home to the largest MRO facility in the U.S. (Delta TechOps) and one of the world's largest military MRO facilities (Warner Robins Air Logistics Complex); and

WHEREAS, Georgia offers world class aerospace education programs and is a leader in research and development through public and private universities conducting over $60 million in specialized aerospace research and over $500 million in general engineering research annually; and

WHEREAS, Georgia's aerospace industry, academia, and state and local governments are collaborating to make Georgia a leader in the rapidly expanding unmanned aerial vehicles

[125] Cherokee County Ordinance Section 42-55, http://www.crpa.net/wp-content/uploads/2016/01/Parks-Ordinance-June-16-2015.pdf.

market, projected to exceed $10 billion annually by 2021, by working with the Federal Aviation Administration to develop the technologies and systems required to integrate unmanned aerial vehicles in the National Air Space System.

NOW, THEREFORE, BE IT RESOLVED BY THE HOUSE OF REPRESENTATIVES that the members of this body recognize the importance of the Georgia Aerospace Industry to the State of Georgia, support efforts to maintain and strengthen the industry, help to foster its continued growth within the State of Georgia for the benefit of all Georgians, and invite Donald Mitchell and Chance McColl to be recognized by the House of Representatives at a date and time designated by the Speaker of the House.

BE IT FURTHER RESOLVED that the Clerk of the House of Representatives is authorized and directed to transmit an appropriate copy of this resolution to the Georgia Aerospace Advisory Council of the Georgia Center of Innovation for Aerospace.[126]

House Resolution (HR) 81: On January 29, 2013, the following HR 81 passed the Georgia House. This resolution reads as follows: Recognizing and commending the Georgia Aerospace Industry for its significant beneficial impact to the people and economy of the State of Georgia; and for other purposes.

WHEREAS, Georgia is a world leader aerospace as the state is home to more than 500 companies employing more than 84,000 Georgians; and

WHEREAS, Delta Air Lines is the world's second largest airline with headquarters in Atlanta and half of its 60,000 member workforce located in Georgia; and

WHEREAS, Gulfstream Aerospace employs more than 7,000 highly skilled workers in Savannah and Brunswick, and Lockheed Martin employs approximately 7,000 workers at its Marietta plant; and

WHEREAS, Thrush and Maule operate aircraft manufacturing facilities in Albany and Moultrie, respectively, while companies throughout the state manufacture key components for major aircraft engine programs throughout the world; and

WHEREAS, Georgia's aerospace industry is a world leader in aerospace exports; between 2002 and 2011, the value of Georgia's aerospace exports grew a remarkable 838% to over $5.75 billion in 2011 (increasing over 30% just between 2010 and 2011), ranking Georgia fourth among U.S. states; and

WHEREAS, Georgia is home to the world's busiest airport, Hartsfield-Jackson Atlanta International Airport, where aviation activities and transportation hubs contribute to the employment of 471,175 people and $62.6 billion in statewide total economic impact; and WHEREAS, Georgia aerospace is a leader in the maintenance, repair, and overhaul (MRO), of military and civil aircraft engines and equipment and is home to the largest MRO facility in the U.S. (Delta TechOps) and one of the world's largest military MRO facilities (Warner Robins Air Logistics Complex); and

[126]HR 80, The Georgia General Assembly, http://www.legis.ga.gov/Legislation/en-US/display/20132014/HR/80.

WHEREAS, Georgia offers world class aerospace education programs and is a leader in research and development through public and private universities conducting over $60 million in specialized aerospace research and over $500 million in general engineering research annually; and

WHEREAS, Georgia's aerospace industry, academia, and state and local governments are collaborating to make Georgia a leader in the rapidly expanding unmanned aerial vehicles market, projected to exceed $10 billion annually by 2021, by working with the Federal Aviation Administration to develop the technologies and systems required to integrate unmanned aerial vehicles in the National Air Space System.

NOW, THEREFORE, BE IT RESOLVED BY THE HOUSE OF REPRESENTATIVES that the members of this body recognize the importance of the Georgia Aerospace Industry to the State of Georgia and support efforts to maintain and strengthen the industry and help to foster its continued growth within the State of Georgia for the benefit of all Georgians.

BE IT FURTHER RESOLVED that the Clerk of the House of Representatives is authorized and directed to transmit an appropriate copy of this resolution to the Georgia Aerospace Advisory Council of the Georgia Center of Innovation for Aerospace.[127]

Senate Resolution (SR) 172: On February 6, 2013, the Georgia Senate read and adopted SR 172, a resolution that reads as follows: Recognizing and commending the Georgia Aerospace Industry for its significant beneficial impact on the people and economy of the State of Georgia; and for other purposes.

WHEREAS, Georgia is a world leader in aerospace as this state is home to more than 500 companies employing more than 84,000 Georgians; and

WHEREAS, Delta Air Lines is the world's second largest airline with headquarters in Atlanta and half of its 60,000 member workforce located in Georgia; and

WHEREAS, Gulfstream Aerospace employs more than 7,000 highly skilled workers in Savannah and Brunswick, and Lockheed Martin employs approximately 7,000 workers at its Marietta plant; and

WHEREAS, Thrush and Maule operate aircraft manufacturing facilities in Albany and Moultrie, respectively, while companies throughout this state manufacture key components for major aircraft engine programs throughout the world; and

WHEREAS, Georgia's aerospace industry is a world leader in aerospace exports; between 2002 and 2011, the value of Georgia's aerospace exports grew a remarkable 838 percent to over $5.75 billion in 2011 (increasing over 30 percent just between 2010 and 2011), ranking Georgia fourth among U.S. states; and

WHEREAS, Georgia is home to the world's busiest airport, Hartsfield-Jackson Atlanta International Airport, where aviation activities and transportation hubs contribute to the employment of 471,175 people and $62.6 billion in statewide total economic impact;

---

[127]HR 81, The Georgia General Assembly, http://www.legis.ga.gov/Legislation/20132014/128994.pdf.

and WHEREAS, Georgia aerospace is a leader in the maintenance, repair, and overhaul (MRO), of military and civil aircraft engines and equipment and is home to the largest MRO facility in the U.S. (Delta TechOps) and one of the world's largest military MRO facilities (Warner Robins Air Logistics Complex); and
WHEREAS, Georgia offers world class aerospace education programs and is a leader in research and development through public and private universities conducting over $60 million in specialized aerospace research and over $500 million in general engineering research annually; and WHEREAS, Georgia's aerospace industry, academia, and state and local governments are collaborating to make Georgia a leader in the rapidly expanding unmanned aerial vehicles market, projected to exceed $10 billion annually by 2021, by working with the Federal Aviation Administration to develop the technologies and systems required to integrate unmanned aerial vehicles into the National Air Space System.
NOW, THEREFORE, BE IT RESOLVED BY THE SENATE that the members of this body recognize the importance of the Georgia Aerospace Industry to the State of Georgia, support efforts to maintain and strengthen the industry, and help to foster its continued growth within the State of Georgia for the benefit of all Georgians.
BE IT FURTHER RESOLVED that the Secretary of the Senate is authorized and directed to transmit an appropriate copy of this resolution to the Georgia Aerospace Advisory Council of the Georgia Center of Innovation for Aerospace.[128]

HAWAII

Hawaii has passed three bills pertaining to UAS, namely, SB 661,[129] SB 1221,[130] and SB 2608.[131]

Additionally, Hawaii has introduced HB 1458 in 2015 to address autonomous motor vehicles;[132] HB 1522 in 2016 to prohibit the use, navigation, piloting, or operation of a UAS less than 500 feet above ground level within the airspace above a residential property without the express consent of the property's owner or lessee;[133] HB 1582 in 2016 to regulate UAS in the state by creating the offenses of violation of privacy through use of a UAS in the first and second degree;[134] HB 1632 in 2016 to amend the offenses of violation of privacy because of UAS;[135] HB 1914 in 2016 relating to the use of UAS

---

[128]SR 172, The Georgia General Assembly, http://www.legis.ga.gov/Legislation/20132014/129816.pdf.
[129]SB 661, Hawaii State Legislature, http://www.capitol.hawaii.gov/session2015/testimony/SB661_SD2_TESTIMONY_EDB_03-24-15_.PDF.
[130]SB 1221, Hawaii State Legislature, https://legiscan.com/HI/text/SB1221/id/743043.
[131]SB 2608, Hawaii State Legislature, https://legiscan.com/HI/bill/SB2608/2014.
[132]HB 1458, Hawaii State Legislature, http://openstates.org/hi/bills/2015%20Regular%20Session/HB1458/.
[133]HB 1522, Hawaii State Legislature, http://www.capitol.hawaii.gov/measure_indiv.aspx?billtype=HB&billnumber=1522.
[134]HB 1582, Hawaii State Legislature, https://legiscan.com/HI/text/HB1582/2016.
[135]HB 1632, Hawaii State Legislature, https://legiscan.com/HI/bill/HB1632/2016.

in water rescues;[136] HB 1979 relating to unwarranted surveillance;[137] HB 2021 in 2016 that increases the fine for conviction of certain offenses when a UAS is used to commit the offense and also provides that the offense of simple trespass can be committed when a person knowingly enters or remains unlawfully in the airspace directly above premises up to 400 feet above ground level;[138] HB 609 in 2015 to prescribe limits on the use of UAS by private entities;[139] HB 632 in 2015 authorizing for testing purposes the operation of autonomous vehicles in the state and requiring the Department of Transportation to establish an application and approval process and report annually to the legislature;[140] HB 636 in 2015 to specify that the purpose of the aerospace advisory committee includes tracking drone development and use and requiring the aerospace advisory committee to appoint a drone regulation temporary working group to study and report to the legislature on recommended legislation, and finally to provide funding for administrative costs associated with the above;[141] HB 637 in 2015 to amend the offenses of violation of privacy in the first and second degrees to specifically address the use of drones in the commission of these offenses;[142] HCR 196 in 2015 requesting the state department of transportation to convene a drone policy task force and recommend a statewide policy for UAS that protects privacy and allows the use of UAS for public and private applications;[143] HR 133 in 2015 requesting the state department of transportation to convene a drone policy task force and recommend a statewide policy for UAS that protects privacy and allows the use of UAS for public and private applications; [144] SB 1329 in 2015 relating to privacy and freedom from unwarranted surveillance;[145] SB 2095 in 2016 relating to drone operator liability insurance;[146] SB 579 in 2015 to ensure that unmanned aircraft technology is used strictly to assist or benefit Hawaii residents and not used in a manner that will impede their right to privacy;[147] SB 630 in 2016 to allow a person who possesses a valid Hawaii driver license to operate an autonomous motor vehicle that employs autonomous technology.[148]

Senate Bill (SB) 661: Hawaii State Legislature adopted SB 661 into Act 208 on July 2, 2015. *Act 208* relates to unmanned aerial systems test sites, and reads as follows:

Be it enacted by the Legislature of the State of Hawaii:

---

[136]HB 1914, Hawaii State Legislature, https://legiscan.com/HI/text/HB1914/2016.

[137]HB 1979, Hawaii State Legislature, https://legiscan.com/HI/text/HB1979/2016.

[138]HB 2021, Hawaii State Legislature, https://legiscan.com/HI/bill/HB2021/2016.

[139]HB 609, Hawaii State Legislature, https://legiscan.com/HI/text/HB609/2016.

[140]HB 632, Hawaii State Legislature, http://www.capitol.hawaii.gov/measure_indiv.aspx?billtype=HB&billnumber=632.

[141]HB 636, Hawaii State Legislature, http://openstates.org/hi/bills/2015%20Regular%20Session/HB636/.

[142]HB 637, Hawaii State Legislature, http://openstates.org/hi/bills/2015%20Regular%20Session/HB637/.

[143]HCR 196, Hawaii State Legislature, http://openstates.org/hi/bills/2015%20Regular%20Session/HCR196/.

[144]HR 133, Hawaii State Legislature, http://openstates.org/hi/bills/2015%20Regular%20Session/HR133/.

[145]SB 1329, Hawaii State Legislature, http://openstates.org/hi/bills/2015%20Regular%20Session/SB1329/#actions.

[146]SB 2095, Hawaii State Legislature, https://legiscan.com/HI/bill/SB2095/2016.

[147]SB 579, Hawaii State Legislature, http://openstates.org/hi/bills/2015%20Regular%20Session/SB579/.

[148]SB 630, Hawaii State Legislature, http://www.capitol.hawaii.gov/measure_indiv.aspx?billtype=SB&billnumber=630.

SECTION 1. The legislature finds that there is a national need to safely integrate unmanned aerial systems into the national air space. The integration of unmanned aerial systems flights into the national air space will generate commercial potential of this technology, estimated to be worth more than $13,600,000,000 during the first three years of integration and more than $82,000,000,000 between 2015 and 2025, and create approximately 103,776 new jobs by 2025.

The legislature further finds that Hawaii offers many unique qualities to support unmanned aerial systems operations including: expansive over-water areas unencumbered by other aviation uses; proximity to the United States Pacific Command, which is projected to be a significant user of future unmanned aerial systems; opportunities for joint operations with Kauai's Pacific missile range facility; and opportunities for long-range point-to-point tests with partner ranges in Alaska and Oregon. Hawaii's proposed test ranges link to military and restricted areas used for current unmanned aerial systems operations. These sites include the Pohakuloa training area on the island of Hawaii, Bradshaw and Wheeler Army airfields, and Kauai's Pacific missile range facility. Test points within the ranges will be used to support shore- and ship-based development, testing and certification of new unmanned aerial systems, training and crew certification of operational unmanned aerial systems, and development of expanded and joint capabilities involving existing communications systems and operations tactics using unmanned aerial systems. The proposed Hawaii ranges have provided an important focus for the development of scientific applications of unmanned aerial systems, including numerous test flights.

The legislature additionally finds existing and potential civilian uses of unmanned aerial systems are wide-ranging, including emergency search and rescue operations; wildfire detection and management; fisheries management; agricultural monitoring; reef health surveys; hazardous spills monitoring; dam and reservoir overflow detection; tsunami damage surveys and assessment; algal bloom detection and mapping; air quality monitoring; motor vehicle traffic management; lava flow monitoring; aerial photography for mapping; disaster management and damage assessment; power line monitoring; flood and pollution control; land use surveys; watershed management; wildlife tracking; geographical, geological, and archaeological surveys; atmospheric monitoring for commercial airline turbulence avoidance; and light detection and ranging mapping of coastal topography to detect beach erosion. In developing these applications, innovative research, business, and education opportunities will emerge, including the development of miniaturized, high performance remote sensing instruments; unmanned aerial systems tracking systems, including command and control hardware and software; training courses and certification programs for unmanned aerial systems operators; and education programs for potential users of unmanned aerial systems technologies.

The legislature also finds that in 2012, the United States Congress directed the Federal Aviation Administration to establish unmanned aerial systems research programs at six national test sites for the development of unmanned aerial systems operating standards and regulations. Hawaii, in partnership with Alaska and Oregon, submitted a tri-state proposal to establish the Pan-Pacific Unmanned Aerial Systems Test Range Complex and was designated by the Federal Aviation Administration as one of the six national test site operators, providing the tri-state team six months to organize and implement the Pan-Pacific Unmanned Aerial Systems Test Range Complex. The University of Alaska has established a board of directors to support performance of the management team and is establishing a Pan-Pacific Unmanned Aerial Systems Test Range Complex management team, to include a chief operating officer from all three states.

The purpose of this Act is to establish a Hawaii unmanned aerial systems test site chief operating officer position to serve on the Pan-Pacific Unmanned Aerial Systems Test Range Complex management team, establish a Hawaii unmanned aerial systems test site advisory board that will formulate an implementation plan and oversee test site development in the State, and appropriate funds for personnel and procurement costs associated with establishing the Hawaii unmanned aerial systems test site.

SECTION 2. Chapter 201, Hawaii Revised Statutes, is amended by adding two new sections to part V to be appropriately designated and to read as follows:

§201-Hawaii unmanned aerial systems test site advisory board; established.

a. There is established a Hawaii unmanned aerial systems test site advisory board, as a subcommittee of the Hawaii aerospace advisory committee, to oversee the planning and operation of the Hawaii unmanned aerial systems test site.
b. The board shall be composed of seven members as follows:
    1. The adjutant general, or the adjutant general's designee;
    2. The director of transportation, or the director's designee;
    3. The director of business, economic development, and tourism, or the director's designee;
    4. The president of the University of Hawaii, or the president's designee;
    5. One member representing the Hawaii business community, to be appointed by the governor pursuant to section 26-34; and
    6. Two members representing the aerospace or aviation industries, to be appointed by the governor pursuant to section 26-34.

    Each board member shall serve for a term of four years; provided that the initial terms shall be staggered, as determined by the governor.
c. Members of the board shall not receive compensation for their services but shall be reimbursed for necessary expenses, including travel expenses, incurred in the performance of their duties under this section.

§201-Hawaii unmanned aerial systems test site chief operating officer.

The department shall hire a chief operating officer who shall manage the operations of Hawaii's unmanned aerial systems test site. The chief operating officer shall:

1. Monitor, oversee, and recommend operations of unmanned aerial systems test site activities;
2. Facilitate opportunities for public and private use of unmanned aerial systems test site facilities;
3. Assist unmanned aerial systems test ranges;
4. Leverage aerospace and related technological capabilities in the State's academic, public, and private sectors to support testing and evaluation at Hawaii's unmanned aerial systems test ranges;
5. Promote innovative education and workforce development programs to enhance public awareness of the benefits and opportunities that unmanned aerial systems technologies and applications can bring to the State;
6. Monitor national and global trends in unmanned aerial systems development and testing, and recommend policies and programs to advance unmanned aerial systems testing in Hawaii;
7. Establish and maintain a public website with updated information on the program and provide information on the Hawaii unmanned aerial systems test site initiative;
8. Contract for services and implement agreements, subject to chapter 103D, as may be necessary to conduct operations at Hawaii's unmanned aerial systems test ranges;
9. Serve as Hawaii's representative on the Pan-Pacific Unmanned Aerial Systems Test Range Complex management team; and
10. Submit an annual report to the legislature no later than twenty days prior to the convening of each regular session, which shall include the status of work, expenditures, and trends regarding Hawaii's unmanned aerial systems test site.

The chief operating officer shall be subject to chapter 84 and may employ one administrative assistant for Hawaii's unmanned aerial systems test site operations and activities. SECTION 3. Section 84-17, Hawaii Revised Statutes, is amended by amending subsection (c) to read as follows:

a. The following persons shall file annually with the state ethics commission a disclosure of financial interests:
    1. The governor, the lieutenant governor, the members of the legislature, and delegates to the constitutional convention; provided that delegates to the constitutional convention shall only be required to file initial disclosures;

2. The directors and their deputies, the division chiefs, the executive directors and the executive secretaries and their deputies, the purchasing agents and the fiscal officers, regardless of the titles by which the foregoing persons are designated, of every state agency and department;
3. The permanent employees of the legislature and its service agencies, other than persons employed in clerical, secretarial, or similar positions;
4. The administrative director of the State, and the assistants in the office of the governor and the lieutenant governor, other than persons employed in clerical, secretarial, or similar positions;
5. The hearings officers of every state agency and department;
6. The president, the vice presidents, assistant vice presidents, the chancellors, and the provosts of the University of Hawaii and its community colleges;
7. The superintendent, the deputy superintendent, the assistant superintendents, the complex area superintendents, the state librarian, and the deputy state librarian of the department of education;
8. The administrative director and the deputy director of the courts;
9. The members of every state board or commission whose original terms of office are for periods exceeding one year and whose functions are not solely advisory;
10. Candidates for state elective offices, including candidates for election to the constitutional convention, provided that candidates shall only be required to file initial disclosures;
11. The administrator and assistant administrator of the office of Hawaiian affairs; and
12. The Hawaii unmanned aerial systems test site chief operating officer.

SECTION 4. There is appropriated out of the general revenues of the State of Hawaii the sum of $150,000 or so much thereof as may be necessary for fiscal year 2015–2016 for the purpose of staffing and operating Hawaii's unmanned aerial systems test site activities.

The sums appropriated shall be expended by the department of business, economic development, and tourism for the purposes of this Act.

SECTION 5. Statutory material to be repealed is bracketed and stricken. New statutory material is underscored.

SECTION 6. This Act shall take effect on July 1, 2015.[149]

Senate Bill (SB) 1221: In 2013, the Hawaii State Legislature passed SB 1221, which reads as follows: A bill for an act relating to higher education. Be it enacted by the legislature of the State of Hawaii:

[149]SB 661, Hawaii State Legislature, http://www.capitol.hawaii.gov/session2015/testimony/SB661_SD2_TESTIMONY_EDB_03-24-15_.PDF.

Section 1. Pursuant to Senate Concurrent Resolution No. 156, S.D. 1, adopted during the regular session of 2012, the legislature urged the creation of an exploratory committee to consider the feasibility of, and necessary resources for, establishing an international aviation training center at the Hilo international airport and an advanced aviation degree training program at the University of Hawaii at Hilo. Tasked with submitting a report to the legislature, the exploratory committee recommended the establishment of three programs and the appropriation of funds by the legislature to hire a program coordinator and a technical support staff member, contracted through the University of Hawaii, who will conduct and draft the necessary planning and studies required for the proposed programs. The proposed professional helicopter pilot program and the proposed professional unmanned aircraft systems pilot program are to be administered through Hawaii community college. The proposed professional airplane pilot program is to be administered by the University of Hawaii at Hilo.

The program coordinator and technical support staff member will work in conjunction with the chancellors of Hawaii community college and the University of Hawaii at Hilo, the exploratory committee, and the state aviation community to complete the necessary studies and planning for the programs' future implementation.

Accordingly, the purpose of this Act is to appropriate funds for the program coordinator and technical support staff member, as recommended by the exploratory committee.

Section 2. There is appropriated out of the general revenues of the State of Hawaii the sum of $100,000 or so much thereof as may be necessary for fiscal year 2013–2014 for a program coordinator and a technical support staff member to complete the necessary planning and studies required for an international flight training center and associated proposed aeronautical training programs at Hawaii community college and the University of Hawaii at Hilo.

The sum appropriated shall be expended by the University of Hawaii for the purposes of this Act.

Section 3. This Act shall take effect on July 1, 2013.[150]

Senate Bill (SB) 2608: In 2014, the Hawaii State Legislature passed SB 2608, a bill for an act relating to unmanned aircrafts, which reads as follows:

Be it enacted by the Legislature of the State of Hawaii:

Section 1. The legislature finds that unmanned aircraft technology is rapidly developing. Public Law 112-95 (2012) requires the Federal Aviation Administration to "accelerate the integration of civil unmanned aircraft systems into the national airspace system." While unmanned aircraft technology offers more efficient and cost saving methods to gather information, this new and innovative technology may also provide opportunities for abuse.

---

[150]SB 1221, Hawaii State Legislature, https://legiscan.com/HI/text/SB1221/id/743043.

In recognizing the progressive opportunities that unmanned aircrafts provide, the legislature also recognizes its duty to protect Hawaii residents from threats to their constitutional right to privacy. The legislature further finds that in 2013, nine other states passed legislation regulating government deployment of unmanned aircraft technology. However, the legislature acknowledges that the Federal Aviation Administration has been mandated by the United States Congress to research and develop a regulatory scheme for the commercial use of unmanned aircrafts. As such, it is the intent of the legislature to wait for the findings from the Federal Aviation Administration before regulating the commercial use of unmanned aircrafts.

The purpose of this Act is to ensure that unmanned aircraft technology is used strictly to assist or benefit Hawaii residents and not used in a manner that will impede their right to privacy.

Section 2. Chapter 803, Hawaii Revised Statutes, is amended by adding a new part to be appropriately designated and to read as follows:

PART _. UNMANNED AIRCRAFTS

§803-A Definitions. As used in this part unless otherwise required by context:

"Law enforcement agency" means any agency of the State, county, or United States that is vested by law with the duty to maintain public order and to enforce the criminal laws.

"Model aircraft" means an unmanned aircraft that is:

1. Capable of sustained flight in the atmosphere;
2. Flown within visual line of sight of the person operating the aircraft; and
3. Flown for hobby or recreational purposes.

"Unmanned aircraft" means an aircraft that is operated without direct human intervention onboard the aircraft.

§803-B Use of unmanned aircrafts prohibited. Except as provided under this part, it shall be unlawful for any law enforcement agency to use an unmanned aircraft to conduct surveillance, including but not limited to capturing images, photographs, or recordings. Nothing in this part shall be construed to limit or prohibit the use of model aircrafts for commercial, hobby, or recreational purposes.

§803-C Exceptions. This part shall not apply to a law enforcement agency under any of the following circumstances:

1. A finding based on credible intelligence is made by the United States Secretary of Homeland Security that the use of an unmanned aircraft is necessary to counter a high-risk terrorist attack by a specific individual or organization;
2. A search warrant is issued pursuant to this chapter that permits the use of an unmanned aircraft; provided that the search warrant shall be valid for a period not

to exceed thirty days but may be extended by the court for no more than thirty additional days upon a showing of good cause;

3. The use of an unmanned aircraft is necessary to assist in search and rescue activities or locate missing persons;
4. The use of an unmanned aircraft is necessary to assist in the operation, exercise, or mission of any branch of the United States military; or
5. The use of an unmanned aircraft is necessary to assist the disaster relief agency under chapter 127 or provide relief pursuant to the governor's declaration of a state disaster under section 209-2.

§803-D Retention of information. If a law enforcement agency uses an unmanned aircraft pursuant to section 803-C, the law enforcement agency shall destroy all information collected by the unmanned aircraft no later than thirty days after such information is collected unless the information is necessary for and relevant to an ongoing criminal investigation or pending criminal trial.

§803-E Disclosure of information. If a law enforcement agency uses an unmanned aircraft pursuant to section 803-C, the law enforcement agency shall not disclose any information collected by the unmanned aircraft unless the collected information is necessary for and relevant to an ongoing criminal investigation or pending criminal trial,

§803-F Remedies for violations. An aggrieved party may initiate a civil action to:

1. Obtain any appropriate relief, including but not limited to equitable relief to prevent or remedy a violation of this part; and
2. Recover monetary damages, including actual and general damages, attorney's fees, and other litigation costs reasonably incurred, in an amount no less than $1,000; or
3. If information collected through the use of an unmanned aircraft was publicly disclosed without the aggrieved party's permission, the aggrieved party is entitled to recover monetary damages, including actual and general damages, attorney's fees, and other litigation costs reasonably incurred, in an amount no less than $10,000.

§803-G Reporting. (a) In January of each year, any law enforcement agency that used an unmanned aircraft during the previous calendar year shall report to the legislature its unmanned aircraft activities and shall provide a copy of the report to the judiciary. The report shall include the following information:

1. The number of times an unmanned aircraft was used, organized by the types of incidents and the types of justification for deployment;
2. The number of crime investigations aided by the use of each unmanned aircraft, along with a description of how the unmanned aircraft was helpful to each investigation; and

3. The total cost of the law enforcement agency's unmanned aircraft program.
   a. In January of each year, any court that has issued a search warrant permitting the use of an unmanned aircraft, issued an extension or modification of such warrant, or denied an application for such warrant during the previous calendar year shall report to the judiciary the number, kind, and duration of search warrants or search warrant extensions applied for, granted, modified, or denied.

Section 3. This Act does not affect rights and duties that matured, penalties that were incurred, and proceedings that were begun before its effective date.
Section 4. In codifying the new sections added by section 2 of this Act, the revisor of statutes shall substitute appropriate section numbers for the letters used in designating the new sections in this Act.
Section 5. This Act shall take effect on July 1, 2050.[151]

IDAHO

Thus far, Idaho has two enacted UAS bills, SB 1213[152] and SB 1134.[153] In 2013, the Idaho senate passed SCR 103, which encouraged Idaho's application to become one of six national unmanned aircraft test sites for the FAA.[154]
Senate Bill (SB) 1213: This bill, which went into effect July 1, 2016, related to fish and game, amends section 36-1101, Idaho Code, to prohibit the use of unmanned aircraft systems for hunting, molesting or locating game animals, game birds and furbearing animals.[155]
Senate Bill (SB) 1134: The Governor of Idaho signed SB 1134 in Session Law Chapter 328 on April 11, 2013. It went into effect on July 1, 2013. This was an act relating to Aeronautics; amending Chapter 2, Title 21, Idaho Code, by the addition of a new section 21-213, Idaho Code, to define a term, to establish provisions relating to restrictions on the use of unmanned aircraft systems, to provide exceptions, to provide for a civil cause of action, to provide for certain damages and to provide that an owner of certain facilities shall not be prohibited from using an unmanned aircraft system to inspect such facilities.
Be it enacted by the Legislature of the State of Idaho:
SECTION 1. That Chapter 2, Title 21, Idaho Code, be, and the same is hereby amended by the addition thereto of a NEW SECTION, to be known and designated as Section 21-213, Idaho Code, and to read as follows:

---

[151]SB 2608, HAWAII STATE LEGISLATURE, https://legiscan.com/HI/bill/SB2608/2014.
[152]SB 1213, IDAHO STATE LEGISLATURE, https://legislature.idaho.gov/legislation/2016/S1213Bookmark.htm.
[153]SB 1134, IDAHO STATE LEGISLATURE, http://www.legislature.idaho.gov/legislation/2013/S1134E2.pdf.
[154]SCR 103, IDAHO STATE LEGISLATURE, http://idahosenate.blogspot.com/2013/02/the-senate-passes-scr-103-which.html.
[155]SB 1213, IDAHO STATE LEGISLATURE, https://legislature.idaho.gov/legislation/2016/S1213Bookmark.htm.

21-213. RESTRICTIONS ON USE OF UNMANNED AIRCRAFT SYSTEMS—DEFINITION—VIOLATION—CAUSE OF ACTION AND DAMAGES.

1. (1)(a) For the purposes of this section, the term "unmanned aircraft system" (UAS) means an unmanned aircraft vehicle, drone, remotely piloted vehicle, remotely piloted aircraft or remotely operated aircraft that is a powered aerial vehicle that does not carry a human operator, can fly autonomously or remotely and can be expendable or recoverable.
   a. Unmanned aircraft system does not include:
      i. Model flying airplanes or rockets including, but not necessarily limited to, those that are radio controlled or otherwise remotely controlled and that are used purely for sport or recreational purposes; and
      ii. An unmanned aircraft system used in mapping or resource management.
2. Absent a warrant, and except for emergency response for safety, search and rescue or controlled substance investigations, no person, entity or state agency shall use an unmanned aircraft system to intentionally conduct surveillance of, gather evidence or collect information about, or photographically or electronically record specifically targeted persons or specifically targeted private property including, but not limited to:
      i. An individual or a dwelling owned by an individual and such dwelling's curtilage, without such individual's written consent;
      ii. A farm, dairy, ranch or other agricultural industry without the written consent of the owner of such farm, dairy, ranch or other agricultural industry.
   a. No person, entity or state agency shall use an unmanned aircraft system to photograph or otherwise record an individual, without such individual's written consent, for the purpose of publishing or otherwise publicly disseminating such photograph or recording. (3) Any person who is the subject of prohibited conduct under subsection (2) of this section shall:
   b. Have a civil cause of action against the person, entity or state agency for such prohibited conduct; and
   c. Be entitled to recover from any such person, entity or state agency damages in the amount of the greater of one thousand dollars ($1,000) or actual and general damages, plus reasonable attorney's fees and other litigation costs reasonably incurred.
3. An owner of facilities located on lands owned by another under a valid easement, permit, license or other right of occupancy is not prohibited in this section from using an unmanned aircraft system to aerially inspect such facilities.[156]

[156] SB 1134, IDAHO STATE LEGISLATURE, http://www.legislature.idaho.gov/legislation/2013/S1134E2.pdf.

ILLINOIS

Illinois has enacted five UAS bills. They are HB 1652,[157] SB 1587,[158] SB 2937,[159] SB 44,[160] and HB 5808.[161]

Additionally, there is HB 1048 in 2015, which is an act concerning wildlife;[162] HB 3136 in 2015 relating to automated motor vehicles;[163] HB 3421 in 2015 to amend the Criminal Code by adding disorderly conduct for a person to knowingly operate any aerial vehicle that does not carry a human operator in flight below the minimum altitudes of flight prescribed by the laws of this state;[164] HB 3996 in 2015 to amend the Criminal Code to provide that a person also commits disorderly conduct when he or she knowingly uses a drone to harass or conduct surveillance of another person without the consent of that other person and with the intent to alarm or disturb that other person;[165] and SB 1371 in 2015 to amend the wildlife code and prohibit the use of a manned or unmanned vehicle to take, pursue, or intentionally harass or disturb wild birds or mammals.[166]

Furthermore, there are a few city ordinances in Illinois. Chicago Ordinance to amend their Chapter 10-36 in 2015, to regulate the use of small UAS in city airspace,[167] Evanston Ordinance Resolution 27-R-13, authorizing the City of Evanston to establish a moratorium on the use of unregulated drone technology,[168] the Village of Manhattan Ordinance regulating UAS within their village,[169] and Schaumburg Ordinance Chapter 109A prohibiting the operation of a drone within 100 feet of the perimeter of any village property or on any village right-of-way during a special event.[170]

House Bill (HB) 1652: On August 16, 2013, the Illinois General Assembly signed HB 1652 into Public Act 98-0402, an act concerning wildlife. The act reads as follows:

Be it enacted by the People of the State of Illinois, represented in the General Assembly:

---

[157]HB 1652, Illinois General Assembly, http://www.ilga.gov/legislation/publicacts/fulltext.asp?Name=098-0402.

[158]SB 1587, Illinois General Assembly, http://www.ilga.gov/legislation/publicacts/fulltext.asp?Name=098-0569.

[159]SB 2937, Illinois General Assembly, http://www.ilga.gov/legislation/publicacts/fulltext.asp?Name=098-0831.

[160]SB 44, Illinois General Assembly, http://www.ilga.gov/legislation/publicacts/fulltext.asp?Name=099-0392.

[161]HB 5808, Illinois General Assembly, http://www.ilga.gov/legislation/publicacts/fulltext.asp?Name=099-0649.

[162]HB 1048, Illinois General Assembly, http://openstates.org/il/bills/99th/HB1048/.

[163]HB 3136, Illinois General Assembly, http://openstates.org/il/bills/99th/HB3136/.

[164]HB 3421, Illinois General Assembly, http://openstates.org/il/bills/99th/HB3421/.

[165]HB 3996, Illinois General Assembly, http://openstates.org/il/bills/99th/HB3996/.

[166]HB 1371, Illinois General Assembly, http://openstates.org/il/bills/99th/SB1371/.

[167]Chicago Ordinance to amend Chapter 10-36, https://chicago.legistar.com/LegislationDetail.aspx?ID=2393877&GUID=8BC7890A-DBA7-4640-B475-7B46DEDBBE68&Options=Advanced&Search=&FullText=1.

[168]Evanston Ordinance 27-R-13, http://uavs.insct.org/wp-content/uploads/2013/06/Evanston-Illinois-Resolution-27_R_13.pdf.

[169]Village of Manhattan Ordinance regulating UAS, http://www.villageofmanhattan.org/vertical/sites/%7BC299AB89-4140-469A-B02A-149D683060CF%7D/uploads/Manhattan_UAS.pdf.

[170]Schaumburg Ordinance Chapter 109A, http://uavs.insct.org/wp-content/uploads/2016/04/Schaumburg-IL-Code-of-Ordinances-Chapter-109A.pdf.

Section 2. The Fish and Aquatic Life Code is amended by changing Section 20-105 as follows: (515 ILCS 5/20-105) (from Ch. 56, par. 20-105)
Sec. 20-105. Revocation and suspension; refusal to issue.

a. Whenever a license or permit is issued to any person under this Code and its holder is found guilty of any misrepresentation in obtaining the license or permit or of a violation of Section 48-3 of the Criminal Code of 2012 or a violation of any of the provisions of this Code, including administrative rules, the license or permit may be revoked by the Department and the Department may refuse to issue any permit or license to that person and may suspend the person from engaging in the activity requiring the permit or license for a period of time not to exceed 5 years following the revocation. Department revocation procedure shall be established by administrative rule.
b. Whenever any person who has not been issued a license or a permit under the provisions of this Code is found guilty of a violation of Section 48-3 of the Criminal Code of 2012 or a violation of the provisions of this Code, including administrative rules, the Department may refuse to issue any permit or license to that person, and suspend that person from engaging in the activity requiring the permit or license for a period of time not to exceed 5 years.
c. Any person who knowingly or intentionally violates any of the provisions of this Code, including administrative rules, during the 5 years following the revocation of his or her license or permit under subsection (a) or during the time he is suspended under subsection (b), shall be guilty of a Class A misdemeanor as provided in Section 20-35. The penalties for a violation of Section 48-3 of the Criminal Code of 2012 shall be as provided in that Section. (d) A person whose license or permit to engage in any activity regulated by this Code has been suspended or revoked may not, during the period of the suspension or revocation or until obtaining such a license or permit,
    i. be in the company of any person engaging in the activity covered by the suspension or revocation or
    ii. serve as a guide, outfitter, or facilitator for a person who is engaged or prepared to engage in the activity covered by the suspension or revocation.
d. No person may be issued or obtain a license or permit or engage in any activity regulated by this Code during the time that the person's privilege to engage in the same or similar activities is suspended or revoked by another state, by a federal agency, or by a province of Canada.

Section 3. The Wildlife Code is amended by changing Section 3.36 as follows:

(520 ILCS 5/3.36) (from Ch. 61, par. 3.36)
Sec. 3.36. Revocation and suspension.

a. Whenever a license or permit is issued to any person under this Act, and the holder thereof is found guilty of any misrepresentation in obtaining such license or permit or of a violation of Section 48-3 of the Criminal Code of 2012 or a violation of any of the provisions of this Act, including administrative rules, his license or permit may be revoked by the Department, and the Department may refuse to issue any permit or license to such person and may suspend the person from engaging in the activity requiring the permit or license for a period of time not to exceed 5 years following such revocation.

   Department revocation procedures shall be established by Administrative rule.
b. Whenever any person who has not been issued a license or a permit under the provisions of this Code is found guilty of a violation of Section 48-3 of the Criminal Code of 2012 or a violation of the provisions of this Code, including administrative rules, the Department may refuse to issue any permit or license to that person, and suspend that person from engaging in the activity requiring the permit or license for a period of time not to exceed 5 years.
c. Any person who knowingly or intentionally violates any of the provisions of this Act, including administrative rules, during such period when his license or permit is revoked or denied by virtue of this Section or during the time he is suspended under subsection (b), shall be guilty of a Class A misdemeanor. The penalties for a violation of Section 48-3 of the Criminal Code of 2012 shall be as provided in that Section.
d. Licenses and permits authorized to be issued under the provisions of this Act shall be prepared by the Department and be in such form as prescribed by the Department. The information required on each license shall be completed thereon by the issuing agent or his sub-agent at the time of issuance and each license shall be signed by the licensee, or initialed by the designated purchaser and then signed immediately upon receipt by the licensee, and countersigned by the issuing agent or his sub-agent at the time of issuance. All such licenses shall be supplied by the Department, subject to such rules and regulations as the Department may prescribe. Any license not properly prepared, obtained and signed as required by this Act shall be void.
e. A person whose license or permit to engage in any activity regulated by this Code has been suspended or revoked may not, during the period of the suspension or revocation or until obtaining such a license or permit,
   i. be in the company of any person engaging in the activity covered by the suspension or revocation or
   ii. serve as a guide, outfitter, or facilitator for a person who is engaged or prepared to engage in the activity covered by the suspension or revocation.

f. No person may be issued or obtain a license or permit or engage in any activity regulated by this Code during the time that the person's privilege to engage in the same or similar activities is suspended or revoked by another state, by a federal agency, or by a province of Canada.

Section 5. The Criminal Code of 2012 is amended by changing Section 48-3 as follows:
(720 ILCS 5/48-3)
Sec. 48-3. Hunter or fisherman interference.

a. Definitions. As used in this Section:

"Aquatic life" means all fish, reptiles, amphibians, crayfish, and mussels the taking of which is authorized by the Fish and Aquatic Life Code.

"Interfere with" means to take any action that physically impedes, hinders, or obstructs the lawful taking of wildlife or aquatic life.

"Taking" means the capture or killing of wildlife or aquatic life and includes travel, camping, and other acts preparatory to taking which occur on lands or waters upon which the affected person has the right or privilege to take such wildlife or aquatic life.

"Wildlife" means any wildlife the taking of which is authorized by the Wildlife Code and includes those species that are lawfully released by properly licensed permittees of the Department of Natural Resources.

b. A person commits hunter or fisherman interference when he or she intentionally or knowingly:
    1. obstructs or interferes with the lawful taking of wildlife or aquatic life by another person with the specific intent to prevent that lawful taking;
    2. drives or disturbs wildlife or aquatic life for the purpose of disrupting a lawful taking of wildlife or aquatic life;
    3. blocks, impedes, or physically harasses another person who is engaged in the process of lawfully taking wildlife or aquatic life;
    4. uses natural or artificial visual, aural, olfactory, gustatory, or physical stimuli to affect wildlife or aquatic life behavior in order to hinder or prevent the lawful taking of wildlife or aquatic life;
    5. erects barriers with the intent to deny ingress or egress to or from areas where the lawful taking of wildlife or aquatic life may occur;
    6. intentionally interjects himself or herself into the line of fire or fishing lines of a person lawfully taking wildlife or aquatic life;
    7. affects the physical condition or placement of personal or public property intended for use in the lawful taking of wildlife or aquatic life in order to impair the usefulness of the property or prevent the use of the property;

8. enters or remains upon or over private lands without the permission of the owner or the owner's agent, with the intent to violate this subsection;
9. fails to obey the order of a peace officer to desist from conduct in violation of this subsection (b) if the officer observes the conduct, or has reasonable grounds to believe that the person has engaged in the conduct that day or that the person plans or intends to engage in the conduct that day on a specific premises; or
10. uses a drone in a way that interferes with another person's lawful taking of wildlife or aquatic life. For the purposes of this paragraph (10), "drone" means any aerial vehicle that does not carry a human operator.

c. Exemptions; defenses.
   1. This Section does not apply to actions performed by authorized employees of the Department of Natural Resources, duly accredited officers of the U.S. Fish and Wildlife Service, sheriffs, deputy sheriffs, or other peace officers if the actions are authorized by law and are necessary for the performance of their official duties.
   2. This Section does not apply to landowners, tenants, or lease holders exercising their legal rights to the enjoyment of land, including, but not limited to, farming and restricting trespass.
   3. It is an affirmative defense to a prosecution for a violation of this Section that the defendant's conduct is protected by his or her right to freedom of speech under the constitution of this State or the United States.
   4. Any interested parties may engage in protests or other free speech activities adjacent to or on the perimeter of the location where the lawful taking of wildlife or aquatic life is taking place, provided that none of the provisions of this Section are being violated.

d. Sentence. A first violation of paragraphs (1) through (8) of subsection (b) is a Class B misdemeanor. A second or subsequent violation of paragraphs (1) through (8) of subsection (b) is a Class A misdemeanor for which imprisonment for not less than 7 days shall be imposed. A person guilty of a second or subsequent violation of paragraphs (1) through (8) of subsection (b) is not eligible for court supervision. A violation of paragraph (9) or (10) of subsection (b) is a Class A misdemeanor. A court shall revoke, for a period of one year to 5 years, any Illinois hunting, fishing, or trapping privilege, license or permit of any person convicted of violating any provision of this Section. For purposes of this subsection, a "second or subsequent violation" means a conviction under paragraphs (1) through (8) of subsection (b) of this Section within 2 years of a prior violation arising from a separate set of circumstances.

e. Injunctions; damages.

1. Any court may enjoin conduct which would be in violation of paragraphs (1) through (8) or (10) of subsection (b) upon petition by a person affected or who reasonably may be affected by the conduct, upon a showing that the conduct is threatened or that it has occurred on a particular premises in the past and that it is not unreasonable to expect that under similar circumstances it will be repeated.
2. A court shall award all resulting costs and damages to any person adversely affected by a violation of paragraphs (1) through (8) or (10) of subsection (b), which may include an award for punitive damages. In addition to other items of special damage, the measure of damages may include expenditures of the affected person for license and permit fees, travel, guides, special equipment and supplies, to the extent that these expenditures were rendered futile by prevention of the taking of wildlife or aquatic life.

Section 99. Effective date. This Act takes effect upon becoming law. This law went into effect on August 16, 2013.[171]

Senate Bill (SB) 1587: On August 27, 2013, the Illinois General Assembly signed SB 1587 into law in Public Act 98-0569. This was an act concerning criminal law, and reads as follows: Be it enacted by the People of the State of Illinois, represented in the General Assembly:

Section 1. Short title.

This Act may be cited as the Freedom from Drone Surveillance Act. Section 5. Definitions. As used in this Act:

"Authority" means the Illinois Criminal Justice Information Authority.

"Drone" means any aerial vehicle that does not carry a human operator.

"Information" means any evidence, images, sounds, data, or other information gathered by a drone.

"Law enforcement agency" means any agency of this State or a political subdivision of this State which is vested by law with the duty to maintain public order and to enforce criminal laws.

Section 10. Prohibited use of drones.

Except as provided in Section 15, a law enforcement agency may not use a drone to gather information.

[171] HB 1652, Illinois General Assembly, http://www.ilga.gov/legislation/publicacts/fulltext.asp?Name=098-0402.

Section 15. Exceptions.

This Act does not prohibit the use of a drone by a law enforcement agency:

1. To counter a high risk of a terrorist attack by a specific individual or organization if the United States Secretary of Homeland Security determines that credible intelligence indicates that there is that risk.
2. If a law enforcement agency first obtains a search warrant based on probable cause issued under Section 108-3 of the Code of Criminal Procedure of 1963. The warrant must be limited to a period of 45 days, renewable by the judge upon a showing of good cause for subsequent periods of 45 days.
3. If a law enforcement agency possesses reasonable suspicion that, under particular circumstances, swift action is needed to prevent imminent harm to life, or to forestall the imminent escape of a suspect or the destruction of evidence. The use of a drone under this paragraph (3) is limited to a period of 48 hours. Within 24 hours of the initiation of the use of a drone under this paragraph (3), the chief executive officer of the law enforcement agency must report in writing the use of a drone to the local State's Attorney.
4. If a law enforcement agency is attempting to locate a missing person, and is not also undertaking a criminal investigation.
5. If a law enforcement agency is using a drone solely for crime scene and traffic crash scene photography. Crime scene and traffic crash photography must be conducted in a geographically confined and time-limited manner to document specific occurrences. The use of a drone under this paragraph (5) on private property requires either a search warrant based on probable cause under Section 108-3 of the Code of Criminal Procedure of 1963 or lawful consent to search. The use of a drone under this paragraph (5) on lands, highways, roadways, or areas belonging to this State or political subdivisions of this State does not require a search warrant or consent to search. Any law enforcement agency operating a drone under this paragraph (5) shall make every reasonable attempt to only photograph the crime scene or traffic crash scene and avoid other areas.

Section 20. Information retention.

If a law enforcement agency uses a drone under Section 15 of this Act, the agency within 30 days shall destroy all information gathered by the drone, except that a supervisor at that agency may retain particular information if:

1. there is reasonable suspicion that the information contains evidence of criminal activity, or
2. the information is relevant to an ongoing investigation or pending criminal trial.

Section 25. Information disclosure.

If a law enforcement agency uses a drone under Section 15 of this Act, the agency shall not disclose any information gathered by the drone, except that a supervisor of that agency may disclose particular information to another government agency, if

1. there is reasonable suspicion that the information contains evidence of criminal activity, or
2. the information is relevant to an ongoing investigation or pending criminal trial.

Section 30. Admissibility.

If the court finds by a preponderance of the evidence that a law enforcement agency used a drone to gather information in violation of the information gathering limits in Sections 10 and 15 of this Act, then the information shall be presumed to be inadmissible in any judicial or administrative proceeding. The State may overcome this presumption by proving the applicability of a judicially recognized exception to the exclusionary rule of the Fourth Amendment to the U.S. Constitution or Article I, Section 6 of the Illinois Constitution to the information. Nothing in this Act shall be deemed to prevent a court from independently reviewing the admissibility of the information for compliance with the aforementioned provisions of the U.S. and Illinois Constitutions.

Section 35. Reporting.

a. If a law enforcement agency owns one or more drones, then subsequent to the effective date of this Act, it shall report in writing annually by April 1 to the Authority the number of drones that it owns.
b. On July 1 of each year, the Authority shall publish on its publicly available website a concise report that lists every law enforcement agency that owns a drone, and for each of those agencies, the number of drones that it owns.

This law went into effect on January 1, 2014.[172]

Senate Bill (SB) 2937: On August 1, 2014, SB 2937 was signed into Public Act 98-0831. This was an act concerning criminal law, which reads as follows:

Be it enacted by the People of the State of Illinois, represented in the General Assembly:

Section 5. The Freedom from Drone Surveillance Act is amended by changing Section 15 and by adding Section 40 as follows:

(725 ILCS 167/15)

---

[172]SB 1587, Illinois General Assembly, http://www.ilga.gov/legislation/publicacts/fulltext.asp?Name=098-0569.

Sec. 15. Exceptions.

This Act does not prohibit the use of a drone by a law enforcement agency:

1. To counter a high risk of a terrorist attack by a specific individual or organization if the United States Secretary of Homeland Security determines that credible intelligence indicates that there is that risk.
2. If a law enforcement agency first obtains a search warrant based on probable cause issued under Section 108-3 of the Code of Criminal Procedure of 1963. The warrant must be limited to a period of 45 days, renewable by the judge upon a showing of good cause for subsequent periods of 45 days.
3. If a law enforcement agency possesses reasonable suspicion that, under particular circumstances, swift action is needed to prevent imminent harm to life, or to forestall the imminent escape of a suspect or the destruction of evidence. The use of a drone under this paragraph (3) is limited to a period of 48 hours. Within 24 hours of the initiation of the use of a drone under this paragraph (3), the chief executive officer of the law enforcement agency must report in writing the use of a drone to the local State's Attorney.
4. If a law enforcement agency is attempting to locate a missing person, and is not also undertaking a criminal investigation.
5. If a law enforcement agency is using a drone solely for crime scene and traffic crash scene photography. Crime scene and traffic crash photography must be conducted in a geographically confined and time-limited manner to document specific occurrences. The use of a drone under this paragraph (5) on private property requires either a search warrant based on probable cause under Section 108-3 of the Code of Criminal Procedure of 1963 or lawful consent to search. The use of a drone under this paragraph (5) on lands, highways, roadways, or areas belonging to this State or political subdivisions of this State does not require a search warrant or consent to search. Any law enforcement agency operating a drone under this paragraph (5) shall make every reasonable attempt to only photograph the crime scene or traffic crash scene and avoid other areas.
6. If a law enforcement agency is using a drone during a disaster or public health emergency, as defined by Section 4 of the Illinois Emergency Management Agency Act. The use of a drone under this paragraph (6) does not require an official declaration of a disaster or public health emergency prior to use. A law enforcement agency may use a drone under this paragraph (6) to obtain information necessary for the determination of whether or not a disaster or public health emergency should be declared, to monitor weather or emergency conditions, to survey damage, or to otherwise coordinate response and recovery efforts. The use of a drone under this paragraph (6) is permissible during the disaster or public health emergency and during subsequent response and recovery efforts.

(725 ILCS 167/40 new)

Sec. 40. Law enforcement use of private drones.

a. Except as provided in Section 15, a law enforcement agency may not acquire information from or direct the acquisition of information through the use of a drone owned by a private third party. In the event that law enforcement acquires information from or directs the acquisition of information through the use of a privately owned drone under Section 15 of this Act, any information so acquired is subject to Sections 20 and 25 of this Act.

b. Nothing in this Act prohibits private third parties from voluntarily submitting information acquired by a privately owned drone to law enforcement. In the event that law enforcement acquires information from the voluntary submission of that information, whether under a request or on a private drone owner's initiative, the information is subject to Sections 20 and 25 of this Act.

This act went into effect on January 1, 2015.[173]

Senate Bill (SB) 44: On August 18, 2015, the Illinois General Assembly signed SB 44 into Public Act 99-0392. This was an act concerning wildlife and reads as follows:

Be it enacted by the People of the State of Illinois, represented in the General Assembly:

Section 1. Short title.

This Act may be cited as the Unmanned Aerial System Oversight Task Force Act.

Section 5. Purpose.

The use of drones is becoming more common in everyday applications both commercially and privately. It is clear that increased drone use creates emerging conflicts and challenges to providing guidance into the safe operation of drones, while not infringing upon the constitutional rights of others. It is necessary to establish a task force to provide oversight and input in creating comprehensive laws and rules for the operation and use of drone technology within this State, subject to federal oversight and regulation.

Section 10. Definitions.

As used in this Act:

"Task Force" means the Unmanned Aerial System Oversight Task Force.

"Unmanned Aerial System" or "UAS" means an unmanned aerial vehicle or drone.

---

[173] SB 2937, Illinois General Assembly, http://www.ilga.gov/legislation/publicacts/fulltext.asp?Name=098-0831.

Section 15. The Unmanned Aerial System Task Force.

a. There is hereby created the Unmanned Aerial System Oversight Task Force to study and make recommendations for the operation, usage, and regulation of Unmanned Aerial Systems, commonly referred to as "drone" technology, within this State.

b. Within 90 days after the effective date of this Act members of the Task Force shall be appointed by the Governor and shall consist of one member from each of the following agencies or interest groups:

1. a member of the Division of Aeronautics of the Department of Transportation, nominated by the Secretary of Transportation;
2. a member of the Department of State Police, nominated by the Director of State Police;
3. a Conservation Police officer of the Department of Natural Resources, nominated by the Director of Natural Resources;
4. a member of the Department of Agriculture, nominated by the Director of Agriculture;
5. a member of the Department of Commerce and Economic Opportunity, nominated by the Director of Commerce and Economic Opportunity;
6. a UAS technical commercial representative;
7. a UAS manufacturing industry representative;
8. a person nominated by the Attorney General;
9. a member of the Illinois Conservation Police Lodge, nominated by the president of the Lodge;
10. a member of a statewide sportsmen's federation, nominated by the president of the federation;
11. a member of a statewide agricultural association, nominated by the president of the association;
12. a member of a statewide commerce association, nominated by the president or executive director of the association;
13. a person nominated by an electric utility company serving retail customers in this State;
14. a member of the Illinois National Guard, nominated by the Adjutant General;
15. a member of a statewide retail association, nominated by the president of the association;
16. a member of a statewide manufacturing trade association, nominated by the president or chief executive officer of the association;
17. a member of a statewide property and casualty insurance association, nominated by the president or chief executive officer of the association;
18. a member of a statewide association representing real estate brokers licensed in this State, nominated by the president of the association;

19. a member of a statewide surveying association, nominated by the president of the association;
20. a law enforcement official from a municipality with a population of 2 million or more inhabitants, nominated by the mayor of the municipality;
21. a law enforcement official from a municipality with a population of less than 2 million inhabitants, nominated by a statewide police chiefs association; and
22. a member of a statewide freight railroad association, nominated by the president of the association.

c. Nominations to the Task Force must be submitted to the Governor within 60 days of the effective date of this Act. The Governor shall make the appointments within 30 days after the close of nominations. The term of the appointment shall be until submission of the report of comprehensive recommendations under subsection (g) of this Section. The member from the Division of Aeronautics of the Department of Transportation shall chair the Task Force and serve as a liaison to the Governor and General Assembly. Meetings of the Task Force shall be held as necessary to complete the duties of the Task Force. Meetings of the Task Force shall be held in the central part of the State.

d. The members of the Task Force shall receive no compensation for serving as members of the Task Force.

e. The Task Force shall consider commercial and private uses of drones, landowner and privacy rights, as well as general rules and regulations for safe operation of drones, and prepare comprehensive recommendations for the safe and lawful operation of UAS in this State.

f. The Department of Transportation shall provide administrative support to the Task Force.

g. The Task Force shall submit a report with recommendations to the Governor and General Assembly no later than July 1, 2016.

Section 20. Expiration.

This Act is repealed on September 1, 2016.

Section 99. Effective date. This Act takes effect upon becoming law.

This act went into effect on August 18, 2015.[174]

House Bill (HB) 5808: On July 28, 2016 this law went into effect expanding the membership of the UAS Oversight Task Force and extending the deadline for the task force to issue a report from July 1, 2016 to July 1, 2017.[175]

---

[174]SB 44, Illinois General Assembly, http://www.ilga.gov/legislation/publicacts/fulltext.asp?Name=099-0392.

[175]HB 5808, Illinois General Assembly, http://www.ilga.gov/legislation/publicacts/fulltext.asp?Name=099-0649.

INDIANA

Indiana has adopted Senate Resolution (SR) 27,[176] House Bill (HB) 1009,[177] HB 1013,[178] and HB 1246.[179] Additionally, there is HB 1013 relating to geolocation information and UAVs.[180]

Senate Resolution (SR) 27: Indiana adopted SR 27 on February 22, 2013. SR 27 was a senate resolution urging the legislative council to establish the interim study committee on the use of aircraft to study the use of unmanned aerial vehicles.

Whereas, With the development of new technological advances, the use of unmanned aerial vehicles, or drones, is increasing;

Whereas, Common applications for unmanned aircraft are numerous, including spreading pesticide on fields, delivering medical supplies in remote areas, or monitoring hundreds of miles of oil pipelines for leaks and borders for security; and

Whereas, Although many worthwhile uses exist for unmanned aerial vehicles, there are also public safety concerns associated with their use, as well as a danger that the use of these vehicles could infringe on the privacy of the citizens of our state: Therefore,

Be it resolved by the Senate of the General Assembly of the State of Indiana:

SECTION 1. That the legislative council is urged to establish the interim study committee on the use of aircraft to study the use of unmanned aerial vehicles.

SECTION 2. That the committee, if established, shall have the following members:

1. The chairperson of the committee and one (1) additional individual, both appointed by the president pro tempore of the senate from among the members of the general assembly.
2. Two (2) members of the general assembly appointed by the minority leader of the senate.
3. Two (2) members of the general assembly appointed by the speaker of the house of representatives.
4. Two (2) members of the general assembly appointed by the minority leader of the senate.
5. One (1) individual who is not a member of the general assembly appointed by the president pro tempore of the senate.
6. One (1) individual who is not a member of the general assembly appointed by the speaker of the house of representatives.
7. One (1) individual who is not a member of the general assembly appointed by the governor.

---

[176]SR 27, Indiana General Assembly, http://www.in.gov/legislative/bills/2013/SRESP/SR0027.html.
[177]HB 1009, Indiana General Assembly, https://iga.in.gov/static-documents/5/3/4/4/5344c8fc/HB1009.06.ENRH.pdf.
[178]HB 1013, Indiana General Assembly, https://iga.in.gov/legislative/2016/bills/house/1013.
[179]HB 1246, Indiana General Assembly, https://iga.in.gov/legislative/2016/bills/house/1246#document-2cf343ef.
[180]HB 1013, Indiana General Assembly, https://iga.in.gov/legislative/2016/bills/house/1013.

SECTION 3. That the committee, if established, shall operate under the direction of the legislative council, and that the committee shall issue a final report when directed to do so by the council.[181]

House Bill (HB) 1009: On March 26, 2014, the Governor of Indiana signed HB 1009 into Public Law 170. Public Law 170 is an act to amend the Indiana Code concerning criminal law and procedure, and it reads as follows:

Be it enacted by the General Assembly of the State of Indiana:

SECTION 1. IC 34-30-2-146.4 IS ADDED TO THE INDIANA CODE AS A NEW SECTION TO READ AS FOLLOWS [EFFECTIVE JULY 1, 2014]: Sec. 146.4. IC 35-33-5-13 (Concerning electronic communications services, remote computing services, and geolocation information services for compliance with warrant laws).

SECTION 2. IC 35-31.5-2-110.5 IS ADDED TO THE INDIANA CODE AS A NEW SECTION TO READ AS FOLLOWS [EFFECTIVE JULY 1, 2014]: Sec. 110.5. "Electronic communication service", for purposes of IC 35-33-5, means a service that provides users with the ability to send or receive wire or electronic communications.

SECTION 3. IC 35-31.5-2-111.5 IS ADDED TO THE INDIANA CODE AS A NEW SECTION TO READ AS FOLLOWS [EFFECTIVE JULY 1, 2014]: Sec. 111.5. "Electronic storage", for purposes of IC 35-33-5, has the meaning set forth in IC 35-33-5-0.5(2).

SECTION 4. IC 35-31.5-2-112.5 IS ADDED TO THE INDIANA CODE AS A NEW SECTION TO READ AS FOLLOWS [EFFECTIVE JULY 1, 2014]: Sec. 112.5. "Electronic user data", for purposes of IC 35-33-5, has the meaning set forth in IC 35-33-5-0.5(3).

SECTION 5. IC 35-31.5-2-143.3 IS ADDED TO THE INDIANA CODE AS A NEW SECTION TO READ AS FOLLOWS [EFFECTIVE JULY 1, 2014]: Sec. 143.3. "Geolocation information" means data generated by an electronic device that can be used to determine the location of the device or the owner of the device. The term includes a cellular telephone, a wireless fidelity (wi-fi) equipped computer, or a GPS navigation or tracking unit. The term does not include the content of a communication.

SECTION 6. IC 35-31.5-2-143.5 IS ADDED TO THE INDIANA CODE AS A NEW SECTION TO READ AS FOLLOWS [EFFECTIVE JULY 1, 2014]: Sec. 143.5. "Geolocation information service" means a person that offers or provides GPS service or other mapping, locational, or directional services to the public by means of an electronic device, including a cellular telephone, a wireless fidelity (wi-fi) equipped computer, or a GPS navigation or tracking unit.

---

[181] SR 27, Indiana General Assembly, http://www.in.gov/legislative/bills/2013/SRESP/SR0027.html.

SECTION 7. IC 35-31.5-2-144, AS ADDED BY P.L.114-2012, SECTION 67, IS AMENDED TO READ AS FOLLOWS [EFFECTIVE JULY 1, 2014]: Sec. 144.

a. "Governmental entity" means:
    1. the United States or any state, county, township, city, town, separate municipal corporation, special taxing district, or public school corporation;
    2. any authority, board, bureau, commission, committee, department, division, hospital, military body, or other instrumentality of any of those entities; or
    3. a state assisted college or state assisted university.
b. For purposes of IC 35-33-5, "governmental entity" also includes a person authorized to act on behalf of a state or local agency.

SECTION 8. IC 35-31.5-2-175.5 IS ADDED TO THE INDIANA CODE AS A NEW SECTION TO READ AS FOLLOWS [EFFECTIVE JULY 1, 2014]: Sec. 175.5. "Intercept", for purposes of IC 35-33-5, has the meaning set forth in IC 35-33-5-0.5(5).

SECTION 9. IC 35-31.5-2-186, AS ADDED BY P.L.114-2012, SECTION 67, IS AMENDED TO READ AS FOLLOWS [EFFECTIVE JULY 1, 2014]: Sec. 186.

a. "Lawful detention" means:
    1. arrest;
    2. custody following surrender in lieu of arrest;
    3. detention in a penal facility;
    4. detention in a facility for custody of persons alleged or found to be delinquent children;
    5. detention under a law authorizing civil commitment in lieu of criminal proceedings or authorizing such detention while criminal proceedings are held in abeyance;
    6. detention for extradition or deportation;
    7. placement in a community corrections program's residential facility;
    8. electronic monitoring;
    9. custody for purposes incident to any of the above including transportation, medical diagnosis or treatment, court appearances, work, or recreation; or
    10. any other detention for law enforcement purposes.
b. Except as provided in subsection (a)(7) and (a)(8), the term does not include supervision of a person on probation or parole or constraint incidental to release with or without bail.
c. The term does not include electronic monitoring through the use of an unmanned aerial vehicle under IC 35-33-5-9.

SECTION 10. IC 35-31.5-2-273.8 IS ADDED TO THE INDIANA CODE AS A NEW SECTION TO READ AS FOLLOWS [EFFECTIVE JULY 1, 2014]: Sec. 273.8. “Remote computing service”, for purposes of IC 35-33-5, has the meaning set forth in IC 35-33-5-0.5(6).

SECTION 11. IC 35-31.5-2-337.5 IS ADDED TO THE INDIANA CODE AS A NEW SECTION TO READ AS FOLLOWS [EFFECTIVE JULY 1, 2014]: Sec. 337.5. “Tracking device”, for purposes of IC 35-33-5 and this chapter, means an electronic or mechanical device that allows a person to remotely determine or track the position or movement of another person or an object. The term includes the following:

1. A device that stores geographic data for subsequent access or analysis.
2. A device that allows real-time monitoring or movement.
3. An unmanned aerial vehicle.
4. A cellular telephone or other wireless or cellular communications device.

SECTION 12. IC 35-31.5-2-342.3 IS ADDED TO THE INDIANA CODE AS A NEW SECTION TO READ AS FOLLOWS [EFFECTIVE JULY 1, 2014]: Sec. 342.3. “Unmanned aerial vehicle”, for purposes of IC 35-33-5, has the meaning set forth in IC 35-33-5-0.5(7).

SECTION 13. IC 35-31.5-2-343.5 IS ADDED TO THE INDIANA CODE AS A NEW SECTION TO READ AS FOLLOWS [EFFECTIVE JULY 1, 2014]: Sec. 343.5. “Use of a tracking device”, for purposes of IC 35-33-5, includes the installation, maintenance, and monitoring of a tracking device. The term does not include:

1. the capture, collection, monitoring, or viewing of images; or
2. the use of a monitoring device with respect to a person required to be tracked or monitored:
   A. as a condition of bail;
   B. as a condition of probation, parole, or community corrections;
   C. as a requirement of sex offender registration; or
   D. as part of a sentence imposed for a crime.

SECTION 14. IC 35-31.5-2-343.7 IS ADDED TO THE INDIANA CODE AS A NEW SECTION TO READ AS FOLLOWS [EFFECTIVE JULY 1, 2014]: Sec. 343.7. “Use of an unmanned aerial vehicle”, for purposes of IC 35-33-5, has the meaning set forth in IC 35-33-5-0.5(8). SECTION 15. IC 35-31.5-2-343.8 IS ADDED TO THE INDIANA CODE AS A NEW SECTION TO READ AS FOLLOWS [EFFECTIVE JULY 1, 2014]: Sec. 343.8. “User”, for purposes of IC 35-33-5, has the meaning set forth in IC 35-33-5-0.5(9).

SECTION 16. IC 35-33-5-0.5 IS ADDED TO THE INDIANA CODE AS A NEW SECTION TO READ AS FOLLOWS [EFFECTIVE JULY 1, 2014]: Sec. 0.5. The following definitions apply throughout this chapter:

1. "Electronic communication service" means a service that provides users with the ability to send or receive wire or electronic communications.
2. "Electronic storage" means any storage of electronic user data on a computer, computer network, or computer system regardless of whether the data is subject to recall, further manipulation, deletion, or transmission. "Electronic storage" includes any storage or electronic communication by an electronic communication service or a remote computing service.
3. "Electronic user data" means any data or records that are in the possession, care, custody, or control of a provider of an electronic communication service, a remote computing service, or any other service or program that stores, uses, collects, or safeguards electronic user data.
4. "Governmental entity" has the meaning set forth in IC 35-31.5-2-144. For purposes of this chapter, "governmental entity" also includes a person authorized to act on behalf of a state or local agency.
5. "Intercept" means to acquire geolocation data through the use of an electronic device, mechanical device, or other device.
6. "Remote computing service" means the provision to the public of computer storage or processing services by means of an electronic communication service.
7. "Unmanned aerial vehicle" means an aircraft that:
   A. does not carry a human operator; and
   B. is capable of flight under remote control or autonomous programming.
8. "Use of an unmanned aerial vehicle" means the use of an unmanned aerial vehicle by a law enforcement officer to obtain evidence relevant to the enforcement of statutes, rules, or regulations. The term includes:
   A. the interception of wire, electronic, or oral communications; and
   B. the capture, collection, monitoring, or viewing of images.
9. "User" means any person who:
   A. uses an electronic communication service, remote computing service, geolocation information service, or an electronic device; and
   B. may or may not be the person or entity having legal title, claim, or right to the electronic device or electronic user data.

SECTION 17. IC 35-33-5-2, AS AMENDED BY P.L.2-2005, SECTION 117, IS AMENDED TO READ AS FOLLOWS [EFFECTIVE JULY 1, 2014]: Sec. 2. (a) Except as provided in section 8 of this chapter, and subject to the requirements of section 11 of this chapter, if applicable, no warrant for search or arrest shall be issued until there is filed with the judge an affidavit:

1. particularly describing:
   A. the house or place to be searched and the things to be searched for; or
   B. particularly describing the person to be arrested;
2. alleging substantially the offense in relation thereto and that the affiant believes and has good cause to believe that:
   A. the things sought are concealed there; or
   B. the person to be arrested committed the offense; and
3. setting forth the facts known to the affiant through personal knowledge or based on hearsay, constituting the probable cause.

B. When based on hearsay, the affidavit must either:
   1. contain reliable information establishing the credibility of the source and of each of the declarants of the hearsay and establishing that there is a factual basis for the information furnished; or
   2. contain information that establishes that the totality of the circumstances corroborates the hearsay.

C. An affidavit for search substantially in the following form shall be treated as sufficient:

STATE OF INDIANA)
) SS:
COUNTY OF ________________________)

A B swears (or affirms, as the case may be) that he believes and has good cause to believe (here set forth the facts and information constituting the probable cause) that (here describe the things to be searched for and the offense in relation thereto) are concealed in or about the (here describe the house or place) of C D, situated in the county of __________________, in said state.

In accordance with Indiana Trial Rule 11, I affirm under the penalties for perjury that the foregoing representations are true.

____

(Signed) Affiant Date

SECTION 18. IC 35-33-5-8 IS AMENDED TO READ AS FOLLOWS [EFFECTIVE JULY 1, 2014]: Sec. 8. (a) A judge may issue a search or arrest warrant without the affidavit required under section 2 of this chapter, if the judge receives testimony subject to the penalties for perjury of the same facts required for an affidavit:

1. in a nonadversarial, recorded hearing before the judge;
2. orally by telephone or radio;

3. in writing by facsimile transmission (FAX); or
4. in writing by electronic mail or other electronic transmission.
    a. If a warrant is issued under subsection (a)(1), the judge shall order the court reporter to type or transcribe the testimony from the hearing for entry in the record. The judge shall then certify the transcript.
    b. After reciting the facts required for an affidavit and verifying the facts recited under penalty of perjury, an applicant for a warrant under subsection (a)(2) shall read to the judge from a warrant form on which the applicant enters the information read by the applicant to the judge. The judge may direct the applicant to modify the warrant. If the judge agrees to issue the warrant, the judge shall direct the applicant to sign the judge's name to the warrant, adding the time of the issuance of the warrant.
    c. After transmitting an affidavit, an applicant for a warrant under subsection (a)(3) or (a)(4) shall transmit to the judge a copy of a warrant form completed by the applicant. The judge may modify the transmitted warrant. If the judge agrees to issue the warrant, the judge shall sign, affix the date and time, and transmit to the applicant a duplicate of the warrant.
    d. If a warrant is issued under subsection (a)(2), the judge shall record the conversation on audio tape and order the court reporter to type or transcribe the recording for entry in the record. The judge shall certify the audio tape, the transcription, and the warrant retained by the judge for entry in the record.
    e. If a warrant is issued under subsection (a)(3), the facsimile copy of the affidavit and warrant sent to the judge shall be retained as if they were the originals. If a warrant is issued under subsection (a)(4), the electronically transmitted copy of the affidavit and warrant sent to the judge shall be printed and retained as if they were the originals.
    f. The court reporter shall notify the applicant who received a warrant under subsection (a)(1) or (a)(2) when the transcription required under this section is entered in the record. The applicant shall sign the transcribed entry upon receiving notice from the court reporter.
    g. The affiant and the judge may use an electronic signature on the affidavit and warrant. An electronic signature may be indicated by "s/Affiant's Name" or "s/Judge's Name" or by any other electronic means that identifies the affiant or judge and indicates that the affiant or judge adopts the contents of the document to which the electronic signature is affixed.

SECTION 19. IC 35-33-5-9 IS ADDED TO THE INDIANA CODE AS A NEW SECTION TO READ AS FOLLOWS [EFFECTIVE JULY 1, 2014]: Sec. 9.

a. Except as provided in subsection (b), a law enforcement officer must obtain a search warrant in order to use an unmanned aerial vehicle.
b. A law enforcement officer or governmental entity may use an unmanned aerial vehicle without obtaining a search warrant if the law enforcement officer determines that the use of the unmanned aerial vehicle:
    1. is required due to:
        A. the existence of exigent circumstances necessitating a warrantless search;
        B. the substantial likelihood of a terrorist attack;
        C. the need to conduct a search and rescue or recovery operation;
        D. the need to conduct efforts: (i) in response to; or (ii) to mitigate; the results of a natural disaster or any other disaster; or
        E. the need to perform a geographical, an environmental, or any other survey for a purpose that is not a criminal justice purpose; or
    2. will be conducted with the consent of any affected property owner.

SECTION 20. IC 35-33-5-10 IS ADDED TO THE INDIANA CODE AS A NEW SECTION TO READ AS FOLLOWS [EFFECTIVE JULY 1, 2014]: Sec. 10. The following are not admissible as evidence in an administrative or judicial proceeding:

1. A communication or an image that is obtained through the use of an unmanned aerial vehicle in violation of section 9 of this chapter.

(2) Evidence derived from a communication or an image described in subdivision (1). SECTION 21. IC 35-33-5-11 IS ADDED TO THE INDIANA CODE AS A NEW SECTION TO READ AS FOLLOWS [EFFECTIVE JULY 1, 2014]: Sec. 11.

a. This subsection does not apply to electronic or video toll collection facilities or activities authorized under any of the following:
    1. IC 8-15-2.
    2. IC 8-15-3.
    3. IC 8-15.5.
    4. IC 8-15.7.
    5. IC 8-16.
    6. IC 9-21-3.5.

    A law enforcement officer may not compel a user to provide a passkey, password, or keycode to any electronic communication service, electronic device, or electronic storage, or any form of stored electronic user data, without a valid search warrant issued by a judge using search warrant procedures.

b. A judge may issue a court order under this section for electronic user data held in electronic storage, including the records and information related to a wire communication or electronic communication held in electronic storage, by a provider of an electronic communication service or a provider of a remote computing service regardless of whether the user data is held at a location in Indiana or at a location in another state.

c. A judge may issue a court order under this section on a service provider that is a corporation or entity that is incorporated or organized under the laws of Indiana or a company or business entity doing business in Indiana under a contract or terms of a service agreement with an Indiana resident. The service provider shall produce all information sought, as required by the court order.

d. Any Indiana corporation that provides electronic communication services or remote computing services to the public shall comply with a valid court order issued in another state that is seeking the information described in this section, if the court order is served on the corporation.

SECTION 22. IC 35-33-5-12 IS ADDED TO THE INDIANA CODE AS A NEW SECTION TO READ AS FOLLOWS [EFFECTIVE JULY 1, 2014]: Sec. 12.

a. A law enforcement officer or law enforcement agency may not use a real time tracking instrument that is capable of obtaining geolocation information concerning a cellular device or a device connected to a cellular network unless:
   1. the law enforcement officer or law enforcement agency has obtained an order issued by a court based upon a finding of probable cause to use the tracking instrument; or
   2. exigent circumstances exist that necessitate using the tracking instrument without first obtaining a court order.

b. If a law enforcement officer or law enforcement agency uses a real time tracking instrument described in subsection (a) based upon the existence of exigent circumstances, the law enforcement officer or law enforcement agency shall seek to obtain an order issued by a court based upon a finding of probable cause not later than seventy-two (72) hours after the initial use of the real time tracking instrument.

SECTION 23. IC 35-33-5-13 IS ADDED TO THE INDIANA CODE AS A NEW SECTION TO READ AS FOLLOWS [EFFECTIVE JULY 1, 2014]: Sec. 13. An electronic communication service, remote computing service, and geolocation information service are immune from civil or criminal liability for providing information or evidence as required by a court order under this chapter.

SECTION 24. IC 35-33-5-14 IS ADDED TO THE INDIANA CODE AS A NEW SECTION TO READ AS FOLLOWS [EFFECTIVE JULY 1, 2014]: Sec. 14.

a. For purposes of IC 34-46-4 (Journalist's Privilege Against Disclosure of Information Source) and subject to subsection (b), if:
    1. a governmental entity requests that a court issue a search warrant to a provider of:
        A. electronic communication service; or
        B. remote computing service; and
    2. the search warrant seeks information or communications concerning a news media entity or a person otherwise described in IC 34-46-4-1; the news media entity or person described in IC 34-46-4-1 shall be given reasonable and timely notice of the search warrant request and shall be given an opportunity to be heard by the court concerning the issuance of the search warrant before the search warrant is issued.
b. If:
    1. the search warrant that would be issued to a provider described in subsection (a)(1) concerns a criminal investigation in which the news media entity or person described in IC 34-46-4-1 is a target of the criminal investigation; and
    2. the notice that would be provided to the news media entity or person described in IC 34-46-4-1 under subsection (a) would pose a clear and substantial threat to the integrity of the criminal investigation; the governmental entity shall certify the threat to the court and notice of the search warrant shall be given to the news media entity or person described in IC 34-46-4-1 as soon as the court determines that the notice no longer poses a clear and substantial threat to the integrity of the criminal investigation.

SECTION 25. IC 35-38-2.5-3, AS AMENDED BY P.L.31-2005, SECTION 2, IS AMENDED TO READ AS FOLLOWS [EFFECTIVE JULY 1, 2014]: Sec. 3.

a. As used in this chapter, "monitoring device" means an electronic device that:
    1. can record or transmit information twenty-four (24) hours each day regarding an offender's:
        A. presence or absence from the offender's home; or
        B. precise location;
    2. is minimally intrusive upon the privacy of the offender or other persons residing in the offender's home;
    3. with the written consent of the offender and with the written consent of other persons residing in the home at the time an order for home detention is entered, may record or transmit:

A. a visual image;

B. an electronic communication or any sound; or

C. information regarding the offender's activities while inside the offender's home; and (4) can notify a probation department, a community corrections program, or a contract agency if the offender violates the terms of a home detention order.

b. The term includes any device that can reliably determine the location of an offender and track the locations where the offender has been, including a device that uses a global positioning system satellite service.

c. The term does not include an unmanned aerial vehicle (as defined in IC 35-31.5-2-342.3).

SECTION 26. IC 35-46-8.5 IS ADDED TO THE INDIANA CODE AS A NEW CHAPTER TO READ AS FOLLOWS [EFFECTIVE JULY 1, 2014]:
Chapter 8.5. Unlawful Photography and Surveillance on Private Property
Sec. 1.

a. This section does not apply to any of the following:
   1. Electronic or video toll collection facilities or activities authorized under any of the following:
      A. IC 8-15-2.
      B. IC 8-15-3.
      C. IC 8-15.5.
      D. IC 8-15.7.
      E. IC 8-16.
      F. IC 9-21-3.5.
   2. A law enforcement officer who has obtained:
      A. a search warrant; or
      B. the consent of the owner or private property; to place a camera or electronic surveillance equipment on private property.

b. A person who knowingly or intentionally places a camera or electronic surveillance equipment that records images or data of any kind while unattended on the private property of another person without the consent of the owner or tenant of the private property commits a Class A misdemeanor.

SECTION 27. [EFFECTIVE JULY 1, 2014] (a) The general assembly urges the legislative council to assign to a study committee during the 2014 legislative interim the topic of digital privacy, including:

1. issues related to:

A. searches of electronic devices;
B. compelling the disclosure of electronic user data;
C. the collection and use of geolocation information; and
D. the collection and use of biometric information; by government agencies; and

2. any other issue concerning digital privacy and related subjects.
3. If a study committee is assigned the topic described in subsection (a), the study committee shall issue to the legislative council a final report containing the study committee's findings and recommendations, including any recommended legislation concerning the topic, in an electronic format under IC 5-14-6 not later than November 1, 2014.
4. (c) This SECTION expires January 1, 2015.[182]

House Bill (HB) 1013: This law allows the use of UAS to photograph or take video of a traffic crash site.[183]
House Bill (HB) 1246: This law prohibits the use of UAS to scout game during hunting season.[184]

IOWA

Iowa only has House File (HF) 2289 on record.[185] Additionally, there is a city ordinance in Iowa adding Chapter 11 entitled "Traffic Cameras, Drones, and License Plate Recognition Systems."[186]
House File (HF) 2289: On May 23, 2014, the Governor of Iowa signed HF 2289 and the act became effective on July 1, 2014. This was an act relating to the regulation and use of unmanned aerial vehicles, which reads as follows:
Be it enacted by the General Assembly of the State of Iowa:

Section 1. New Section.

321.492B Use of unmanned aerial vehicle—prohibition—traffic law enforcement.
The state or a political subdivision of the state shall not use an unmanned aerial vehicle for traffic law enforcement.

Section 2. New Section.

808.15 Unmanned aerial vehicle—information—admissibility.

---

[182]HB 1009, Indiana General Assembly, https://iga.in.gov/static-documents/5/3/4/4/5344c8fc/HB1009.06.ENRH.pdf.
[183]HB 1013, Indiana General Assembly, https://iga.in.gov/legislative/2016/bills/house/1013.
[184]HB 1246, Indiana General Assembly, https://iga.in.gov/legislative/2016/bills/house/1246#document-2cf343ef.
[185]HF 2289, Iowa General Assembly, https://legiscan.com/IA/text/HF2289/2013.
[186]Iowa City Ordinance, http://www.thenewspaper.com/rlc/docs/2013/ia-drones.pdf.

Information obtained as a result of the use of an unmanned aerial vehicle is not admissible as evidence in a criminal or civil proceeding, unless the information is obtained pursuant to the authority of a search warrant, or unless the information is otherwise obtained in a manner that is consistent with state and federal law.

Section 3. Unmanned Aerial Vehicle—Report.

The department of public safety, in consultation with the attorney general, state and local agencies, and other interested organizations, including but not limited to organizations with expertise in unmanned aerial vehicle technology, shall examine whether the Iowa criminal code should be modified to regulate the use of unmanned aerial vehicles, shall develop model guidelines for the use of unmanned aerial vehicles, and shall report such guidelines to the general assembly no later than December 31, 2014.[187]

KANSAS

From 2014, there was SB 409 restricting the use of images obtained from the use of unmanned vehicles or aircraft, which died in chamber,[188] and in 2015, HB 2397 was introduced, relating to UAV regulation and privacy,[189] and HB 2683 regulating UAVs, which died in committee.[190] Furthermore, Wichita City Ordinance 9.35.210 was passed in 2016.[191]

The state of Kansas has two enacted bills: SB 319[192] and SB 249.[193]

Senate Bill (SB) 319: This law expands the definition of harassment in the Protection from Stalking Act to include certain uses of UAS.[194]

Senate Bill (SB) 249: This law appropriates funds that can be used to focus on research and development efforts related to UAS by state educational institutions. The law specifies a number of focuses for the research, including the use of UAS for inspection and surveillance by the department of transportation, highway patrol and state bureau of investigation. It requires that the director of UAS make recommendations regarding state laws and rules that balance privacy concerns and the need for robust UAS economic development in the state.[195]

---

[187]HF 2289, Iowa General Assembly, https://legiscan.com/IA/text/HF2289/2013.

[188]SB 409, Kansas Legislative Sessions, http://kslegislature.org/li_2014/b2013_14/measures/sb409/.

[189]HB 2397, Kansas Legislative Sessions, http://www.kslegislature.org/li/b2015_16/measures/hb2397/.

[190]HB 2683, Kansas Legislative Sessions, http://kslegislature.org/li_2014/b2013_14/measures/hb2683/.

[191]Wichita City Code Section 9.35.210, http://uavs.insct.org/wp-content/uploads/2016/04/Wichita-KS-Code-of-Ordinances-9.35.210.pdf.

[192]SB 319, Kentucky State Legislature, http://www.kslegislature.org/li/b2015_16/measures/documents/sb319_enrolled.pdf.

[193]SB 249, Kentucky State Legislature, http://www.kslegislature.org/li/b2015_16/measures/documents/sb249_enrolled.pdf.

[194]SB 319, Kentucky State Legislature, http://www.kslegislature.org/li/b2015_16/measures/documents/sb319_enrolled.pdf.

[195]SB 249, Kentucky State Legislature, http://www.kslegislature.org/li/b2015_16/measures/documents/sb249_enrolled.pdf.

KENTUCKY

A number of bills have been introduced. HB 12 in 2015 that was to prohibit a law enforcement agency from using a drone to gather evidence or other information;[196] HB 22 in 2016 also relating to drone surveillance;[197] HB 67 in 2016 relating to making drone harassment a violation for the first offense, a Class B misdemeanor for the second, and a Class A misdemeanor for a third or subsequent offense;[198] and HB 120 to prohibit the use of a drone for purposes of harassment, to commit acts of voyeurism, or facilitation burglary.[199]

LOUISIANA

Louisiana has enacted seven bills pertaining to UAS, namely, HB 1029,[200] SB 183,[201] HB 19,[202] SB 141,[203] HB 635,[204] HB 335,[205] and SB 73.[206]

House Bill (HB) 1029: The governor signed HB 1029 into law in Act 661. This act then became effective on August 1, 2014. Act 661 enacted R.S. 14:336 and 337, relative to offenses against the public; to create the crimes of unlawful aiming of a laser at an aircraft and unlawful use of an unmanned aircraft system; to provide for definitions; to provide for criminal penalties; to provide for exceptions; and to provide for related matters.

Be it enacted by the Legislature of Louisiana:

Section 1. R.S. 14:336 and 337 are hereby enacted to read as follows:

§336. Unlawful aiming of a laser at an aircraft

A. "Unlawful aiming of a laser at an aircraft" is the intentional projection of a laser on or at an aircraft or at the flight path of an aircraft in the aircraft jurisdiction of the state of Louisiana.

B. For purposes of this Section, the following terms have the following meanings:

   1. "Laser" means any device that projects a beam or point of light by means of light amplification by stimulated emission of radiation or any device that emits light which simulates the appearance of a laser.
   2. "Police officer" shall include commissioned police officers, sheriffs, deputy sheriffs, marshals, deputy marshals, correctional officers, constables, wildlife enforcement agents, and probation and parole officers.

---

[196]HB 12, Kentucky State Legislature, https://legiscan.com/KY/bill/HB12/2015.
[197]HB 22, Kentucky State Legislature, https://legiscan.com/KY/bill/HB22/2016.
[198]HB 67, Kentucky State Legislature, https://legiscan.com/KY/bill/HB67/2016.
[199]HB 120, Kentucky State Legislature, https://legiscan.com/KY/bill/HB120/2016.
[200]HB 1029, Louisiana State Legislature, https://www.legis.la.gov/legis/ViewDocument.aspx?d=915322.
[201]SB 183, Louisiana State Legislature, http://www.legis.la.gov/Legis/ViewDocument.aspx?d=935697.
[202]HB 19, Louisiana State Legislature, https://legiscan.com/LA/comments/HB19/2016.
[203]SB 141, Louisiana State Legislature, http://www.legis.la.gov/legis/ViewDocument.aspx?d=1012711.
[204]HB 635, Louisiana State Legislature, http://www.legis.la.gov/legis/ViewDocument.aspx?d=1013224.
[205]HB 335, Louisiana State Legislature, http://www.legis.la.gov/legis/ViewDocument.aspx?d=1012772.
[206]SB 73, Louisiana State Legislature, https://www.legis.la.gov/legis/ViewDocument.aspx?d=1008337.

C. The provisions of this Section shall not prohibit aiming of a laser at an aircraft by any of the following:
   1. An authorized individual in the conduct of research and development or flight test operations conducted by an aircraft manufacturer, the Federal Aviation Administration, or any other person authorized by the Federal Aviation Administration to conduct research and development or flight test operations.
   2. Members or employees of the United States Department of Defense, United States Department of Homeland Security, or police officers acting in the course and scope of their official duties for the purpose of research, development, operations, testing, or training.
   3. A person using a laser emergency signaling device to send an emergency distress signal. D. (1) Whoever commits the crime of unlawful aiming of a laser at an aircraft shall be imprisoned with or without hard labor for not less than one year nor more than five years and shall be fined two thousand dollars.

D. On a conviction for a second or subsequent offense, the offender shall be imprisoned with or without hard labor for not less than two years nor more than ten years and shall be fined four thousand dollars.

§337. Unlawful use of an unmanned aircraft system

A. Unlawful use of an unmanned aircraft system is the intentional use of an unmanned aircraft system to conduct surveillance of, gather evidence or collect information about, or photographically or electronically record a targeted facility without the prior written consent of the owner of the targeted facility.

B. As used in this Section, the following definitions shall apply:
   1. "Unmanned aircraft system" means an unmanned, powered aircraft that does not carry a human operator, can be autonomous or remotely piloted or operated, and can be expendable or recoverable. "Unmanned aircraft system" does not include any of the following:
      a. A satellite orbiting the earth.
      b. An unmanned aircraft system used by the federal government or a person who is acting pursuant to contract with the federal government to conduct surveillance of specific activities.
      c. An unmanned aircraft system used by the state government or a person who is acting pursuant to a contract with the state government to conduct surveillance of specific activities.
      d. An unmanned aircraft system used by a local government law enforcement agency or fire department.

2. "Targeted facility" means the following systems:
   a. Petroleum and alumina refineries.
   b. Chemical and rubber manufacturing facilities.
   c. Nuclear power electric generation facilities.
3. "Federal government" means the United States of America and any department, agency, or instrumentality thereof.
4. "State government" means the state of Louisiana and any department, agency, or instrumentality thereof.

C. Nothing in this Section shall prohibit a person from using an unmanned aircraft system to conduct surveillance of, gather evidence or collect information about, or photographically or electronically record his own property that is either of the following:
   a. Located on his own immovable property.
   b. Located on immovable property owned by another under a valid lease, servitude, right-of-way, right of use, permit, license, or other right.
   c. Third persons retained by the owner of the property described in Paragraph (1) of this Subsection shall not be prohibited under this Section from using an unmanned aircraft system to conduct activities described in Paragraph (1) of this Subsection.

D. The provisions of this Section shall apply unless preempted by applicable federal law or by regulations adopted by the Federal Aviation Administration.

E. Whoever commits the crime of unlawful use of an unmanned aircraft system shall be fined not more than five hundred dollars, or imprisoned for not more than six months, or both.
   1. On a conviction for a second or subsequent offense, the offender shall be fined not less than five hundred dollars nor more than two thousand dollars, or imprisoned, with or without hard labor, for not less than six months nor more than one year, or both.

F. The provisions of this Section shall not apply to unmanned aircraft systems used for motion picture, television, or similar production where the filming is authorized by the property owner.[207]

Senate Bill (SB) 183: On June 23, 2015 the Louisiana Governor signed SB 183 into law in Act 166 which reads: AIRCRAFT/AVIATION and provides for the regulation of unmanned aerial systems in agricultural commercial operations.

To enact Chapter 1-A of Title 3 of the Louisiana Revised Statutes of 1950, to be comprised of R.S. 3:41 through 47, relative to the regulation of unmanned aerial systems in agricultural

[207]HB 1029, Louisiana State Legislature, https://www.legis.la.gov/legis/ViewDocument.aspx?d=915322.

commercial operations; to provide for definitions; to authorize the commissioner to adopt rules; to provide for license and registration requirements; to provide relative to the operation of unmanned aerial systems; to provide for violations; to authorize the commissioner to issue stop orders; to provide for penalties; and to provide for related matters.
Be it enacted by the Legislature of Louisiana:
Section 1. Chapter 1-A of Title 3 of the Louisiana Revised Statutes of 1950, comprised of R.S. 3:41 through 47 is hereby enacted to read as follows:

CHAPTER 1-A. UNMANNED AERIAL SYSTEMS

§41. Definitions

For the purposes of this Chapter, the following definitions shall apply:

1. "Agricultural commercial operation" means the use of any agricultural facility or agricultural land for agricultural production or agricultural processing and includes the production and processing of crops or products, livestock or livestock products, farm-raised fish and fish products, wood, timber or forest products, fowl or plants for breeding or sale, and poultry or poultry products for commercial or industrial purposes. "Agricultural commercial operation" also includes the use of farm machinery, equipment, devices, chemicals, products for agricultural use, materials, and structures designed for agricultural use and used in accordance with traditional farm practices.
2. "Aircraft" means an unmanned aircraft operated as a part of an unmanned aerial system.
3. "Commissioner" means the Louisiana commissioner of agriculture and forestry.
4. "Data" means any information, pictures, images, samples, or other such information obtained through an unmanned aerial system.
5. "Department" means the Louisiana Department of Agriculture and Forestry.
6. "Unmanned aerial system" means an unmanned aircraft and all associated support equipment, control station, data links, telemetry, communications, and navigation equipment necessary to operate the unmanned aircraft. Such system may include drones, remote controlled aircraft, unmanned aircraft, or any other such craft that is controlled autonomously by computer or remote control from the ground.

§42. Powers of the commissioner

A. The commissioner shall adopt rules and regulations to implement the provisions of this Chapter, including but not limited to rules and regulations governing the registration of unmanned aerial systems used in the course of agricultural commercial operations. All rules and regulations shall be adopted in accordance with the Administrative Procedure Act.

B. The commissioner shall administer and enforce the provisions of this Chapter and the rules and regulations adopted pursuant to the provisions of this Chapter.

C. The commissioner may appoint committees to advise on the implementation of the provisions of this Chapter.

§43. Licenses and registration

A. Each person operating an unmanned aerial system in the course of an agricultural commercial operation shall obtain a license from the department which shall be issued upon the applicant meeting the following requirements:

1. Applicants must submit a written application to the department as provided by the commissioner by rule.
2. Prior to applying for a license, each applicant shall complete an educational and safety training course conducted by the Louisiana State University Agricultural Center and comply with any other licensing requirements established by the commissioner by rule.

B. Each unmanned aerial system operated in the course of an agricultural commercial operation shall be registered with the department.

C. Licenses and registrations shall be valid for three years and may be renewed for additional three-year periods in accordance with rules adopted by the commissioner.

§44. Unmanned aerial systems; operation

A. Unmanned aerial systems may operate in agricultural commercial operations in accordance with this Chapter and the rules and regulations established by the commissioner, except as prohibited by federal law.

B. Private landowners engaged in agricultural commercial operations on their private property may use unmanned aerial systems within the geographical confines of such property.

1. Producers, tenants, lessees, or other contracted or hired personnel working on private property who are engaged in agricultural commercial operations may use unmanned aerial systems within the geographical confines of the property, only with written permission of the landowner or entity controlling the agricultural commercial use of the property.
2. Data obtained through the use of an unmanned aerial system shall be used solely in the course of conducting a generally accepted agricultural commercial operation, or in conjunction with an agricultural research, extension program, or initiative conducted by a Louisiana public postsecondary educational institution.

3. All data obtained through the use of an unmanned aerial system shall remain the property of the legal owner of the property where the data was collected, unless written approval is given by the property owner for other uses.

§45. Violations

Violations of this Part shall include but shall not be limited to the following:

1. The failure or refusal to obtain an unmanned aerial system license in accordance with R.S. 3:43.
2. The failure or refusal to register an unmanned aerial system operated in the course of an agricultural commercial operation in accordance with R.S. 3:43.
3. Any violation of this Chapter or of any rule or regulation adopted pursuant to the provisions of this Chapter.
4. Any interference with the commissioner and the department or its representatives in the performance of their duties as prescribed in this Chapter.
5. Failure to timely pay any penalties or costs due pursuant to the provisions of this Chapter or any rule or regulation of the commissioner adopted pursuant to this Chapter.

§46. Stop orders

A. When the commissioner has reason to believe that a violation has occurred, the commissioner may issue a stop order prohibiting the continued use of an unmanned aerial system.

B. Any violation of a stop order shall constitute a separate violation.

C. The stop order may be released by the commissioner upon a determination by the commissioner that the cause for issuing the stop order has been remedied.

D. Any person aggrieved by a stop order may petition the commissioner for a hearing to contest the validity of the stop order by making a written request within five calendar days after issuance of the stop order. The hearing shall be held within ten calendar days of receipt of the written request for a hearing. The commissioner may appoint a hearing officer to preside over the matter. The commissioner shall issue a ruling in the matter. The hearing and any subsequent appeal shall be held in accordance with the provisions of the Administrative Procedure Act.

E. Based upon the results of the hearing, or a consent agreement, the commissioner may take one or more of the following actions:
   1. Release the stop order.
   2. Require the cause for the stop order to be remedied prior to releasing the stop order.
   3. Amend the stop order.

§47. Penalties

A. If the commissioner determines that a violation of the provisions of this Chapter or of the rules and regulations adopted pursuant to the provisions of this Chapter has occurred, an adjudicatory hearing shall be held to make a determination with respect to the suspected violation.

B. The commissioner shall give written notice to the alleged violator, which shall comply with the requirements of the Administrative Procedure Act, at least five days prior to the date set for such adjudicatory hearing.

C. The commissioner shall designate a hearing officer to preside at all adjudicatory proceedings.

D. At any such adjudicatory hearing, the alleged violator shall be accorded all of the rights set forth in the Administrative Procedure Act.

E. Whenever the commissioner makes a determination from the proceedings of the adjudicatory hearing that a violation of this Chapter or these regulations has occurred, the commissioner may impose a civil penalty not to exceed five hundred dollars for each violation. Civil penalties shall be imposed only on the basis of an adjudication of violations pursuant to an adjudicatory hearing held in accordance with the Administrative Procedure Act.

F. Each separate day on which a violation occurs shall be considered a separate violation. Section 2. This Act shall become effective upon signature by the governor or, if not signed by the governor, upon expiration of the time for bills to become law without signature by the governor, as provided by Article III, Section 18 of the Constitution of Louisiana. If vetoed by the governor and subsequently approved by the legislature, this Act shall become effective on the day following such approval.[208]

House Bill (HB) 19: This law, effective August 1, 2016, prohibits using a drone to conduct surveillance of, gather evidence or collect information about, or take photo or video of a school, school premises, or correctional facilities. The law also establishes a penalty of a fine of up to $2,000 and up to six months in jail.[209]

Senate Bill (SB) 141: This law, effective June 17, 2016, specifies that surveillance by an unmanned aircraft constitutes criminal trespass under certain circumstances.[210]

House Bill (HB) 635: This law, effective August 1, 2016, adds the use of UAS to the crimes of voyeurism, video voyeurism and peeping tom.[211]

---

[208]SB 183, Louisiana State Legislature, http://www.legis.la.gov/Legis/ViewDocument.aspx?d=935697.

[209]HB 19, Louisiana State Legislature, https://legiscan.com/LA/comments/HB19/2016.

[210]SB 141, Louisiana State Legislature, http://www.legis.la.gov/legis/ViewDocument.aspx?d=1012711.

[211]HB 635, Louisiana State Legislature, http://www.legis.la.gov/legis/ViewDocument.aspx?d=1013224.

House Bill (HB) 335: This law, effective August 1, 2016, authorizes the establishment of registration and licensing fees for UAS, with a limit of $100.[212]
Senate Bill (SB) 73: This law, effective August 1, 2016, adds intentionally crossing a police cordon using a drone to the crime of obstructing an officer. Furthermore, it allows law enforcement or fire department personnel to disable the UAS if it endangers the public or an officer's safety.[213]

MAINE

Maine has a sole Legislative Document (LD) 25 on its books.[214]
Legislative Document (LD) 25: On July 2, 2015, LD 25 was enacted into Public Law Chapter 307, an act to regulate domestic unmanned aerial vehicle use. Chapter 307 reads as follows: Be it enacted by the People of the State of Maine as follows:
Sec. 1. 25 MRSA Pt. 12 is enacted to read:

PART 12 UNMANNED AERIAL VEHICLES CHAPTER 551

REGULATION OF UNMANNED AERIAL VEHICLES

§4501. Regulation of unmanned aerial vehicles

1. Findings.

   The Legislature finds that evolving technology regarding unmanned aerial vehicles presents a potential economic driver for the State, an opportunity for research and development and a very real benefit for security, for search and rescue efforts and for disaster prevention and relief, as well as a tool for the investigation of serious crimes, but the technology also presents a potential threat to the privacy of citizens of this State if used by law enforcement in the conduct of criminal investigations without appropriate guidelines and supervision.
2. Definitions.

   As used in this section, unless the context otherwise indicates, the following terms have the following meanings.

   A. "Law enforcement agency" has the same meaning as in section 3701, subsection 1.

   B. "Unmanned aerial vehicle" means an aircraft operated without a physical human presence within or on the aircraft that, in the manner in which the

[212] HB 335, LOUISIANA STATE LEGISLATURE, http://www.legis.la.gov/legis/ViewDocument.aspx?d=1012772.
[213] SB 73, LOUISIANA STATE LEGISLATURE, https://www.legis.la.gov/legis/ViewDocument.aspx?d=1008337.
[214] LD 25, MAINE LEGISLATURE, http://www.mainelegislature.org/legis/bills/getPDF.asp?paper=HP0024&item=3&snum=127.

aircraft is used or the manner in which it is equipped, is capable of performing audio or visual surveillance.

3. Acquisition of unmanned aerial vehicles.
   A. The acquisition of an unmanned aerial vehicle by a law enforcement agency must be approved by the governing body of the governmental unit overseeing the law enforcement agency seeking to make such an acquisition or, in the case of a state agency, by the commissioner of that agency.
4. Law enforcement agency operation of unmanned aerial vehicles.

   A law enforcement agency's operation of an unmanned aerial vehicle must fully comply with all Federal Aviation Administration requirements and guidelines, including the acquisition of a certificate of authorization or waiver from the Federal Aviation Administration. Additionally, a law enforcement agency's use of an unmanned aerial vehicle is governed by the following provisions.
   A. A law enforcement agency may not use an unmanned aerial vehicle before adopting standards that meet, at a minimum, the standards set forth in subsection 5.
   B. Except as permitted by a recognized exception to the requirement for a warrant under the Constitution of Maine or the United States Constitution, a law enforcement agency may not use an unmanned aerial vehicle for criminal investigations without a warrant.
   C. Notwithstanding paragraph A, a law enforcement agency may use an unmanned aerial vehicle for the purpose of a search and rescue operation when the law enforcement agency determines that use of an unmanned aerial vehicle is necessary to alleviate an immediate danger to any person or for training exercises related to such uses.
   D. Notwithstanding paragraph A, a law enforcement agency may use an unmanned aerial vehicle for purposes other than the investigation of crime, including, but not limited to, aerial photography for the assessment of accidents, forest fires and other fire scenes, flood stages and storm damage.
   E. In no case may a weaponized unmanned aerial vehicle be used or its use facilitated by a state or local law enforcement agency in this State.
   F. A law enforcement agency may not use an unmanned aerial vehicle to conduct surveillance of private citizens peacefully exercising their constitutional rights of free speech and assembly.
   G. Notwithstanding paragraph A, a law enforcement agency may use an unmanned aerial vehicle for an emergency use approved by the chief administrative officer of the agency or the Governor.

5. Minimum standards for law enforcement. The Board of Trustees of the Maine Criminal Justice Academy, in consultation with the Office of the Attorney General, shall establish minimum standards for written policies and protocols for use of unmanned aerial vehicles by law enforcement agencies. The standards must include at a minimum:
   A. Training and certification requirements for a person operating an unmanned aerial vehicle;
   B. Requirements for prior authorization for the use of an unmanned aerial vehicle by the chief administrative officer of the law enforcement agency seeking to use such a vehicle;
   C. Approval by the Attorney General or chief prosecuting attorney for the appropriate jurisdiction for the deployment of an unmanned aerial vehicle for criminal investigation purposes;
   D. Restrictions on the use of night vision technology, high-powered zoom lenses, video analytics, facial recognition technology, thermal imaging and other such enhancement technology;
   E. Procedures to minimize the inadvertent audio or visual recording of private spaces of 3rd parties who are not under investigation;
   F. Procedures for destroying any unnecessary audio or visual recordings without further duplication or dissemination;
   G. Recommended minimum altitudes and speeds at which an unmanned aerial vehicle may be flown in order to minimize the invasion of privacy of 3rd parties who are not under investigation;
   H. Methods to minimize the number of unmanned aerial vehicles deployed at any one time in any one area or at any one event;
   I. Procedures to avoid hazards to persons and property on land and in the air due to the operation of unmanned aerial vehicles;
   J. Methods for tracking and recording the flight of each unmanned aerial vehicle;
   K. Requirements for regular statistical reporting of all uses of unmanned aerial vehicles, including the purposes, the results and the duration of such uses, to the appropriate governmental bodies; and
   L. Accountability of a law enforcement agency for any mistake in deployment or misuse of an unmanned aerial vehicle, including sanctions as provided in section 2803-C or section 2806-A, as applicable.
6. Data collection. On or before July 1, 2016 and July 1st of each subsequent year, the Commissioner of Public Safety shall submit to the Legislature a report containing the number of instances in which an unmanned aerial vehicle has been deployed by any law enforcement agency in the State with summary descriptions

of the number of deployments for investigative purposes, the general nature of those investigations and the number of search warrants sought and the number of search warrants obtained for the deployment of unmanned aerial vehicles.[215]

MARYLAND

Maryland has enacted two bills pertaining to UAS. They are HB 620[216] and SB 370.[217]

In 2013, HB 100 appropriated $500,000 for the state's unmanned aerial system test site.[218] Also introduced were HB 351 requiring the use of a drone by an agent of the state or a political subdivision of the state to fully comply with all specified applicable federal requirements and guidelines,[219] HB 8 establishing a task force to study issues related to the use of self-driving vehicles,[220] HB 88 prohibiting an individual from operating an automated motor vehicle in automated mode on a highway,[221] and SB 126 establishing the task force to study issues relating to the use of self-driving issues.[222]

House Bill (HB) 620: HB 620 was introduced and read for the first time on February 12, 2015. It is a bill entitled, *An Act Concerning Crimes—Unmanned Aircraft Systems—Unauthorized Surveillance*, and reads as follows:

FOR the purpose of prohibiting, except under certain circumstances, a person from using a certain unmanned aircraft system to intentionally conduct surveillance of, gather evidence or collect information about, or photographically or electronically record a certain individual or property; providing penalties for a violation of this Act; defining a certain term; and generally relating to unmanned aircraft systems.

BY adding to Article—Criminal Law—*Section 3–908*—Annotated Code of Maryland (2012 Replacement Volume and 2014 Supplement)

SECTION 1.

Be it enacted by the General Assembly of Maryland, that the Laws of Maryland read as follows:

---

[215] LD 25, Maine Legislature, http://www.mainelegislature.org/legis/bills/getPDF.asp?paper=HP0024&item=3&snum=127.

[216] HB 620, Maryland General Assembly, http://mgaleg.maryland.gov/2015RS/bills/hb/hb0620F.pdf.

[217] SB 370, Maryland General Assembly, http://mgaleg.maryland.gov/2015RS/bills/sb/sb0370f.pdf.

[218] HB 100, Maryland General Assembly, http://mgaleg.maryland.gov/webmga/frmMain.aspx?pid=billpage&stab=01&id=hb0100&tab=subject3&ys=2013RS.

[219] HB 351, Maryland General Assembly, http://mgaleg.maryland.gov/webmga/frmMain.aspx?id=HB0351&stab=01&pid=billpage&tab=subject3&ys=2016RS.

[220] HB 8, Maryland General Assembly, http://mgaleg.maryland.gov/webmga/frmMain.aspx?id=HB0008&stab=01&pid=billpage&tab=subject3&ys=2016RS.

[221] HB 88, Maryland General Assembly, http://mgaleg.maryland.gov/webmga/frmMain.aspx?id=HB0088&stab=01&pid=billpage&tab=subject3&ys=2016RS.

[222] SB 126, Maryland General Assembly, http://mgaleg.maryland.gov/webmga/frmMain.aspx?id=SB0126&stab=01&pid=billpage&tab=subject3&ys=2016rs.

Article—Criminal Law 3-908

A. (1) In this Section, "Unmanned Aircraft System" means a powered aerial vehicle that:
   i. does not carry a human operator;
   ii. uses aerodynamic forces to provide vehicle lift;
   iii. can fly autonomously or be piloted remotely;
   iv. is expendable or recoverable; and
   v. carries equipment that can be used for surveillance purposes.

   (2) "Unmanned aircraft system" does not include:
   i. a model flying airplane or rocket, including one that is radio-controlled or otherwise remotely controlled and used purely for sport or recreational purposes; or
   ii. an unmanned aircraft system used in mapping or resource management.

B. A person may not use an unmanned aircraft system to intentionally conduct surveillance of, gather evidence or collect information about, or photographically or electronically record specifically targeted individuals or private property.

C. This section does not prohibit the use of an unmanned aircraft system:
   1. In accordance with a valid search warrant;
   2. In connection with an emergency response for safety, search and rescue, or controlled dangerous substance investigation;
   3. With the written consent of the individual being observed; or
   4. By an owner of a facility located on property owned by another person under an easement, permit, license, or other right of occupancy using an unmanned aircraft system to inspect the facility.

D. A person who violates this section is guilty of a misdemeanor and on conviction is subject to imprisonment not exceeding 1 year or a fine not exceeding $2,500 or both.

SECTION 2.

And be it further enacted, that this Act shall take effect October 1, 2015.[223]

Senate Bill (SB) 370: The governor of Maryland signed SB 370 into law, Chapter 164, which went into effect on July 1, 2015. This is a bill entitled, *An Act concerning Unmanned Aircraft Systems Research, Development, Regulation, and Privacy Act of 2015*. It reads as follows:

FOR the purpose of providing that only the State may enact a law or take other action to prohibit, restrict, or regulate the testing or operation of unmanned aircraft systems in the State; preempting the authority of a county or municipality to prohibit, restrict, or regulate the testing or operation of unmanned aircraft systems and providing that certain laws or ordinances are superseded; specifying that federal preemption of State law is not

[223]HB 620, Maryland General Assembly, http://mgaleg.maryland.gov/2015RS/bills/hb/hb0620F.pdf.

affected by this Act; requiring the Department of Business and Economic Development, in consultation with the University of Maryland and other interested parties, to study the benefits of and the concerns regarding the use of certain unmanned aircraft systems in the State; requiring the Department to report its findings and recommendations to the General Assembly on or before a certain date; defining certain terms; and generally relating to unmanned aircraft systems.

BY adding to Article—Economic Development—Section 14–301 to be under the new subtitle "Subtitle 3. Unmanned Aircraft Systems" Annotated Code of Maryland (2008 Volume and 2014 Supplement)

Preamble

WHEREAS, The Federal Aviation Act of 1958 established the Federal Aviation Administration (FAA) and made the FAA responsible for the control and use of navigable airspace within the United States; and

WHEREAS, The FAA regulates unmanned aircraft systems, also known as drones; and

WHEREAS, Public entities, including law enforcement agencies, firefighter units, border and port patrols, disaster relief workers, search and rescue personnel, qualifying public universities, military training facilities, and other government operational mission units, may operate unmanned aircraft systems by applying for a Certificate of Waiver or Authorization from the FAA; and

WHEREAS, In December 2013, the FAA selected six public entities around the nation, including the Mid-Atlantic Aviation Partnership, to develop test sites for unmanned aircraft systems research; and

WHEREAS, The FAA is in the process of developing rules that are intended to safely integrate small unmanned aircraft systems into the national airspace system and that, following issuance of a Notice of Proposed Rulemaking and public comment period, are expected to be released in late 2016 or 2017; and

WHEREAS, The small unmanned aircraft systems rules are anticipated to be similar to the current hobbyist rules for operations of model unmanned aircraft systems that limit the area of operation to low-risk and controlled environments and the size of the systems to less than 55 pounds; and

WHEREAS, While the FAA is developing the small unmanned aircraft systems rules, private commercial entities on a company-by-company basis may apply to the FAA for a Section 333 exemption from the current rules for manned aircraft by showing that the entity operates with at least an equivalent level of safety; and WHEREAS, In September 2014, six companies in the film production industry were granted Section 333 exemptions and the FAA has stated that many more exemptions are in the processing stage; and

WHEREAS, In December 2014, as part of the Mid-Atlantic Aviation Partnership, the University of Maryland's unmanned aircraft systems testing operation opened a test site near Naval Air Station Patuxent River, allowing commercial companies to partner with the University to develop their projects; and

WHEREAS, In addition to being used in military training and operations and public safety areas, unmanned aircraft systems may be used in a wide variety of activities, including: oil and natural gas pipeline inspection; transportation; natural disaster aid; precision agriculture; natural resource and environmental protection; bridge and infrastructure inspection; public utility support; construction and building inspection; surveying; golf course marketing; wind turbine inspection; realtor marketing and photography; and prison monitoring; and

WHEREAS, Other countries have advanced in the testing and operation of unmanned aircraft systems; and

WHEREAS, While the public has expressed concerns with the operation of unmanned aircraft systems, including privacy and public safety issues, there are benefits that may be realized by the State, including the State's various industry sectors, from conducting research on unmanned aircraft systems in the State and developing, manufacturing, and operating unmanned aircraft systems in the State; now, therefore,

SECTION 1.

Be it enacted by the General Assembly of Maryland, that the Laws of Maryland read as follows:

Article—Economic Development

SUBTITLE 3. Unmanned Aircraft Systems.
14–301.

A. In this section the following words have the meanings indicated.
   1. "Unmanned aircraft" means the flying portion of an unmanned aircraft system, flown by a pilot via a ground control system, or autonomously through use of an onboard computer, a communication link, and any additional equipment that is necessary for the unmanned aircraft to operate safely.
   2. "Unmanned aircraft system" means an unmanned aircraft and all the associated support equipment, control stations, data links, telemetry, communications and navigation equipment, and other equipment necessary to operate the unmanned aircraft.

B. Only the State may enact a law or take any other action to prohibit, restrict, or regulate the testing or operation of unmanned aircraft systems in the State.

C. Subsection (B) of this section:
   1. Preempts the authority of a county or municipality to prohibit, restrict, or regulate the testing or operation of unmanned aircraft systems; and

2. Supersedes any existing law or ordinance of a county or municipality that prohibits, restricts, or regulates the testing or operation of unmanned aircraft systems.

D. This section does not affect federal preemption of state law.

SECTION 2.

And be it further enacted, that:

a. The Department of Business and Economic Development, in consultation with the University of Maryland in its role as a member of the Mid-Atlantic Aviation Partnership and with other interested parties, shall:
   1. identify the benefits, including job creation, a cleaner environment, positive economic impacts, increased public safety, and enhanced efficiencies, that may be realized by the State from conducting research on unmanned aircraft systems in the State and developing, manufacturing, and operating unmanned aircraft systems in the State;
   2. determine the benefits that may be realized within each of the various industry sectors in the State that have the potential for the use of unmanned aircraft systems and evaluate the extent to which the use of unmanned aircraft systems in the State is in the public interest;
   3. identify ways in which the benefits and other positive aspects associated with the use of unmanned aircraft systems may be conveyed to the public in order to achieve public support;
   4. identify policies that should be implemented to ensure that any concerns, including privacy and public safety issues, are addressed; and
   5. as a way to provide certainty to companies that want to use unmanned aircraft systems in the State, develop qualification guidelines for companies to follow when applying to the Federal Aviation Administration for a Section 333 exemption.

b. On or before December 31, 2015, the Department shall report, in accordance with § 2–1246 of the State Government Article, its findings and recommendations to the General Assembly.

SECTION 3.

And be it further enacted, that this Act shall take effect July 1, 2015.[224]

---

[224] SB 370, Maryland General Assembly, http://mgaleg.maryland.gov/2015RS/bills/sb/sb0370f.pdf.

MASSACHUSETTS

The Commonwealth of Massachusetts has introduced HB 1322 to regulate the use of UAVs;[225] HB 2977 for legislation to authorize the operation of autonomous vehicles without active control or monitoring by a human operator;[226] SB 1835 for legislation to regulate the use of UAVs;[227] SB 1841 relating to autonomous vehicles;[228] SB 447 for legislation relative to hunting with UAS;[229] HB 2169 relative to military equipment procurement by local law enforcement departments, to include unmanned aerial vehicles;[230] and HB 1357 relative to the use of information derived from UAVs as evidence in judicial, regulatory, or other government proceedings.[231]

The city of Northampton has passed an Ordinance,[232] and in April 2016, the Town of Barnstable banned drones on Barnstable's public beaches.[233]

City of Northampton Ordinance: On June 27, 2013, the city of Northampton, in Massachusetts passed the following resolution on drone aircraft that reads:

III. I. Navigable airspace and drone aircraft.

WHEREAS, the former City Councilor and Mayor of Northampton, Calvin Coolidge, as President of the United States, signed into law the Air Commerce Act of 1926, establishing the national airspace system in the United States; and

WHEREAS, this act declared that the airspace above the minimum safe altitudes of flight, generally understood to be about 500 feet or more above the surface, is 'navigable airspace'; and

WHEREAS, aircraft, including drone aircraft, have the 'public right of transit' through navigable airspace; and

WHEREAS, navigable airspace is preempted by federal laws and therefore not generally subject to state and local control; and

WHEREAS Congress and the FAA now appear to be expanding navigable airspace down to the ground in order to accommodate low-flying drone aircraft; and

WHEREAS, such an expansion of navigable airspace threatens long-standing property rights, expectations of privacy, and state and local sovereignty; and

---

[225]HB 1322, Commonwealth of Massachusetts, https://legiscan.com/MA/research/H1322/2015.
[226]HB 2977, Commonwealth of Massachusetts, https://legiscan.com/MA/research/H2977/2015.
[227]SB 1835, Commonwealth of Massachusetts, https://legiscan.com/MA/research/S1835/2015.
[228]SB 1841, Commonwealth of Massachusetts, https://legiscan.com/MA/bill/S1841/2015.
[229]SB 447, Commonwealth of Massachusetts, https://legiscan.com/MA/research/S447/2015.
[230]HB 2169, Commonwealth of Massachusetts, https://malegislature.gov/Bills/189/House/H2169.
[231]HB 1357, Commonwealth of Massachusetts, https://legiscan.com/MA/bill/H1357/2013.
[232]Resolution on Drone Aircraft, City of Northampton, http://www.northamptonma.gov/DocumentCenter/View/1103.
[233]*Drones Banned on Barnstable's Public Beaches*, Town of Barnstable, http://www.townofbarnstable.us/townmanager/deskoftm/enews.aspx?filename=FridayApril222016.html.

WHEREAS, such an expansion of navigable airspace could give the FAA unprecedented authority to restrict any landowner activity that could potentially interfere with low-flying drone aircraft; and

WHEREAS, such an expansion of navigable airspace is contrary to the findings of the landmark 1946 Supreme Court case, *United States v. Causby*, which affirmed that the landowner 'must have exclusive control over the immediate reaches of the enveloping atmosphere' and that 'the landowner owns at least as much of the airspace as they can occupy or use in connection with the land'; and

WHEREAS, such an expansion of navigable airspace is contrary to the aviation statutes in Title 49 of the United States Code, which have been in force, largely unchanged, since President Coolidge first signed them into law in 1926; and

IV. II. Surveillance and Weaponized Drone Aircraft

WHEREAS, drone aircraft are poised to gain unprecedented access to private property at any altitude and for any purpose, including but not limited to purposes of advertising, news reporting, environmental monitoring, package delivery, recreation, and private investigations; and

WHEREAS, police departments in the United States have begun using drone aircraft in the absence of clear guidance from lawmakers; and

WHEREAS, some drone aircraft being marketed to law enforcement agencies are designed to carry weapons including tear gas, rubber buckshot, and firearms and the introduction of such technology sends a chilling message to the American people; and

WHEREAS, the rapid development of drone aircraft technology, driven largely by manufacturers producing drone aircraft for military use, poses a threat to the privacy and constitutional rights of the American people, including the residents of Northampton; and

WHEREAS, the federal use of weaponized drone aircraft overseas is a poor precedent for their domestic use, in that the extrajudicial use of drones has turned public opinion against the US government in Pakistan, Yemen, Somalia, and Afghanistan; and

WHEREAS, drone aircraft strikes have killed more non-targeted people than those targeted, including men, women, and children, some known by name and others unidentified;

BE IT RESOLVED, that the City of Northampton calls on the US government to immediately end its practice of extrajudicial killing by armed drone aircraft; and

BE IT FURTHER RESOLVED, that the City of Northampton affirms that, within the city limits, the navigable airspace for drone aircraft shall not be expanded below the long-established airspace for manned aircraft; and

BE IT FURTHER RESOLVED, that the City of Northampton affirms that within the city limits, landowners subject to state laws and local ordinances have exclusive control of the immediate reaches of the airspace and that no drone aircraft shall have the 'public right of transit' through this private property; and

BE IT FURTHER RESOLVED, that the City of Northampton calls on the United States Congress, the Federal Aviation Administration, and the General Assembly of the Commonwealth of Massachusetts to respect legal precedent and constitutional guarantees of privacy, property rights, and local sovereignty in all matters pertaining to drone aircraft and navigable airspace.[234]

MICHIGAN

In Michigan, there are currently two bills that have been enacted, SB 54[235] and SB 55.[236] In 2013, HR 280 sought to support and endorse the proposal submitted by the Michigan Advanced Aerial Systems Consortium requesting that Michigan be named one of six federally identified UAS test sites.[237] Also introduced were HB 4866 calling for prohibition against use of UAVs within 1,500 feet of a correctional facility;[238] HB 4867 calling for sentencing guidelines for operating UAVs within 1,500 feet of a correctional facility;[239] HB 4868 calling for the criminalization of certain uses of UAVs;[240] HB 5026 relating to law enforcement, investigations, use of UAVs, and collection of evidence;[241] HB 5027 calling for criminal sentencing guidelines for UAV statute violations;[242] HB 5148 prohibiting the operation of drones in airspace above the capitol or the capitol grounds without authorization from the Michigan State Capitol commission;[243] SB 432 prohibiting operation of UAVs near the Mackinac Bridge;[244] SB 433 adding the definition of drone to the Aeronautics Code of the State of Michigan;[245] SB 487 prohibiting and providing for penalties for the operation of UAVs within 1,000 feet of a correctional facility;[246] SB 488 requiring sentencing guidelines for operating UAVs within 1,000 feet of a correctional

---

[234] *Resolution on drone aircraft*, City of Northampton, http://www.northamptonma.gov/DocumentCenter/View/1103.

[235] SB 54, Michigan Legislature, http://www.legislature.mi.gov/documents/2015-2016/publicact/pdf/2015-PA-0012.pdf.

[236] SB 55, Michigan Legislature, http://www.legislature.mi.gov/documents/2015-2016/publicact/pdf/2015-PA-0013.pdf.

[237] HR 280, Michigan Legislature, http://www.legislature.mi.gov/(S(cr1ee5j5tvwdl12wvloj0102))/mileg.aspx?page=getobject&objectName=2013-HR-0280.

[238] HB 4866, Michigan Legislature, http://www.legislature.mi.gov/(S(pty45bjvt0xx0ga01erclaoj))/mileg.aspx?page=GetObject&objectname=2015-HB-4866.

[239] HB 4867, Michigan Legislature, http://www.legislature.mi.gov/(S(wldnvadnv3rf3gjhryqsvzix))/mileg.aspx?page=GetObject&objectName=2015-HB-4867.

[240] HB 4868, Michigan Legislature, http://www.legislature.mi.gov/(S(rkqeson45pgtlzgc1n0eubch))/mileg.aspx?page=BillStatus&objectname=2015-HB-4868.

[241] HB 5026, Michigan Legislature, http://www.legislature.mi.gov/(S(fu1ylbvahpaulabfjjzkklrg))/mileg.aspx?page=GetObject&objectname=2015-HB-5026.

[242] HB 5027, Michigan Legislature, http://www.legislature.mi.gov/(S(hwspczbnuhkdxe125qdrkssr))/mileg.aspx?page=GetObject&objectname=2015-HB-5027.

[243] HB 5148, Michigan Legislature, https://www.legislature.mi.gov/(S(xcsimz5yqcwfvlmhtbmdvgww))/mileg.aspx/mileg.aspx?page=getobject&objectName=2015-HB-5148.

[244] SB 432, Michigan Legislature, https://www.legislature.mi.gov/(S(ll4v53jxe4ot3tpaycrh2yqy))/mileg.aspx?page=GetObject&objectName=2015-SB-0432.

[245] SB 433, Michigan Legislature, http://openstates.org/mi/bills/2015-2016/SB433/.

[246] SB 487, Michigan Legislature, http://www.legislature.mi.gov/(S(pgnxafw00a35hbokurpt0jmb))/mileg.aspx?page=GetObject&objectName=2015-SB-0487.

facility;[247] and SB 549 prohibiting the operation of UAVs above Michigan State Capitol Historic Site or its grounds.[248] Finally, East Lansing City Council passed Ordinance No. 1369 to regulate the use of UAVs adding section 26-67 to Division 2, Disorderly Conduct, of Chapter 26.[249]

Senate Bill (SB) 54: On April 14, 2015, the Governor of Michigan signed SB 54 into Public Act 12 of 2015, effective July 13, 2015. *Public Law 12* is an act to amend 1994 PA 451, entitled "An act to protect the environment and natural resources of the state; to codify, revise, consolidate, and classify laws relating to the environment and natural resources of the state; to regulate the discharge of certain substances into the environment; to regulate the use of certain lands, waters, and other natural resources of the state; to protect the people's right to hunt and fish; to prescribe the powers and duties of certain state and local agencies and officials; to provide for certain charges, fees, assessments, and donations; to provide certain appropriations; to prescribe penalties and provide remedies; and to repeal acts and parts of acts," by amending sections 40112 and 48703a (MCL 324.40112 and 324.48703a), section 40112 as amended by 1996 PA 316 and section 48703a as amended by 2014 PA 281.

The People of the State of Michigan enact*:*

Sec. 40112.

1. An individual shall not obstruct or interfere in the lawful taking of animals or fish by another individual.
2. An individual violates this section when the individual intentionally or knowingly does any of the following:
    a. Drives or disturbs animals or fish for the purpose of disrupting a lawful taking.
    b. Blocks, impedes, or harasses another individual who is engaged in the process of lawfully taking an animal or fish.
    c. Uses a natural or artificial visual, aural, olfactory, gustatory, or physical stimulus or an unmanned vehicle or unmanned device that uses aerodynamic forces to achieve flight or that operates on the surface of the water or underwater, to affect animal or fish behavior in order to hinder or prevent the lawful taking of an animal or a fish.
    d. Erects barriers to deny ingress or egress to areas where the lawful taking of animals or fish may occur. This subdivision does not apply to an individual who erects barriers to prevent trespassing on his or her property.

[247]SB 488, Michigan Legislature, http://openstates.org/mi/bills/2015-2016/SB488/.
[248]SB 549, Michigan Legislature, http://openstates.org/mi/bills/2015-2016/SB549/.
[249]Ordinance No. 1369, City of East Lansing, http://eastlansing.granicus.com/MetaViewer.php?view_id=2&clip_id=567&meta_id=35066.

e. Interjects himself or herself into the line of fire of an individual lawfully taking wildlife. (f) Affects the condition or placement of personal or public property intended for use in the lawful taking of an animal or a fish in order to impair the usefulness of the property or prevent the use of the property.

f. Enters or remains upon private lands without the permission of the owner or the owner's agent, for the purpose of violating this section.

g. Engages in any other act or behavior for the purpose of violating this section.

3. Upon petition of an aggrieved person or an individual who reasonably may be aggrieved by a violation of this section, a court of competent jurisdiction, upon a showing that an individual was engaged in and threatens to continue to engage in illegal conduct under this section, may enjoin that conduct.
4. An individual who violates this section is guilty of a misdemeanor punishable by imprisonment for not more than 93 days or a fine of not less than $500.00 or more than $1,000.00, or both, and the costs of prosecution. An individual who violates this section a second or subsequent time is guilty of a misdemeanor punishable by imprisonment for not more than 1 year or a fine of not less than $1,000.00 or more than $2,500.00, or both, and the costs of prosecution. In addition to the penalties provided for in this subsection, any permit or license issued by the department authorizing the individual to take animals or fish shall be revoked. A prosecution under this section does not preclude prosecution or other action under any other criminal or civil statute.
5. This section does not apply to a peace officer while the peace officer performs his or her lawful duties.

Sec. 48703a.

1. The legislature finds and declares that aquatic invasive species, including Asian carp, represent a significant threat to the state's fisheries, aquatic resources, outdoor recreation and tourism economies, and public safety.
2. The commission has the exclusive authority to regulate sport fishing under this part. The commission shall, to the greatest extent practicable, utilize principles of sound scientific management in making decisions regarding the regulation of sport fishing under this part. The commission may take testimony from department personnel, independent experts, and others, and review scientific literature and data, among other sources, in support of the commission's duty to use principles of sound scientific management. The commission shall issue orders regarding the regulation of sport fishing under this part following a public meeting and an opportunity for public input. Not less than 30 days before issuing an order, the commission shall provide a copy of the order to each of the following:

   a. Each member of each standing committee of the senate or house of representatives that considers legislation pertaining to conservation, the environment, natural resources, recreation, tourism, or agriculture.
   b. The chairperson of the senate appropriations committee and the chairperson of the house of representatives appropriations committee.
   c. The members of the subcommittee of the senate appropriations committee and the subcommittee of the house of representatives appropriations committee that consider the budget of the department of natural resources.
3. For the fiscal year ending September 30, 2015, there is appropriated for the department the sum of $1,000,000.00 to implement management practices necessary for rapid response, prevention, control, or elimination of aquatic invasive species, including Asian carp. Any portion of the amount under this section that is not expended in the fiscal year ending September 30, 2015 shall not lapse to the general fund but shall be carried forward in a work project account that is in compliance with section 451a of the management and budget act, 1984 PA 431, MCL 18.1451a.

Enacting section 1. This amendatory act takes effect 90 days after the date it is enacted into law. This act is ordered to take immediate effect.[250]

Senate Bill (SB) 55: On April 14, 2015, the governor of Michigan signed SB 55 into Public Law 13 of 2015, which went into effect on July 13, 2015. *Public Law 13* is an act to amend 1994 PA 451, entitled "An act to protect the environment and natural resources of the state; to codify, revise, consolidate, and classify laws relating to the environment and natural resources of the state; to regulate the discharge of certain substances into the environment; to regulate the use of certain lands, waters, and other natural resources of the state; to protect the people's right to hunt and fish; to prescribe the powers and duties of certain state and local agencies and officials; to provide for certain charges, fees, assessments, and donations; to provide certain appropriations; to prescribe penalties and provide remedies; and to repeal acts and parts of acts," by amending section 40111c (MCL 324.40111c), as added by 2008 PA 301.

The People of the State of Michigan enact:

Sec. 40111c.

1. A person other than the department shall not take game using a tranquilizer propelled from a bow or firearm.

---

[250]SB 54, MICHIGAN LEGISLATURE, http://www.legislature.mi.gov/documents/2015-2016/publicact/pdf/2015-PA-0012.pdf.

2. An individual shall not take game or fish using an unmanned vehicle or unmanned device that uses aerodynamic forces to achieve flight or using an unmanned vehicle or unmanned device that operates on the surface of water or underwater.

Enacting section 1. This amendatory act takes effect 90 days after the date it is enacted into law.

Enacting section 2. This amendatory act does not take effect unless Senate Bill No. 54 of the 98th Legislature is enacted into law. This act is ordered to take immediate effect.[251]

MINNESOTA

There have been a number of bills introduced. HB 1194 to require a search warrant for the use of UAVs by law enforcement as well as to secure search warrants to receive electronic device location information;[252] HB 1197 to regulate the use of UAVs by law enforcement;[253] HB 1491 also to regulate the use of UAVs by law enforcement;[254] HB 786 relating to data practices and classifying law enforcement surveillance technology;[255] HB 986 prohibiting unmanned aerial camera usage to gather evidence by law enforcement agencies;[256] SB 1299 relating to public safety, regulating the use of UAVs by law enforcement agencies;[257] SB 1792 prohibiting the use of UAVs by law enforcement agencies to gather evidence;[258] SB 685 regulating the use of UAVs by law enforcement;[259] SB 686 also regulating the use of UAVs by law enforcement as well as the requirement for search warrant to gather electronic location information;[260] and SF 1299 adding a new section to Minnesota Statutes, chapter 626 addressing the use of drones by law enforcement.[261] Additionally, the City of St. Bonifacius has passed Resolution 2013-8 restricting the use of drones in the City of St. Bonifacius airspace.[262]

---

[251]SB 55, MICHIGAN LEGISLATURE, http://www.legislature.mi.gov/documents/2015-2016/publicact/pdf/2015-PA-0013.pdf.
[252]HB 1194, MINNESOTA STATE LEGISLATURE, https://legiscan.com/MN/bill/HF1194/2015.
[253]HB 1197, MINNESOTA STATE LEGISLATURE, https://legiscan.com/MN/bill/HF1197/2015.
[254]HB 1491, MINNESOTA STATE LEGISLATURE, https://legiscan.com/MN/bill/HF1491/2015.
[255]HB 786, MINNESOTA STATE LEGISLATURE, https://legiscan.com/MN/bill/HF786/2015.
[256]HB 986, MINNESOTA STATE LEGISLATURE, https://legiscan.com/MN/bill/HF986/2015.
[257]SB 1299, MINNESOTA STATE LEGISLATURE, https://legiscan.com/MN/bill/SF1299/2015.
[258]SB 1792, MINNESOTA STATE LEGISLATURE, https://legiscan.com/MN/bill/SF1792/2015.
[259]SB 685, MINNESOTA STATE LEGISLATURE, https://legiscan.com/MN/bill/SF685/2015.
[260]SB 686, MINNESOTA STATE LEGISLATURE, https://legiscan.com/MN/drafts/SF686/2015.
[261]SB 1299, MINNESOTA STATE LEGISLATURE, http://www.senate.leg.state.mn.us/departments/scr/billsumm/summary_display_from_db.php?ls=&id=2689.
[262]Resolution 2013-8, CITY OF ST. BONIFACIUS, http://blogs.mprnews.org/statewide/files/legacy//St.%20Bonifacius.pdf.

MISSISSIPPI

Mississippi has introduced HB 347 to regulate the use of drones, to prohibit any person from using a drone to capture unauthorized images, with exceptions, and to provide penalties for violations of this act,[263] and has enacted bill SB 2022.[264]

Senate Bill (SB) 2022: The governor of Mississippi signed SB 2022 into Public Law, Chapter 489 that went into effect on July 1, 2015. *Chapter 489* is an act to amend Section 97-29-61, Mississippi Code of 1972, to prohibit "peeping tom" activities that do not amount to felonious trespass; to amend section 97-29-63, Mississippi Code of 1972, to prohibit surreptitious photography of a person's body or underclothing under circumstances in which a person would reasonably expect to not be the subject of such photography; and for related purposes.

Be it enacted by the legislature of the State of Mississippi:

SECTION 1. Section 97-29-61, Mississippi Code of 1972, is amended as follows:

97-29-61. (1)(a) Any person who enters upon real property, whether the original entry is legal or not, and thereafter pries or peeps through a window or other opening in a dwelling or other building structure for the lewd, licentious and indecent purpose of spying upon the occupants thereof, shall be guilty of a felonious trespass.

b. Any person who looks through a window, hole or opening, or otherwise views by means of any instrumentality, including, but not limited to, a periscope, telescope, binoculars, drones, camera, motion-picture camera, camcorder or mobile phone, into the interior of a bedroom, bathroom, changing room, fitting room, dressing room, spa, massage room or therapy room or tanning booth, or the interior of any other area in which the occupant has a reasonable expectation of privacy, with the intent to invade the privacy of a person or persons inside and without the consent or knowledge of every person present, for the lewd, licentious and indecent purpose of spying upon the occupant or occupants thereof, shall be guilty of a felony.

2(a). Except as provided in paragraph (b) of this subsection, a person who was over the age of twenty-one (21) at the time of the offense who is convicted of a violation of subsection (1) of this section shall be imprisoned in the custody of the Department of Corrections not more than five (5) years.

b. When one or more occupants spied upon is a child under sixteen (16) years of age, a person who was over the age of twenty-one (21) at the time of the offense who is convicted of a violation of subsection (1) of this section shall be imprisoned in the custody of the Department of Corrections not more than ten (10) years.

---

[263]HB 347, MISSISSIPPI LEGISLATURE, http://billstatus.ls.state.ms.us/documents/2015/pdf/HB/0300-0399/HB0347IN.pdf.

[264]SB 2022, MISSISSIPPI LEGISLATURE, http://billstatus.ls.state.ms.us/documents/2015/pdf/SB/2001-2099/SB2022SG.pdf.

SECTION 2. Section 97-29-63, Mississippi Code of 1972, is amended as follows: 97-29-63.

1a. It is a felony for any person with lewd, licentious or indecent intent to photograph, film, videotape, record or otherwise reproduces the image of another person without the permission of the other person when the other person is located in a place where a person would intend to be in a state of undress and have a reasonable expectation of privacy, including, but not limited to, private dwellings or any facility, public or private, used as a restroom, bathroom, shower room, tanning booth, locker room, fitting room, dressing room or bedroom shall be guilty of a felony.

b. It is a felony for any person to invade the privacy of another person and with lewd, licentious or indecent intent to photograph, film, videotape, record or otherwise reproduce the image of another, identifiable person under or through the clothing being worn by that other person for the purpose of viewing the body of, or the undergarments worn by, the other person without the consent or knowledge of the other person and under circumstances in which the other person has a reasonable expectation that the other person's body or undergarments would not be viewed or would not be the subject of a reproduced image.

2a. Except as provided in paragraph (b) of this subsection, a person who was over the age of twenty-one (21) at the time of the offense who is convicted of a violation of subsection (1) of this section shall be punished by a fine of Five Thousand Dollars ($5,000.00) or by imprisonment of not more than five (5) years in the custody of the Department of Corrections, or both.

b. Where the person who is secretly photographed, filmed, videotaped or otherwise reproduced is a child under sixteen (16) years of age, a person who was over the age of twenty-one (21) at the time of the offense who is convicted of a violation of subsection (1) of this section shall be punished by a fine of Five Thousand Dollars ($5,000.00) or by imprisonment of not more than ten (10) years in the custody of the Department of Corrections, or both.

SECTION 3. This act shall take effect and be in force from and after July 1, 2015.[265]

MONTANA

Montana has one enacted bill, SB 196.[266]

Senate Bill (SB) 196: On May 1, 2013, the 63rd Legislature of Montana signed SB 196 into law as Chapter 377. This was an act limiting the use of unmanned aerial vehicles by law enforcement; and prohibiting the use of unlawfully obtained information as evidence in court, and it reads as follows:

---

[265]SB 2022, Mississippi Legislature, http://billstatus.ls.state.ms.us/documents/2015/pdf/SB/2001-2099/SB2022SG.pdf.
[266]SB 196, Montana Legislature, http://leg.mt.gov/bills/2013/billpdf/SB0196.pdf.

Be it enacted by the legislature of the State of Montana:
Section 1. Limitations on unmanned aerial vehicles.

1. In any prosecution or proceeding within the state of Montana, information from an unmanned aerial vehicle is not admissible as evidence unless the information was obtained:
   a. pursuant to the authority of a search warrant; or
   b. in accordance with judicially recognized exceptions to the warrant requirement.
2. Information obtained from the operation of an unmanned aerial vehicle may not be used in an affidavit of probable cause in an effort to obtain a search warrant unless the information was obtained under the circumstances described in subsection (1)(a) or (1)(b) or was obtained through the monitoring of public lands or international borders.
3. For the purposes of this section, "unmanned aerial vehicle" means an aircraft that is operated without direct human intervention from on or within the aircraft. The term does not include satellites.

Section 2. Codification instruction.

[Section 1] is intended to be codified as an integral part of Title 46, chapter 5, part 1, and the provisions of Title 46, chapter 5, part 1, apply to [section 1].[267]

NEVADA

Nevada has passed Senate Concurrent Resolution (SCR) 7,[268] and enacted Assembly Bill (AB) 507,[269] and AB 239.[270]

Senate Concurrent Resolution (SCR) 7: On May 8, 2013, SCR 7 was enrolled and delivered to the Secretary of State as File No. 30. This was a resolution that became effective upon passage, and reads as follows:

SENATE CONCURRENT RESOLUTION—Expressing the determination of the Nevada Legislature that Nevada is the optimal location for unmanned aircraft system development and testing.

WHEREAS, The State of Nevada has a long and distinguished history of directly contributing to the national security and defense of the United States of America; and WHEREAS, This State is indebted to the dedicated and selfless men and women assigned to Fallon Naval Air Station, Creech Air Force Base, Nellis Air Force Base and auxiliary fields, and the Reserve and National Guard units across the State; and

---

[267]SB 196, Montana Legislature, http://leg.mt.gov/bills/2013/billpdf/SB0196.pdf.

[268]SCR 7, Nevada Legislature, http://www.leg.state.nv.us/Session/77th2013/Bills/SCR/SCR7_EN.pdf.

[269]AB 507, Nevada Legislature, http://www.leg.state.nv.us/Session/77th2013/Bills/AB/AB507_EN.pdf.

[270]AB 239, Nevada Legislature, available at https://legiscan.com/NV/text/AB239/id/1241015/Nevada-2015-AB239-Enrolled.pdf.

WHEREAS, The State of Nevada's military operations have attracted the aerospace, defense and national security industry to Nevada, and have led to increased technological advances and important supporting relationships with our universities, research laboratories and technical colleges; and

WHEREAS, National security and defense sectors are increasingly important to Nevada's economy, as they provide and support approximately 190,000 jobs and generate more than $18 billion in revenue; and

WHEREAS, The aforesaid economic impact of military operations would not have been possible without the past and continuing congressional, state and local support; and

WHEREAS, Our military investments, military expertise, technically trained workforce and geographic and climatic diversity have positioned the State of Nevada well to retain its military operations and compete for next-generation technology projects; and

WHEREAS, Remotely piloted aircraft, cyberspace and intelligence, surveillance and reconnaissance mission areas will be high-priority areas of defense and national security budgets; now, therefore, be it

RESOLVED BY THE SENATE OF THE STATE OF NEVADA, THE ASSEMBLY CONCURRING, That the members of the 77th Session of the Nevada Legislature hereby endorse the promotion of efforts to support the establishment of Nevada as the "Silicon Valley" of unmanned aircraft systems education, testing, research and manufacture; and be it further RESOLVED, That:

1. As military spending continues to be downsized in the foreseeable future, the State of Nevada must look to find ways to sustain and expand its direct support of the national security and defense sectors;
2. This State should specifically leverage its Congressional and State Leadership, current military expertise, technical workforce, geographical, climate diversity and vast aerospace to provide a concerted effort to aggressively pursue the high-priority requirements of the United States Department of Defense and the Federal Aviation Administration;
3. The State of Nevada must compete to be selected as one or more of the six national test ranges under the 2012 National Defense Authorization Act that directs the Federal Aviation Administration in coordination with the United States Department of Defense, and the National Aeronautics and Space Administration to determine how to safely operate unmanned aircraft systems in shared national airspace; and
4. The creation of these test ranges will mark the first step in what will undoubtedly be a long-range process, leading to a common-day occurrence in which manned and unmanned aircraft fly safely and seamlessly together in the national airspace;

> and be it further RESOLVED, That Nevada is an optimal location for unmanned aircraft system development and testing, and would be honored to be one of the sites selected for such testing and development.[271]

Assembly Bill (AB) 507: In 2013, Nevada, through AB 507, appropriated $4,000,000 to the Interim Finance Committee for allocation to the Governor's Office of Economic Development for the Unmanned Aerial Vehicle (UAV) Program. The funds could only be appropriated if Nevada is selected as a Federal Aviation Administration (FAA) Test Site.[272]

Assembly Bill (AB) 239: On June 2, 2015, the governor of Nevada approved AB 239 into law as Chapter 327. This is an act relating to aircraft; regulating operators of unmanned aerial vehicles in this State; revising the definition of "aircraft" to include unmanned aerial vehicles; prohibiting the operation or use of an unmanned aerial vehicle under certain circumstances or for certain purposes; authorizing a law enforcement agency to operate an unmanned aerial vehicle at certain locations without a warrant under certain circumstances and for any other lawful purpose; prohibiting a law enforcement agency from operating an unmanned aerial vehicle without first obtaining a warrant under certain circumstances; authorizing a public agency to operate an unmanned aerial vehicle only under certain circumstances; requiring the Department of Public Safety, to the extent that money is available, to establish and maintain a registry of unmanned aerial vehicles that are operated by public agencies in this State; requiring the Department to report certain information to the Legislature with respect to the operation of unmanned aerial vehicles by public agencies in this State; requiring the Department to adopt regulations prescribing the public purposes for which a public agency may operate an unmanned aerial vehicle in this State; providing certain criminal and civil penalties for the unlawful operation or use of an unmanned aerial vehicle in this State; and providing other matters properly relating thereto. This act reads as follows:

The People of the State of Nevada, represented in Senate and Assembly, do enact as follows:

Section 1. Chapter 493 of NRS is hereby amended by adding thereto the provisions set forth as sections 2 to 22, inclusive, of this act.

Secs. 2-17. (Deleted by amendment.)

Sec. 18.

1. A person shall not weaponize an unmanned aerial vehicle or operate a weaponized unmanned aerial vehicle. A person who violates this section is guilty of a category D felony and shall be punished as provided in NRS 193.130.

---

[271]SCR 7, NEVADA LEGISLATURE, http://www.leg.state.nv.us/Session/77th2013/Bills/SCR/SCR7_EN.pdf.
[272]AB 507, NEVADA LEGISLATURE, http://www.leg.state.nv.us/Session/77th2013/Bills/AB/AB507_EN.pdf.

2. A person who weaponizes an unmanned aerial vehicle in violation of subsection 1 and who discharges the weapon is guilty of a category C felony and shall be punished as provided in NRS 193.130.

Sec. 18.5.

1. A person shall not operate an unmanned aerial vehicle within:
   a. A horizontal distance of 500 feet or a vertical distance of 250 feet from a critical facility without the written consent of the owner of the critical facility.
   b. Except as otherwise provided in subsection 2, 5 miles of an airport.
2. A person may operate an unmanned aerial vehicle within 5 miles of an airport only if the person obtains the consent of the airport authority or the operator of the airport, or if the person has otherwise obtained a waiver, exemption or other authorization for such operation pursuant to any rule or regulation of the Federal Aviation Administration. A person who is authorized to operate an unmanned aerial vehicle within 5 miles of an airport pursuant to this subsection shall, at all times during such operation, maintain on his or her person documentation of any waiver, exemption, authorization or consent permitting such operation.
3. A person who violates this section is guilty of a misdemeanor.
4. As used in this section, "airport" means any area of land or water owned, operated or maintained by or on behalf of a city, county, town, municipal corporation or airport authority that is designed and set aside for the landing and taking off of aircraft and that is utilized in the interest of the public for such purposes.

Sec. 19.

1. Except as otherwise provided in subsection 2, a person who owns or lawfully occupies real property in this State may bring an action for trespass against the owner or operator of an unmanned aerial vehicle that is flown at a height of less than 250 feet over the property if:
   a. The owner or operator of the unmanned aerial vehicle has flown the unmanned aerial vehicle over the property at a height of less than 250 feet on at least one previous occasion; and
   b. The person who owns or occupies the real property notified the owner or operator of the unmanned aerial vehicle that the person did not authorize the flight of the unmanned aerial vehicle over the property at a height of less than 250 feet. For the purposes of this paragraph, a person may place the owner or operator of an unmanned aerial vehicle on notice in the manner prescribed in subsection 2 of NRS 207.200.

2. A person may not bring an action pursuant to subsection 1 if:
   a. The unmanned aerial vehicle is lawfully in the flight path for landing at an airport, airfield or runway.
   b. The unmanned aerial vehicle is in the process of taking off or landing.
   c. The unmanned aerial vehicle was under the lawful operation of:
      1. A law enforcement agency in accordance with section 20 of this act.
      2. A public agency in accordance with section 21 of this act.
   d. The unmanned aerial vehicle was under the lawful operation of a business licensed in this State or a land surveyor if:
      1. The operator is licensed or otherwise approved to operate the unmanned aerial vehicle by the Federal Aviation Administration;
      2. The unmanned aerial vehicle is being operated within the scope of the lawful activities of the business or surveyor; and
      3. The operation of the unmanned aerial vehicle does not unreasonably interfere with the existing use of the real property.
3. A plaintiff who prevails in an action for trespass brought pursuant to subsection 1 is entitled to recover treble damages for any injury to the person or the real property as the result of the trespass. In addition to the recovery of damages pursuant to this subsection, a plaintiff may be awarded reasonable attorney's fees and costs and injunctive relief.

Sec. 20.

1. Except as otherwise provided in this section, nothing in this section shall be deemed to otherwise prohibit the operation of an unmanned aerial vehicle by a law enforcement agency for any lawful purpose in this State.
2. Except as otherwise provided in subsection 3, a law enforcement agency shall not operate an unmanned aerial vehicle for the purpose of gathering evidence or other information within the curtilage of a residence or at any other location or upon any property in this State at which a person has a reasonable expectation of privacy, unless the law enforcement agency first obtains a warrant from a court of competent jurisdiction authorizing the use of the unmanned aerial vehicle for that purpose. A warrant authorizing the use of an unmanned aerial vehicle must specify the period for which operation of the unmanned aerial vehicle is authorized. A warrant must not authorize the use of an unmanned aerial vehicle for a period of more than 10 days. Upon motion and a showing of probable cause, a court may renew a warrant after the expiration of the period for which the warrant was initially issued.

3. A law enforcement agency may operate an unmanned aerial vehicle without obtaining a warrant issued pursuant to subsection 2:
    a. If the law enforcement agency has probable cause to believe that a person has committed a crime, is committing a crime or is about to commit a crime, and exigent circumstances exist that make it unreasonable for the law enforcement agency to obtain a warrant authorizing the use of the unmanned aerial vehicle.
    b. If a person provides written consent to the law enforcement agency authorizing the law enforcement agency to acquire information about the person or the real or personal property of the person. The written consent must specify the information to be gathered and the time, place and manner in which the information is to be gathered by the law enforcement agency.
    c. For the purpose of conducting search and rescue operations for persons and property in distress.
    d. Under circumstances in which the law enforcement agency believes that an imminent threat exists to the life and safety of an individual person or to the public at large, including, without limitation, the threat of an act of terrorism. A law enforcement agency that operates an unmanned aerial vehicle pursuant to this paragraph shall document the factual basis for its belief that such an imminent threat exists and shall, not later than 2 business days after initiating operation, file a sworn statement with a court of competent jurisdiction describing the nature of the imminent threat and the need for the operation of the unmanned aerial vehicle.
    e. Upon the declaration of a state of emergency or disaster by the Governor. A law enforcement agency that operates an unmanned aerial vehicle pursuant to this paragraph shall not use the unmanned aerial vehicle outside of the geographic area specified in the declaration or for any purpose other than the preservation of public safety, the protection of property, or the assessment and evaluation of environmental or weather-related damage, erosion or contamination.
4. Any photograph, image, recording or other information that is acquired by a law enforcement agency through the operation of an unmanned aerial vehicle in violation of this section, or that is acquired from any other person or governmental entity, including, without limitation, a public agency and any department or agency of the Federal Government, that obtained the photograph, image, recording or other information in a manner inconsistent with the requirements of this section, and any evidence that is derived therefrom:
    a. Is not admissible in and must not be disclosed in a judicial, administrative or other adjudicatory proceeding; and

b. May not be used to establish reasonable suspicion or probable cause as the basis for investigating or prosecuting a crime or offense.

Sec. 21.

1. A public agency:
   a. May operate an unmanned aerial vehicle only if:
      1. Before the operation of the unmanned aerial vehicle, the public agency registers the unmanned aerial vehicle with the Department pursuant to subsection 2 of section 22 of this act.
      2. The public agency operates the unmanned aerial vehicle in accordance with the regulations adopted by the Department pursuant to subsection 4 of section 22 of this act.
   b. Must not operate an unmanned aerial vehicle for the purposes of assisting a law enforcement agency with law enforcement or conducting a criminal prosecution.
2. Any photograph, image, recording or other information that is acquired by a public agency through the operation of an unmanned aerial vehicle in violation of this section, and any evidence that is derived therefrom:
   a. Is not admissible in, and must not be disclosed in, a judicial, administrative or other adjudicatory proceeding; and
   b. May not be used to establish reasonable suspicion or probable cause as the basis for investigating or prosecuting a crime or offense.

Sec. 22.

1. The Department shall, to the extent that money is available for this purpose, establish and maintain a registry of unmanned aerial vehicles that are operated by public agencies in this State. The Department shall include on its Internet website the information that is maintained in the registry.
2. A public agency shall, for each unmanned aerial vehicle the public agency intends to operate, submit to the Department, on a form provided by the Department, for inclusion in the registry:
   a. The name of the public agency;
   b. The name and contact information of each operator of the unmanned aerial vehicle;
   c. Sufficient information to identify the unmanned aerial vehicle; and
   d. A statement describing the use of the unmanned aerial vehicle by the public agency.

3. The Department shall, on or before February 1 of each year, prepare and submit to the Director of the Legislative Counsel Bureau for submission to the Legislature, or to the Legislative Commission when the Legislature is not in regular session, a report outlining the activities of public agencies with respect to the operation of unmanned aerial vehicles in this State.
4. The Department shall adopt regulations prescribing the public purposes for which a public agency may operate an unmanned aerial vehicle that is registered with the Department pursuant to this section, including, without limitation:
   a. The provision of fire services.
   b. The provision of emergency medical services.
   c. The protection of a critical facility that is public property. (d) Search and rescue operations conducted for persons and property in distress.

Sec. 22.5.
NRS 493.010 is hereby amended to read as follows:
493.010 NRS 493.010 to 493.120, inclusive, and sections 18 to 22, inclusive, of this act may be cited as the Uniform State Law for Aeronautics.
Sec. 23.
NRS 493.020 is hereby amended to read as follows:
493.020 As used in NRS 493.010 to 493.120, inclusive, and sections 18 to 22, inclusive, of this act, unless the context otherwise requires:

1. "Aircraft" includes a balloon, airplane, hydroplane, unmanned aerial vehicle and any other vehicle used for navigation through the air. A hydroplane, while at rest on water and while being operated on or immediately above water, is governed by the rules regarding water navigation. A hydroplane while being operated through the air other than immediately above water, is an aircraft.
2. "Critical facility" means a petroleum refinery, a petroleum or chemical production, transportation, storage or processing facility, a chemical manufacturing facility, a pipeline and any appurtenance thereto, a wastewater treatment facility, a water treatment facility, a mine as that term is defined in NRS 512.006, a power generating station, plant or substation and any appurtenances thereto, any transmission line that is owned in whole or in part by an electric utility as that term is defined in subsection 5 of NRS 704.187, a county, city or town jail or detention facility and any prison, facility or institution under the control of the Department of Corrections. The term does not include any facility or infrastructure of a utility that is located underground.
3. "Department" means the Department of Public Safety.
4. "Law enforcement agency" means an agency, office, bureau, board, commission, department or division of this State or a political subdivision of this State, the primary duty of which is to enforce the law.

5. "Operator" includes aviator, pilot, balloonist and any other person having any part in the operation of aircraft while in flight.
6. "Passenger" includes any person riding in an aircraft, but having no part in its operation.
7. "Public agency" means an agency, office, bureau, board, commission, department or division of this State or a political subdivision of this State other than a law enforcement agency.
8. "Unmanned aerial vehicle" means a powered aircraft of any size without a human operator aboard the vehicle and that is operated remotely or autonomously.

Sec. 24. (Deleted by amendment.)
Sec. 24.2.
NRS 493.050 is hereby amended to read as follows:
493.050

1. Flight of an aircraft over the lands and waters of this state is lawful:
    a. Unless at such a low altitude as to interfere with the then existing use to which the land or water, or the space over the land or water, is put by the owner.
    b. Unless so conducted as to be imminently dangerous to persons or property lawfully on the land or water beneath.
    c. Unless specifically prohibited by the provisions of NRS 493.010 to 493.120, inclusive, and sections 18 to 22, inclusive, of this act, or any regulations adopted pursuant thereto.
2. The landing of an aircraft on the lands or waters of another, without his or her consent, is unlawful, except in the case of a forced landing. For damages caused by a forced landing, the owner, lessee or operator of the aircraft is liable as provided in NRS 493.060.

Sec. 24.4.
NRS 493.100 is hereby amended to read as follows:
493.100

1. Any operator or passenger, while an aircraft is in flight over a heavily populated area or over a public gathering within this state, who:
    a. Except as otherwise provided in subsection 2, engages in trick or acrobatic flying, or in any acrobatic feat;
    b. Except while in landing or taking off, flies at such a low level as to endanger the persons on the surface beneath; or
    c. (c) Drops any object with reckless disregard for the safety of other persons and willful indifference to injuries that could reasonably result from dropping the object, is guilty of a misdemeanor.

2. The provisions of paragraph (a) of subsection 1 do not apply to the operator of an unmanned aerial vehicle in a park unless the operator is operating the unmanned aerial vehicle with reckless disregard for the safety of other persons and with willful indifference to injuries that could reasonably result from such operation.

Sec. 24.6.
NRS 493.120 is hereby amended to read as follows:
493.120
NRS 493.010 to 493.120, inclusive, and sections 18 to 22, inclusive, of this act shall be so interpreted and construed as to effectuate their general purpose to make uniform the law of those states which enact them, and to harmonize, as far as possible, with federal laws and regulations on the subject of aeronautics. They shall not be interpreted or construed to apply in any manner to aircraft owned and operated by the Federal Government.
Sec. 24.8.
NRS 493.130 is hereby amended to read as follows:
493.130

1. Any person operating an aircraft in the air, or on the ground or water:
    a. While under the influence of intoxicating liquor or a controlled substance, unless in accordance with a lawfully issued prescription; or
    b. In a careless or reckless manner so as to endanger the life or property of another, is guilty of a gross misdemeanor.
2. As used in this section:
    a. "Aircraft" includes an unmanned aerial vehicle as that term is defined in subsection 8 of NRS 493.020.
    b. "Controlled substance" has the meaning ascribed to it in 21 U.S.C. § 802(6).
    c. "Prescription" has the meaning ascribed to it in NRS 453.128.

Sec. 25.
The provisions of subsection 1 of NRS 218D.380 do not apply to any provision of this act which adds or revises a requirement to submit a report to the Legislature.
Sec. 26.
This act becomes effective:

1. Upon passage and approval for the purpose of adopting regulations and performing any other preparatory administrative tasks necessary to carry out the provisions of this act; and
2. On October 1, 2015, for all other purposes.[273]

---

[273]AB 239, Nevada Legislature, https://legiscan.com/NV/text/AB239/id/1241015/Nevada-2015-AB239-Enrolled.pdf.

NEW HAMPSHIRE

New Hampshire has enacted a sole bill, SB 222,[274] and has introduced HB 1654 relating to flying a drone above a correctional facility;[275] HB 240 prohibiting law enforcement agencies from using a drone to collect evidence;[276] HB 602 regulating the use of drones by government agencies and individuals and establishing criminal penalties and civil remedies for violations of the law;[277] and SB 459 establishing requirements for the operations of drones.[278]

Senate Bill (SB) 222: On May 7, 2015, the governor of New Hampshire signed SB 222 into law as Chapter 38, which became effective on January 1, 2016. This is an act relative to harassment of hunting, fishing, or trapping. This bill extends the prohibition on harassment of persons hunting, fishing, or trapping and adds the use of unmanned aerial vehicles to the activities prohibited. This bill is a request of the fish and game department, and reads as follows:

Be it enacted by the Senate and House of Representatives in General Court convened:

38:1 Fish and Game; Harassment. Amend RSA 207:57 to read as follows:

207:57 Harassment.

- I. No person shall purposely obstruct or impede the participation of any individual in the lawful activity of hunting, fishing or trapping. No person shall purposely obstruct or impede the participation of any individual in the lawful activity of hunting, fishing or trapping. No person shall purposely engage in an activity that will tend to disturb wild animals, with intent to prevent their lawful taking. No person shall use a drone or UAV with the intent to conduct video surveillance of private citizens who are lawfully hunting, fishing, or trapping without obtaining the written consent of the persons being surveilled prior to conducting the surveillance.
- I-a. For purposes of this section, "unmanned aerial vehicle" or "UAV" means any device capable of flying in the air which is remotely, automatically, or otherwise piloted without an occupant, including but not limited to, drones.
- II. The provisions of paragraph I shall not apply to the actions of law enforcement officers and personnel of the department of fish and game in the performance of their official duties. The provisions of paragraph I shall not apply to any incidental interference arising from the lawful and normal activities of public land users.

---

[274]SB 222, New Hampshire Legislature, https://legiscan.com/NH/text/SB222/id/1224996/New_Hampshire-2015-SB222-Chaptered.html.

[275]HB 1654, New Hampshire Legislature, https://legiscan.com/NH/bill/HB1654/2016.

[276]HB 240, New Hampshire Legislature, https://legiscan.com/NH/research/HB240/2016.

[277]HB 602, New Hampshire Legislature, https://legiscan.com/NH/text/HB602/id/1155470.

[278]SB 459, New Hampshire Legislature, https://legiscan.com/NH/bill/SB459/2016.

III. The executive director shall adopt rules, pursuant to RSA 541-A, to administer this section.

IV. Any person violating the provisions of this section shall be guilty of a violation.

38:2 Effective Date. This act shall take effect January 1, 2016.[279]

NEW JERSEY

New Jersey Legislature has introduced a number of bills. AB 1039 setting forth certain standards to be followed by law enforcement agencies and fire departments when utilizing drones;[280] AB 4344 criminalizing the use of drones to collect surveillance of or fly over critical infrastructures, as well as requiring certain drones to be registered and insured;[281] AB 4806 creating the fourth degree crime of operating a drone equipped with a weapon;[282] AB 4807 permitting municipalities to enact ordinances prohibiting the operation of drones under certain circumstances;[283] AB 4837 requiring certain drones to contain geo-fencing technology and making a violation a fourth degree crime;[284] AB 4838 requiring certain retailers to provide notice of FAA safety guidelines for flying drones to customers;[285] SB 2310 setting forth certain standards to be followed by law enforcement agencies and fire departments when utilizing drones;[286] SB 3174 requiring certain retailers to provide notice of FAA safety guidelines for flying drones to consumers;[287] SB 3175 requiring certain drones to contain geo-fencing technology and making a violation a fourth degree crime;[288] SB 3183 criminalizing the use of drones to conduct surveillance of or fly over critical infrastructures, and requires certain drones to be registered and insured;[289] SB 379 requiring certain retailers to provide notice of FAA safety guidelines for flying drones to consumers;[290] SB 380 requiring certain drones to contain geo-fencing technology and making violation a fourth degree crime;[291] SB 465 setting forth certain standards to be followed by law enforcement agencies and fire departments when utilizing drones;[292] and SR 141 urging the FAA to expand restrictions on flights of UAVs.[293]

---

[279]SB 222, New Hampshire Legislature, https://legiscan.com/NH/text/SB222/id/1224996/New_Hampshire-2015-SB222-Chaptered.html.
[280]AB 1039, New Jersey Legislation, https://legiscan.com/NJ/bill/A1039/2014.
[281]AB 4344, New Jersey Legislation, https://legiscan.com/NJ/bill/A4344/2014.
[282]AB 4806, New Jersey Legislation, https://legiscan.com/NJ/text/A4806/id/1275956.
[283]AB 4807, New Jersey Legislation, https://legiscan.com/NJ/text/A4807/id/1275971.
[284]AB 4837, New Jersey Legislation, https://legiscan.com/NJ/text/A4837/id/1277114.
[285]AB 4838, New Jersey Legislation, https://legiscan.com/NJ/text/A4838/id/1277155.
[286]SB 2310, New Jersey Legislation, https://legiscan.com/NJ/text/S2310/id/1290049.
[287]SB 3174, New Jersey Legislation, https://legiscan.com/NJ/text/S3174/id/1266097.
[288]SB 3175, New Jersey Legislation, https://legiscan.com/NJ/text/S3175/id/1266090.
[289]SB 3183, New Jersey Legislation, https://legiscan.com/NJ/text/S3183/id/1266096.
[290]SB 379, New Jersey Legislation, https://legiscan.com/NJ/bill/S379/2016.
[291]SB 380, New Jersey Legislation, https://legiscan.com/NJ/bill/S380/2016.
[292]SB 465, New Jersey Legislation, https://legiscan.com/NJ/bill/S465/2016.
[293]SR 141, New Jersey Legislation, https://legiscan.com/NJ/bill/SR141/2014.

Additionally, Bernards Township passed Ordinance 2328 prohibiting the flying and/or launching of unmanned aircraft by the public within any park or recreation facility,[294] and Chatham passed Ordinance 2015-16 to add certain and specific regulations of drone technology and unmanned aircraft in the Township to protect the public health, welfare, and safety.[295]

NEW MEXICO

New Mexico has introduced SB 303 relating to privacy and enacting the freedom from unwarranted surveillance act as well as providing penalties for violations;[296] SB 82 relating to wildlife and prohibiting the use of drones in hunting as well as providing penalties for violations;[297] and SB 556 relating to privacy and enacting the freedom from unwarranted surveillance act as well as providing penalties for violations.[298]

NEW YORK

New York has introduced AB 1247 to impose limitations on the use of drones within the state;[299] AB 2683 to amend the criminal procedure law in relation to protecting against unwarranted surveillance by UAVs;[300] AB 31 providing for and regulating the operation and testing of motor vehicles with autonomous technology;[301] AB 3597 criminalizing the unlawful use of an unmanned aircraft;[302] AB 6713 regulating the use of unmanned aerial vehicles by the state and political subdivisions thereof;[303] AB 8505 to amend civil rights law relating to a private right of action for unwarranted surveillance of a neighbor to include drones;[304] AB 8605 relating to the unlawful use of a model aircraft and prohibiting the use of a model aircraft in restricted areas, out of the user's line of sight, near stadiums or above 400 feet;[305] AB 8625 relating to prohibiting civilian drone use within 1,000 feet of a correctional facility;[306] Senate Bill (SB) 1841 relating to protection against unwarranted surveillance;[307] SB 411 imposing limitations on the use of drones within the state;[308] SB 6222 relating to prohibiting civilian drone use within 1,000 feet

[294]Ordinance 2328, Township of Barnards, http://www.bernards.org/Parks%20and%20Recreation/Document/Parks%20and%20Facilities%20Docs/Current%20Municipal%20Parks%20Ordinance%20.pdf.

[295]Ordinance 2015-16, Township of Chatham, http://www.chathamtownship-nj.gov/ordinances/2015/Ordinance-2015-16.pdf.

[296]SB 303, New Mexico Legislature, https://www.nmlegis.gov/Sessions/15%20Regular/bills/senate/SB0303.pdf.

[297]SB 82, New Mexico Legislature, https://www.nmlegis.gov/Sessions/15%20regular/bills/senate/SB0082.PDF.

[298]SB 556, New Mexico Legislature, https://www.nmlegis.gov/Sessions/13%20regular/bills/senate/SB0556.pdf.

[299]AB 1247, New York State Legislature, https://legiscan.com/NY/bill/A01247/2015.

[300]AB 2683, New York State Legislature, https://legiscan.com/NY/bill/A02683/2015.

[301]AB 31, New York State Legislature, https://legiscan.com/NY/bill/A00031/2015.

[302]AB 3597, New York State Legislature, https://legiscan.com/NY/bill/A03597/2015.

[303]AB 6713, New York State Legislature, https://legiscan.com/NY/bill/A06713/2015.

[304]AB 8505, New York State Legislature, https://legiscan.com/NY/bill/A08505/2015.

[305]AB 8605, New York State Legislature, https://legiscan.com/NY/bill/A08605/2015.

[306]AB 8625, New York State Legislature, https://legiscan.com/NY/bill/A08625/2015.

[307]SB 1841, New York State Legislature, https://legiscan.com/NY/bill/S01841/2015.

[308]SB 411, New York State Legislature, https://legiscan.com/NY/bill/S00411/2015.

of a correctional facility;[309] SB 6455 relating to the registration of drones;[310] SB 6456 relating to prohibiting the unlawful use of a drone;[311] SB 6306 criminalizing the unlawful use of an unmanned aircraft;[312] and SB 6335 to include the possession of an unmanned aircraft containing a deadly weapon within the Class C felony of criminal possession of a weapon in the second degree.[313]

Additionally, New York City passed File No 614, a local law requiring the registration and insurance of unmanned aerial vehicles,[314] the Town of Orchard Park passed Ordinance 99-3 making it unlawful for any person to operate an unmanned aircraft or unmanned aircraft system upon or in the airspace within one mile of an open-air event in this town, as well as within three miles of the Ralph Wilson Stadium Complex during events,[315] and the City of Syracuse passed a resolution urging federal and state officials to adopt related safeguards which ensure that the use of unmanned aircraft systems (UAVs or drones) do not infringe with the civil liberties and civil rights of all people in Syracuse, New York.[316]

NORTH CAROLINA

North Carolina has enacted three Senate bills: SB 402,[317] SB 744,[318] and SB 446.[319] Additionally, the state has introduced HB 4 to clarify that agents or agencies of the state or a political subdivision of the state have the authority to procure and operate unmanned aircraft systems upon approval of the state CIO and would modify the regulation of unmanned aircraft systems to comply with federal guidelines;[320] HB 782 to direct the division of motor vehicles to study how to implement autonomous vehicle technology on the roads and highways of the state;[321] SB 600 also to direct the division of motor vehicles to study how to implement autonomous vehicle technology on the roads and highways

---

309 SB 6222, New York State Legislature, https://legiscan.com/NY/bill/S06222/2015.
310 SB 6455, New York State Legislature, https://legiscan.com/NY/bill/S06455/2015.
311 SB 6456, New York State Legislature, https://legiscan.com/NY/bill/S06456/2015.
312 SB 6306, New York State Legislature, https://legiscan.com/NY/bill/S06306/2015.
313 SB 6335, New York State Legislature, https://legiscan.com/NY/bill/S06335/2015.
314 File No. 614, New York City Council, http://legistar.council.nyc.gov/LegislationDetail.aspx?ID=2119765&GUID=85AA1161-E2D5-42A5-AE4C-D2F1CC1BFF6F&Options=ID%7c&Search=.
315 Chapter 99. Peace and Good Order, Town of Orchard Park, http://uavs.insct.org/wp-content/uploads/2016/04/Town-of-Orchard-Park-NY.pdf.
316 Syracuse Resolution, City of Syracuse, available at http://uavs.insct.org/wp-content/uploads/2016/04/Syracuse-City-Municipal-Ordinance.pdf.
317 SB 402, North Carolina General Assembly, http://www.ncga.state.nc.us/Sessions/2013/Bills/Senate/HTML/S402v7.html.
318 SB 744, North Carolina General Assembly, http://www.ncleg.net/Sessions/2013/Bills/Senate/PDF/S744v8.pdf.
319 SB 446, North Carolina General Assembly, http://www.ncga.state.nc.us/Sessions/2015/Bills/Senate/HTML/S446v6.html.
320 HB 4, North Carolina General Assembly, http://www.ncleg.net/gascripts/BillLookUp/BillLookUp.pl?Session=2015&BillID=H0004.
321 HB 782, North Carolina General Assembly, http://www.ncleg.net/gascripts/BillLookUp/BillLookUp.pl?Session=2015&BillID=h782.

of the state, as recommended by the department of transportation;[322] and SB 622 to prohibit the use of unmanned aircraft systems by law enforcement agencies to photograph gatherings of persons on private property.[323]

Furthermore, the Town of Nags Head has adopted an ordinance regulating the use of unmanned aircraft,[324] and Kannapolis City has passed Section 12-21, making it unlawful for any person within the confines of any park to voluntarily take off, land, or cause to descend or take off any airplane, flying machine (to include radio-controlled airplanes, helicopters, rockets and drones), balloon, parachute, or other apparatus of aviation, except as expressly permitted by the city manager.[325]

Senate Bill (SB) 402: SB 402 was added to Chapter SL 2013-360 on July 26, 2013. The parts applicable to UAS regulation read as follows:

SECTION 7.16. (e)

Until July 1, 2015, no State or local governmental entity or officer may procure or operate an unmanned aircraft system or disclose personal information about any person acquired through the operation of an unmanned aircraft system unless the State CIO approves an exception specifically granting disclosure, use, or purchase. Any exceptions to the prohibition in this subsection shall be reported immediately to the Joint Legislative Oversight Committee on Information Technology and the Fiscal Research Division. The following definitions apply in this section:

1. "Unmanned aircraft" means an aircraft that is operated without the possibility of human intervention from within or on the aircraft.
2. "Unmanned aircraft system" means an unmanned aircraft and associated elements, including communication links and components that control the unmanned aircraft that are required for the pilot in command to operate safely and efficiently in the national airspace system.

SECTION 7.16. (f)

If the State Chief Information Officer determines that there is a requirement for unmanned aircraft systems for use by State or local agencies, planning may begin for the possible development, implementation, and operation of an unmanned aircraft system program within the State of North Carolina. This planning effort shall be accomplished in coordination with the Chief Information Officer for the Department of Transportation and the

---

[322]SB 600, North Carolina General Assembly, http://www.ncleg.net/gascripts/BillLookUp/BillLookUp.pl?Session=2015&BillID=S0600.

[323]SB 622, North Carolina General Assembly, http://www.ncleg.net/gascripts/BillLookUp/BillLookUp.pl?Session=2015&BillID=S622.

[324]Ordinance no. 16-03-XXX, Town of Nags Head, https://lintvwavy.files.wordpress.com/2016/03/3-j1-ph-town-of-nags-head-drones-ord.pdf.

[325]Code of Ordinances Section 12-31, Kannapolis, http://uavs.insct.org/wp-content/uploads/2016/04/Kannapolis-NC-Code-of-Ordinances-12-31.pdf.

DOT Aviation Division Director. If the State CIO decides to plan for an unmanned aircraft system program, a proposal for the implementation of the program shall be provided by March 1, 2014, to the Joint Legislative Oversight Committee on Information Technology, the Joint Transportation Legislative Oversight Committee, and the Fiscal Research Division. At a minimum, the proposal shall include the following:

1. Governance structure to include the appropriate use at each level of government.
2. Guidelines for program implementation to include limitations on unmanned aircraft system use.
3. Potential participants.
4. Costs associated with establishing a program.
5. Potential sources of funding.
6. Issues associated with establishing a program to include limitations on entities that may already have purchased unmanned aircraft systems.
7. Recommendations for legislative proposals.[326]

Senate Bill (SB) 744: SB 744 was added to Chapter SL 2013-360 on May 15, 2014. The parts applicable to UAS regulation read as follows:

VEHICLE MANAGEMENT

Section 7.11 (a) Section 7.16 (e) of S.L. 2013-360 reads as rewritten:
Section 7.16 (e): Until December 31, 2015, no State or local governmental entity or officer may procure or operate an unmanned aircraft system or disclose personal information about any person acquired through the operation of an unmanned aircraft system unless the State CIO approves an exception specifically granting disclosure, use, or purchase. Any exceptions to the prohibition in this subsection shall be reported immediately to the Joint Legislative Oversight Committee on Information Technology and the Fiscal Research Division. The following definitions apply in this section:

1. "Unmanned aircraft" means an aircraft that is operated without the possibility of human intervention from within or on the aircraft.
2. "Unmanned aircraft system" means an unmanned aircraft and associated elements, including communication links and components that control the unmanned aircraft that are required for the pilot in command to operate safely and efficiently in the national airspace system."

---

[326]SB 402, North Carolina General Assembly, http://www.ncga.state.nc.us/Sessions/2013/Bills/Senate/HTML/S402v7.html.

SECTION 7.11. (b)

If Section 7.16(e) of S.L. 2013-360 is repealed during the 2014 Session of the 2013 General Assembly, then Section 7.16 of S.L. 2013-360 is amended by adding the following new subsection:

"SECTION 7.16.(g) Until December 31, 2015, no State or local governmental entity or officer may procure or operate an unmanned aircraft system or disclose personal information about any person acquired through the operation of an unmanned aircraft system unless the State CIO approves an exception specifically granting disclosure, use, or purchase. Any exceptions to the prohibition in this subsection shall be reported immediately to the Joint Legislative Oversight Committee on Information Technology and the Fiscal Research Division. The following definitions apply in this section:

1. "Unmanned aircraft" means an aircraft that is operated without the possibility of human intervention from within or on the aircraft.
2. "Unmanned aircraft system" means an unmanned aircraft and associated elements, including communication links and components that control the unmanned aircraft that are required for the pilot in command to operate safely and efficiently in the national airspace system."

REGULATION OF UNMANNED AIRCRAFT SYSTEMS

SECTION 34.30. (a) Chapter 15A of the General Statutes is amended by adding a new Article to read:

Article 16B. Use of Unmanned Aircraft Systems.

§ 15A-300.1. Restrictions on use of unmanned aircraft systems.

a. Definitions.—The following definitions apply to this Article:
    1. Manned aircraft.—An aircraft, as defined in G.S. 63-1, that is operated with a person in or on the aircraft.
    2. Model aircraft.—An aircraft, as defined in G.S. 63-1, that is mechanically driven or launched into flight and that meets all of the following requirements:
        a. Is flown solely for hobby or recreational purposes.
        b. Is not used for payment, consideration, gratuity, or benefit, directly or indirectly charged, demanded, received, or collected, by any person for the use of the aircraft or any photographic or video image produced by the aircraft.
    3. Unmanned aircraft.—An aircraft, as defined in G.S. 63-1, that is operated without the possibility of human intervention from within or on the aircraft and that does not meet the definition of model aircraft.

4. Unmanned aircraft system.—An unmanned aircraft and associated elements, including communication links and components that control the unmanned aircraft that are required for the pilot in command to operate safely and efficiently in the national airspace system.

b. General Prohibitions.—Except as otherwise provided in this section, no person, entity, or State agency shall use an unmanned aircraft system to do any of the following:
    1. Conduct surveillance of:
        a. A person or a dwelling occupied by a person and that dwelling's curtilage without the person's consent.
        b. Private real property without the consent of the owner, easement holder, or lessee of the property.
    2. Photograph an individual, without the individual's consent, for the purpose of publishing or otherwise publicly disseminating the photograph. This subdivision shall not apply to newsgathering, newsworthy events, or events or places to which the general public is invited.

c. Law Enforcement Exceptions.—Notwithstanding the provisions of subsection (b) of this section, the use of unmanned aircraft systems by law enforcement agencies of the State or a political subdivision of the State is not prohibited in the following instances:
    1. To counter a high risk of a terrorist attack by a specific individual or organization if the United States Secretary of Homeland Security or the Secretary of the North Carolina Department of Public Safety determines that credible intelligence indicates that such a risk exists.
    2. To conduct surveillance in an area that is within a law enforcement officer's plain view when the officer is in a location the officer has a legal right to be.
    3. If the law enforcement agency first obtains a search warrant authorizing the use of an unmanned aircraft system.
    4. If the law enforcement agency possesses reasonable suspicion that, under particular circumstances, swift action is needed to prevent imminent danger to life or serious damage to property, to forestall the imminent escape of a suspect or the destruction of evidence, to conduct pursuit of an escapee or suspect, or to facilitate the search for a missing person.
    5. To photograph gatherings to which the general public is invited on public or private land.

d. Limitations on Use of Special Imaging Technology.—Commercial and private unmanned aircraft systems may be equipped with infrared or other thermal imaging technology subject to the provisions of this subsection. Infrared or other similar thermal imaging technology equipment shall be for the sole purpose of scientific investigation; scientific research; mapping and evaluating the earth's surface, including terrain and surface water bodies and other features; investigation or evaluation of crops, livestock, or farming operations; investigation of forests and forest management; and other similar investigations of vegetation or wildlife.

e. Any person who is the subject of unwarranted surveillance, or whose photograph is taken in violation of the provisions of this section, shall have a civil cause of action against the person, entity, or State agency that conducts the surveillance or that uses an unmanned aircraft system to photograph for the purpose of publishing or otherwise disseminating the photograph. In lieu of actual damages, the person whose photograph is taken may elect to recover five thousand dollars ($5,000) for each photograph or video that is published or otherwise disseminated, as well as reasonable costs and attorneys' fees and injunctive or other relief as determined by the court.

f. Evidence obtained or collected in violation of this section is not admissible as evidence in a criminal prosecution in any court of law in this State except when obtained or collected under the objectively reasonable, good-faith belief that the actions were lawful.

§ 15A-300.2. Regulation of launch and recovery sites.

a. No unmanned aircraft system may be launched or recovered from any State or private property without consent.

b. A unit of local government may adopt an ordinance to regulate the use of the local government's property for the launch or recovery of unmanned aircraft systems."

SECTION 34.30. (b) Chapter 14 of the General Statutes is amended by adding a new Article to read:

Article 2F. Crimes by Unmanned Aircraft Systems.

§ 14-7.45. Crimes committed by use of unmanned aircraft systems.

All crimes committed by use of an unmanned aircraft system, as defined in G.S. 15A-300.1, while in flight over this State shall be governed by the laws of this State, and the question of whether the conduct by an unmanned aircraft system while in flight over this State constitutes a crime by the owner of the unmanned aircraft system shall be determined by the laws of this State."

SECTION 34.30. (c) Article 36 of Chapter 14 of the General Statutes is amended by adding a new section to read:

§ 14-280.3. Interference with manned aircraft by unmanned aircraft systems.

a. Any person who willfully damages, disrupts the operation of, or otherwise interferes with a manned aircraft through use of an unmanned aircraft system, while the manned aircraft is taking off, landing, in flight, or otherwise in motion, is guilty of a Class H felony.

b. The following definitions apply to this section:
   1. Manned aircraft.—As defined in G.S. 15A-300.1.
   2. Unmanned aircraft system.—As defined in G.S. 15A-300.1."

SECTION 34.30. (d) Article 52 of Chapter 14 of the General Statutes is amended by adding a new section to read:
§ 14-401.24. Unlawful possession and use of unmanned aircraft systems.

a. It shall be a Class E felony for any person to possess or use an unmanned aircraft or unmanned aircraft system that has a weapon attached.

b. It shall be a Class 1 misdemeanor for any person to fish or to hunt using an unmanned aircraft system.

c. The following definitions apply to this section:
   1. To fish.—As defined in G.S. 113-130.
   2. To hunt.—As defined in G.S. 113-130.
   3. Unmanned aircraft.—As defined in G.S. 15A-300.1.
   4. Unmanned aircraft system.—As defined in G.S. 15A-300.1.
   5. Weapon.—Those weapons specified in G.S. 14-269, 14-269.2, 14-284.1, or 14-288.8 and any other object capable of inflicting serious bodily injury or death when used as a weapon.

d. This section shall not prohibit possession or usage of an unmanned aircraft or unmanned aircraft system that is authorized by federal law or regulation."

SECTION 34.30. (e) Article 52 of Chapter 14 of the General Statutes is amended by adding a new section to read:
§ 14-401.25. Unlawful distribution of images.

It shall be a Class A1 misdemeanor to publish or disseminate, for any purpose, recorded images taken by a person or non-law enforcement entity through the use of infrared or other similar thermal imaging technology attached to an unmanned aircraft system, as defined in G.S. 15A-300.1, and revealing individuals, materials, or activities inside of a structure without the consent of the property owner."
SECTION 34.30. (f) G.S. 113-295 reads as rewritten:

§ 113-295. Unlawful harassment of persons taking wildlife resources.

a. It is unlawful for a person to interfere intentionally with the lawful taking of wildlife resources or to drive, harass, or intentionally disturb any wildlife resources for the purpose of disrupting the lawful taking of wildlife resources. It is unlawful to take or abuse property, equipment, or hunting dogs that are being used for the lawful taking of wildlife resources. This subsection does not apply to a person who incidentally interferes with the taking of wildlife resources while using the land for other lawful activity such as agriculture, mining, or recreation. This subsection also does not apply to activity by a person on land he owns or leases.

   Violation of this subsection is a Class 2 misdemeanor for a first conviction and a Class 1 misdemeanor for a second or subsequent conviction.

   1. It is unlawful to use an unmanned aircraft system, as defined in G.S. 15A-300.1, to violate subsection (a) of this section. Violation of this subsection is a Class 1 misdemeanor.

b. The Wildlife Resources Commission may, either before or after the institution of any other action or proceeding authorized by this section, institute a civil action for injunctive relief to restrain a violation or threatened violation of subsection (a) of this section pursuant to G.S. 113-131. The action shall be brought in the superior court of the county in which the violation or threatened violation is occurring or about to occur and shall be in the name of the State upon the relation of the Wildlife Resources Commission. The court, in issuing any final order in any action brought pursuant to this subsection may, in its discretion, award costs of litigation including reasonable attorney and expert-witness fees to any party."

SECTION 34.30. (g) Chapter 63 of the General Statutes is amended by adding a new Article to read:
Article 10. Operation of Unmanned Aircraft Systems.
§ 63-95. Training required for operation of unmanned aircraft systems.

a. As used in this Article, the term "Division" means the Division of Aviation of the Department of Transportation.

b. The Division shall develop a knowledge and skills test for operating an unmanned aircraft system that complies with all applicable State and federal regulations and shall provide for administration of the test. The Division may permit a person, including an agency of this State, an agency of a political subdivision of this State, an employer, or a private training facility, to administer the test developed pursuant to this subsection, provided the test is the same as that administered by the Division and complies with all applicable State and federal regulations.

c. No agent or agency of the State, or agent or agency of a political subdivision of the State, may operate an unmanned aircraft system within the State without completion of the test set forth in subsection (b) of this section.

§ 63-96. License required for commercial operation of unmanned aircraft systems.

a. No person shall operate an unmanned aircraft system, as defined in G.S. 15A-300.1, in this State for commercial purposes unless the person is in possession of a license issued by the Division valid for the unmanned aircraft system being operated. Application for such license shall be made in the manner provided by the Division. Unless suspended or revoked, the license shall be effective for a period to be established by the Division not exceeding eight years.
b. No person shall be issued a license under this section unless all of the following apply:
   1. The person is at least 18 years of age.
   2. The person possesses a valid drivers license issued by any state or territory of the United States or the District of Columbia.
   3. (The person has passed the knowledge and skills test for operating an unmanned aircraft system as prescribed in G.S. 63-95(b).
   4. The person has satisfied all other applicable requirements of this Article or federal regulation.
c. A license to operate an unmanned aircraft system for commercial purposes shall not be issued to a person while the person's license to operate an unmanned aircraft system is suspended, revoked, or cancelled in any state.
d. The Division shall develop and administer a program to license operators of unmanned aircraft systems for commercial purposes. The program must include the following components:
   1. A system for classifying unmanned aircraft systems based on characteristics determined to be appropriate by the Division.
   2. A fee structure for licenses.
   3. A license application process.
   4. Technical guidance for complying with program requirements.
   5. Criteria under which the Division may suspend or revoke a license.
   6. Criteria under which the Division may waive licensure requirements for applicants currently holding a valid license to operate unmanned aircraft systems issued by another state or territory of the United States, the District of Columbia, or the United States.

7. A designation of the geographic area within which a licensee shall be authorized to operate an unmanned aircraft system.
8. Requirements pertaining to the collection, use, and retention of data by licensees obtained through the operation of unmanned aircraft systems, to be established in consultation with the State Chief Information Officer.
9. Requirements for the marking of each unmanned aircraft system operated pursuant to a license issued under this section sufficient to permit identification of the owner of the system and the person licensed to operate it.
10. A system for providing agencies that conduct other operations within regulated airspace with the identity and contact information of licensees and the geographic areas within which the licensee is permitted to operate an unmanned aircraft system.

e. A person who operates an unmanned aircraft system for commercial purposes other than as permitted under this section shall be guilty of a Class 1 misdemeanor.

f. The Division may issue rules and regulations to implement the provisions of this section."

SECTION 34.30. (h) The Division of Aviation of the Department of Transportation shall develop and implement the knowledge and skills test required by G.S. 63-95, as enacted in subsection (g) of this section, no later than May 31, 2015, and shall report to the Joint Legislative Transportation Oversight Committee on the status of implementation by June 15, 2015.

SECTION 34.30 (i) The Division of Aviation of the Department of Transportation shall immediately begin developing the licensing system for commercial operation required by G.S. 63-96, as enacted in subsection (g) of this section, and shall ensure that the system complies with Federal Aviation Administration (FAA) guidelines on commercial operation, as those guidelines become available. Within 60 days of issuance of the FAA guidelines and authorization by the FAA for commercial operations to begin, the Division shall implement the licensing system required by G.S. 63-96, as enacted in subsection (g) of this section.

SECTION 34.30. (j) No operation of unmanned aircraft systems by agents or agencies of the State, or agents or agencies of a political subdivision of the State, shall be authorized in this State until the knowledge and skills test required by G.S. 63-95, as enacted in subsection (g) of this section, has been implemented.

No operation of unmanned aircraft systems for commercial purposes shall be authorized in this State until the FAA has authorized commercial operations and the licensing system required by G.S. 63-96, as enacted in subsection (g) of this section, has been implemented.

SECTION 34.30. (k) The Division of Aviation of the Department of Transportation shall use funds appropriated in this act to the Division to cover the administration costs incurred

from developing and implementing the knowledge and skills test and licensing system for commercial operation required by this section.
SECTION 34.30. (l) Subsection (a) of this section becomes effective October 1, 2014, and applies to acts occurring on or after that date. Subsections (b), (c), (d), (e), and (f) of this section become effective December 1, 2014, and apply to offenses committed on or after that date. The remainder of this section is effective when it becomes law.[327]
Senate Bill (SB) 446: On August 25, 2015, the Governor of North Carolina signed into law, SB 446 that expands the authority of the state's Chief Information Officer to approve the purchase and operation of UAS by the state and modifies the state regulation of UAS to conform to FAA guidelines.[328]

NORTH DAKOTA

North Dakota has enacted House Bill (HB) 1328[329] and HB 1018.[330] In 2013, SB 2018 granted $1 million from the state general fund to pursue designation as an FAA UAS test site. Had the state been selected, the law would grant an additional $4 million to operate the site.[331] Also in 2013, North Dakota adopted House Concurrent Resolution (HCR) 3012 to recognize the benefits of a thriving UAS industry in the state.[332] Furthermore, Grand Forks passed Ordinance 9-0202 prohibiting landing and taking off of aircraft (including unmanned vehicles weighing more than 55 pounds) at unauthorized airports, helipads, or other unauthorized locations unless approved by the city council.[333]
House Bill (HB) 1328: On April 2, 2015, the North Dakota Senate passed HB 1328, a Bill for an Act to provide for limitations on the use of unmanned aircraft for surveillance, and reads as follows:
Be it enacted by the Legislative Assembly of North Dakota:

SECTION 1.

Definitions.
As used in this Act:

1. "Law enforcement agency" means a person authorized by law, or funded by the state, to investigate or prosecute offenses against the state.

---

[327]SB 744, North Carolina General Assembly, http://www.ncleg.net/Sessions/2013/Bills/Senate/PDF/S744v8.pdf.
[328]SB 446, North Carolina General Assembly, http://www.ncga.state.nc.us/Sessions/2015/Bills/Senate/HTML/S446v6.html.
[329]HB 1328, North Dakota General Assembly, http://www.legis.nd.gov/assembly/64-2015/documents/15-0259-02000.pdf?20151004134739.
[330]HB 1018, North Dakota General Assembly, https://legiscan.com/ND/bill/1018/2015.
[331]SB 2018, North Dakota General Assembly, http://www.legis.nd.gov/assembly/63-2013/documents/13-8168-01000.pdf?20130513175955.
[332]HCR 3012, North Dakota General Assembly, http://www.legis.nd.gov/files/resource/63-2013/library/hcr3012.pdf.
[333]Ordinance 9-0202, Grand Forks, http://uavs.insct.org/wp-content/uploads/2016/04/Grand-Forks-ND-Code-of-Ordinances-9-0202.pdf.

2. "Unmanned aircraft" means any aircraft that is operated without the possibility of direct human intervention within or on the aircraft.
3. "Unmanned aircraft system" means an unmanned aircraft and associated elements, including communication links and the components that control the unmanned aircraft, which are required for the pilot in command to operate safely and efficiently in state airspace.

SECTION 2.

Prohibited use of unmanned aircraft system.

1. Except as provided in section 4 of this Act, a law enforcement agency may not use an unmanned aircraft for surveillance of a person within the state or for the surveillance of personal or business property located within the borders of the state to gather evidence or other information pertaining to criminal conduct, or conduct in violation of a statute or regulation except to the extent authorized in a warrant issued by a court which satisfies the requirements of the Constitution of North Dakota.
2. Warrants to conduct surveillance with an unmanned aircraft may be issued only in the investigation of a felony. Unmanned aircraft may not be used to conduct investigations of misdemeanors, traffic infractions, or other non-felony violations of law.

SECTION 3.

Warrant requirements.

A warrant for the use of an unmanned aircraft must satisfy the requirements of the Constitution of North Dakota. In addition, the warrant must contain a data collection statement that includes:

1. The persons that will have the power to authorize the use of the unmanned aircraft;
2. The locations in which the unmanned aircraft system will operate;
3. The maximum period for which the unmanned aircraft system will operate in each flight; and
4. Whether the unmanned aircraft system will collect information or data about individuals or groups of individuals, and if so:
    a. The circumstances under which the unmanned aircraft system will be used; and
    b. The specific kinds of information or data the unmanned aircraft system will collect about individuals and how that information or data, as well as conclusions drawn from that information or data, will be used, disclosed, and otherwise handled, including:

1. The period for which the information or data will be retained; and
2. Whether the information or data will be destroyed, and if so, when and how the information or data will be destroyed.

SECTION 4.

Exceptions.

This Act does not prohibit any use of an unmanned aircraft for surveillance during the course of:

1. Patrol of national borders. The use of an unmanned aircraft to patrol within twenty-five miles [40.23 kilometers] of a national border, for purposes of policing that border to prevent or deter the illegal entry of any individual, illegal substance, or contraband.
2. Exigent circumstances. The use of an unmanned aircraft by a law enforcement agency is permitted when exigent circumstances exist. For the purposes of this subsection, exigent circumstances exist when a law enforcement agency possesses reasonable suspicion that absent swift preventative action, there is an imminent danger to life or bodily harm.
3. An environmental or weather-related catastrophe. The use of an unmanned aircraft by state authorities to preserve public safety, protect property, survey environmental damage to determine if a state of emergency should be declared, or conduct surveillance for the assessment and evaluation of environmental or weather-related damage, erosion, flood, or contamination.
4. Testing, training, education, and research of unmanned aircraft systems.

SECTION 5.

Prohibited surveillance.

1. Use of force. A state agency may not authorize the use of, including granting a permit to use, an unmanned aircraft armed with any lethal or nonlethal weapons, including firearms, pepper spray, bean bag guns, mace, and sound-based weapons.
2. This Act prohibits any use of an unmanned aircraft for:
   a. Domestic use in private surveillance. A state agency may not authorize the use of, including granting a permit to use, an unmanned aircraft to permit any private person to conduct surveillance on any other private person without the express, informed consent of that other person or the owner of any real property on which that other private person is present.

b. Surveillance of the exercise of constitutional rights. A state agency may not authorize the use, including granting a permit to use, of an unmanned aircraft for the purpose of the surveillance of persons engaged in the lawful exercise of the constitutional right of freedom of speech and freedom of assembly.

SECTION 6.

Remedies for violation.

Any aggrieved party may obtain in a civil action all appropriate relief to prevent or remedy a violation of this Act.

SECTION 7.

Prohibitions on the conduct of unmanned aircraft surveillance and the use of acquired surveillance as evidence.

1. Evidence obtained or collected in violation of this Act is not admissible as evidence in a criminal prosecution, including use during trial, at sentencing, before a grand jury, as rebuttal evidence, or during administrative hearings in any court of law in the state.
2. Any imaging or other forms of observational data gathered by unmanned aircraft surveillance from or concerning the parties or places subjected to surveillance in violation of this Act may not be preserved by law enforcement for any purpose unless required by a court of law.
3. Any imaging or any other forms of data lawfully obtained under this Act for which there is not a reasonable and articulable suspicion that those images or data contain evidence of a crime, or are relevant to an ongoing investigation or trial, may not be retained for more than ninety days, unless the retention is attendant to general agency guidelines regarding the retention of evidence in criminal cases. In those cases, the imaging or other data may not be distributed to agencies, entities, or individuals if the distribution is not necessary to meet general agency guidelines regarding the retention of evidence in criminal cases.
4. An unmanned aircraft may not conduct any type of surveillance that would violate state laws regarding the interception of aural communications, electronic messaging, personal location data, or acquire video or still images of any individual within a home or place without first obtaining all required warrants in compliance with state statutes relevant to the interception of such voice communications, digital communications, physical surveillance data, or to capture the still or video images of any individual or interior or a place for which a search warrant is prerequisite to its lawful search.

SECTION 8.

Documentation of unmanned aircraft surveillance.

1. The person authorized to conduct the surveillance under this Act shall document all use of an unmanned aircraft for surveillance. The person shall document all surveillance flights as to duration, flight path, and mission objectives, including the names of place or persons authorized to be subject to surveillance.
2. The flight information must be certified as accurate and complete by the supervising person authorized by a court to conduct the surveillance.
3. The flight information required under this section must be retained for five years.
4. Except for the operational capabilities of the unmanned aircraft system and other operational information strictly related to the technical conduct and physical security of the surveillance operation, a person accused of a crime that includes evidence gathered through the use of an unmanned aircraft system surveillance may obtain all information relating to the person acquired in the course of the surveillance through subpoena and discovery proceedings available in criminal proceedings.
5. Any other person that has an interest in obtaining the documentation required by this section may obtain that documentation pursuant to chapter 44-04.
6. The documentation required by this section applies to all uses of unmanned aircraft systems, including testing, training, education, and research.

SECTION 9.

Application to federal law.

This Act may not be construed to limit, constrain, or adversely impact testing and operations of a state test range under the FAA Modernization and Reform Act of 2012 [Pub. L. 112-95; 126 Stat. 11].[334]

House Bill (HB) 1018: On June 16, 2015, HB 1018 was signed by the Governor. This was an act to provide an appropriation for defraying the expenses of the department of commerce; to provide exemptions; to provide for transfers; to amend and reenact section 54-60-17 of the North Dakota Century Code, relating to workforce development; and to provide for a legislative management study.[335]

OHIO

Ohio has enacted HB 292.[336] Ohio has introduced HB 218 to enact sections 4561.50 and 4561.51 of the Revised Code to regulate the operation of drones near airports and

---

[334]HB 1328, North Dakota General Assembly, http://www.legis.nd.gov/assembly/64-2015/documents/15-0259-02000.pdf?20151004134739.

[335]HB 1018, North Dakota General Assembly, https://legiscan.com/ND/bill/1018/2015.

[336]HB 292, Ohio Legislature, http://archives.legislature.state.oh.us/bills.cfm?ID=130_HB_292.

to impose certain record-keeping requirements on retail sellers of drones,[337] HB 228 to amend sections 2981.01 and 2981.02 and to enact section 2917.48 of the Revised Code to prohibit the commission of any of a list of specified offenses by using a drone or by directing or supervising another person's use of a drone,[338] and SB 251 to enact sections 2933.67, 2933.68, 2933.69, and 2933.70 of the Revised Code to regulate the use of drones for gathering evidence and information by law enforcement officers in Ohio.[339]

Furthermore, the City of Celina has introduced Ordinance 52-15-O, an ordinance restricting the use of UAVs by individuals and businesses over the airspace above any city-owned property, above any city right-of-way, and above any easement granted in favor of the city with certain exemptions,[340] and Cleveland has passed legislation to restrict the use of UAS in certain areas, and to permit authorities to assist the FAA in regulating aerial drones.[341]

House Bill (HB) 292: The General Assembly passed HB 292 to enact section 122.98 of the Revised Code to create the Ohio Aerospace and Aviation Technology Committee. The law that went into effect on September 17, 2014, reads as follows:
Be it enacted by the General Assembly of the State of Ohio:
SECTION 1. That section 122.98 of the Revised Code be enacted to read as follows:
Sec. 122.98.

A. There is hereby created the Ohio aerospace and aviation technology committee, consisting of the following members:

1. Three members of the senate, appointed by the president of the senate, not more than two of whom may be members of the same political party;
2. Three members of the house of representatives, appointed by the speaker of the house of representatives, not more than two of whom may be members of the same political party;
3. Fifteen members representing the aviation, aerospace, or technology industry, the military, or academia. One such member shall be appointed by the governor, and fourteen such members shall be appointed by majority vote of the six members representing the senate and house of representatives.

The legislative members of the committee shall be appointed not later than September 1, 2014, and the remaining members shall be appointed within ten days thereafter. The initial term of all members shall end on December 31, 2016. Thereafter, the term of all

---

[337]HB 218, Ohio Legislature, https://www.legislature.ohio.gov/legislation/legislation-summary?id=GA131-HB-218.
[338]HB 228, Ohio Legislature, https://www.legislature.ohio.gov/legislation/legislation-summary?id=GA131-HB-228.
[339]HB 251, Ohio Legislature, https://www.legislature.ohio.gov/legislation/legislation-summary?id=GA131-SB-251.
[340]Ordinance 57-15-O, City of Celina, http://www.ci.celina.oh.us/documents/57-15-O.Celinadroneordinancedraft4.pdf.
[341]Ordinance 330-16, City of Cleveland, https://www.docdroid.net/kR9KX0D/cleveland-drone-ordinance.pdf.html.

members shall end on the thirty-first day of December of the year following the year of appointment. Vacancies shall be filled in the manner of the original appointment.

The first legislator appointed to the committee by the speaker of the house of representatives after the effective date of H.B. 292 of the 130th general assembly shall serve as the first chairperson of the committee and shall serve until December 31, 2016. Every general assembly thereafter, the chairperson shall alternate between the first legislator appointed by the president of the senate and the first legislator appointed by the speaker of the house of representatives.

B. The duties of the committee shall include, but are not limited to, all of the following:
   1. Studying and developing comprehensive strategies to promote the aviation, aerospace, and technology industry throughout the state, including through the commercialization of aviation, aerospace, and technology products and ideas;
   2. Encouraging communication and resource-sharing among individuals and organizations involved in the aviation, aerospace, and technology industry, including business, the military, and academia;
   3. Promoting research and development in the aviation, aerospace, and technology industry, including research and development of unmanned aerial vehicles;
   4. Providing assistance related to military base realignment and closure.

C. The Ohio aerospace and aviation council shall serve as an advisory council to the committee.

D. The committee shall compile an annual report of its activities, findings, and recommendations and shall furnish a copy of the report to the governor, president of the senate, and speaker of the house of representatives not later than July 1, 2015, and the first day of July of each year thereafter.[342]

OKLAHOMA

Oklahoma has introduced HB 1295 creating the Oklahoma Unmanned Aerial Surveillance Act, defining terms, requiring use of unmanned aerial vehicles to comply with FAA rules, prohibiting agents of state from using UAV, proscribing information disclosure, providing exceptions, limiting use in exceptions provided, barring facial recognition, and other matching technology on non-targets, excluding equipping UAVs with weapons, blocking use of information gathered in violation of the act of evidence among others;[343] HB 2337

---

[342]HB 292, Ohio Legislature, http://archives.legislature.state.oh.us/bills.cfm?ID=130_HB_292.
[343]HB 1295, Oklahoma Legislature, https://legiscan.com/OK/bill/HB1295/2016.

creating the Oklahoma Unmanned Surveillance Act;[344] HB 2368 prohibiting operation of unmanned aircraft over private agricultural property;[345] HB 2591 requiring contact information be permanently affixed to aircraft and prohibiting operation over private property below specified altitude;[346] HB 3001 creating the Oklahoma Use of Unmanned Aerial Vehicles Act;[347] SB 492 providing civil immunity for damage or destruction of a drone on personal property;[348] and SB 503 restricting the use of drones.[349]

House Bill (HB) 2599: On May 18, 2016, the governor of Oklahoma signed this bill into law, prohibiting operation of an unmanned aircraft over a critical infrastructure facility.[350]

OREGON

Oregon passed one senate bill; SB 5702[351] and four house bills, HB 4066,[352] HB 2710,[353] HB 2534,[354] and HB 2354.[355]

Senate Bill (SB) 5702: Effective April 4, 2016, this law specifies the fees for registration of public UAS.[356]

House Bill (HB) 4066: Signed by the governor in 2016, this law makes a Class A misdemeanor if any person intentionally, knowingly, or recklessly operates a drone that is capable of firing a bullet or projectile. It creates a new violation of reckless interference with aircraft. Additionally, the law requires public bodies using drones in Oregon to establish policies and procedures for use, storage, accessing, sharing, and retention of data collected. Probable cause may be established through public data without a warrant. This law creates a violation if a person knowingly or intentionally operates a drone over a critical facility at an altitude of 400 feet or lower or makes contact with the critical facility. Critical facility includes prisons, petroleum refineries, chemical manufacturing facilities and water treatment facilities.[357]

---

[344]HB 2337, Oklahoma Legislature, https://legiscan.com/OK/bill/HB2337/2016.
[345]HB 2368, Oklahoma Legislature, https://legiscan.com/OK/bill/HB2368/2016.
[346]HB 2591, Oklahoma Legislature, https://legiscan.com/OK/bill/HB2591/2016.
[347]HB 3001, Oklahoma Legislature, https://legiscan.com/OK/bill/HB3001/2016.
[348]SB 492, Oklahoma Legislature, https://legiscan.com/OK/bill/SB492/2016.
[349]SB 503, Oklahoma Legislature, https://legiscan.com/OK/bill/SB503/2016.
[350]HB 2599, Oklahoma Legislature, https://legiscan.com/OK/bill/HB2599/2016.
[351]SB 5702, Oregon Legislative Assembly, https://olis.leg.state.or.us/liz/2016R1/Downloads/MeasureDocument/SB5702/Enrolled.
[352]HB 4066, Oregon Legislative Assembly, https://olis.leg.state.or.us/liz/2016R1/Downloads/MeasureDocument/HB4066/Enrolled.
[353]HB 2710, Oregon Legislative Assembly, https://olis.leg.state.or.us/liz/2013R1/Downloads/MeasureDocument/HB2710/Enrolled.
[354]HB 2534, Oregon Legislative Assembly, https://olis.leg.state.or.us/liz/2015R1/Downloads/MeasureDocument/HB2534/Enrolled.
[355]HB 2354, Oregon Legislative Assembly, https://olis.leg.state.or.us/liz/2015R1/Downloads/MeasureDocument/HB2354/Enrolled.
[356]SB 5702, Oregon Legislative Assembly, https://olis.leg.state.or.us/liz/2016R1/Downloads/MeasureDocument/SB5702/Enrolled.
[357]HB 4066, Oregon Legislative Assembly, https://olis.leg.state.or.us/liz/2016R1/Downloads/MeasureDocument/HB4066/Enrolled.

House Bill (HB) 2710: On June 24, 2013, the Oregon Legislative Assembly adopted HB 2710. This is an act relating to drones and declaring an emergency, and reads as follows: Be it enacted by the People of the State of Oregon:
DEFINITIONS
SECTION 1. As used in sections 1 to 17 of this 2013 Act:

1. "Drone" means an unmanned flying machine. "Drone" does not include a model aircraft as defined in section 336 of the FAA Modernization and Reform Act of 2012 (P.L. 112-95) as in effect on the effective date of this 2013 Act.
2. "Law enforcement agency" means an agency that employs police officers, as defined in ORS 133.525, or that prosecutes offenses.
3. "Public body" has the meaning given that term in ORS 174.109.
4. "Warrant" means a warrant issued under ORS 133.525 to 133.703.

USE OF DRONES BY LAW ENFORCEMENT AGENCIES

SECTION 2.

1. Except as otherwise provided in sections 2 to 7 of this 2013 Act, a law enforcement agency may not operate a drone, acquire information through the operation of a drone or disclose information acquired through the operation of a drone.
2. Any image or other information that is acquired through the use of a drone by a law enforcement agency in violation of sections 2 to 7 of this 2013 Act, and any evidence derived from that image or information:
    a. Is not admissible in, and may not be disclosed in, a judicial proceeding, administrative proceeding, arbitration proceeding or other adjudicatory proceeding; and
    b. May not be used to establish reasonable suspicion or probable cause to believe that an offense has been committed.

SECTION 3.

1. A law enforcement agency may operate a drone, acquire information through the operation of a drone, or disclose information acquired through the operation of a drone, if: (a) A warrant is issued authorizing use of a drone; or
    a. The law enforcement agency has probable cause to believe that a person has committed a crime, is committing a crime or is about to commit a crime,

and exigent circumstances exist that make it unreasonable for the law enforcement agency to obtain a warrant authorizing use of a drone.

2. A warrant authorizing the use of a drone must specify the period for which operation of the drone is authorized. In no event may a warrant provide for the operation of a drone for a period of more than 30 days. Upon motion and good cause shown, a court may renew a warrant after the expiration of the 30-day period.

SECTION 4.

A law enforcement agency may operate a drone for the purpose of acquiring information about an individual, or about the individual's property, if the individual has given written consent to the use of a drone for those purposes.

SECTION 5.

1. A law enforcement agency may operate a drone, acquire information through the operation of a drone, or disclose information acquired through the operation of a drone, for the purpose of search and rescue activities, as defined in ORS 404.200.
2. A law enforcement agency may operate a drone, acquire information through the operation of a drone, or disclose information acquired through the operation of a drone, for the purpose of assisting an individual in an emergency if:
    a. The law enforcement agency reasonably believes that there is an imminent threat to the life or safety of the individual, and documents the factual basis for that belief; and
    b. Not more than 48 hours after the emergency operation begins, an official of the law enforcement agency files a sworn statement with the circuit court that describes the nature of the emergency and the need for use of a drone.
3. A law enforcement agency may operate a drone, acquire information through the operation of a drone, or disclose information acquired through the operation of a drone, during a state of emergency that is declared by the Governor under ORS chapter 401 if:
    a. The drone is used only for the purposes of preserving public safety, protecting property or conducting surveillance for the assessment and evaluation of environmental or weather related damage, erosion or contamination; and
    b. The drone is operated only in the geographical area specified in a proclamation pursuant to ORS 401.165 (5).

SECTION 6.

1. A law enforcement agency may operate a drone, acquire information through the operation of a drone, or disclose information acquired through the operation of a drone, for the purpose of reconstruction of a specific crime scene, or similar physical assessment, related to a specific criminal investigation.
2. The period that a law enforcement agency may operate a drone under this section may not exceed five days for the purpose of reconstruction of a specific crime scene, or similar physical assessment, related to a specific criminal investigation.

SECTION 7.

1. A law enforcement agency may operate a drone for the purpose of training in:
   a. The use of drones; and
   b. The acquisition of information through the operation of a drone.
2. Any image or other information that is acquired through the use of a drone by a law enforcement agency under this section, and any evidence derived from that image or information:
   a. Is not admissible in, and may not be disclosed in, a judicial proceeding, administrative proceeding, arbitration proceeding or other adjudicatory proceeding; and
   b. May not be used to establish reasonable suspicion or probable cause to believe that an offense has been committed.

REGISTRATION OF DRONES USED BY PUBLIC BODIES

SECTION 8.

1. A public body may not operate a drone in the airspace over this state without registering the drone with the Oregon Department of Aviation.
2. The Oregon Department of Aviation may impose a civil penalty of up to $10,000 against a public body that violates subsection (1) of this section.
3. Evidence obtained by a public body through the use of a drone in violation of subsection (1) of this section is not admissible in any judicial or administrative proceeding and may not be used to establish reasonable suspicion or probable cause to believe that an offense has been committed.
4. The Oregon Department of Aviation shall establish a registry of drones operated by public bodies and may charge a fee sufficient to reimburse the department for the maintenance of the registry.

5. The Oregon Department of Aviation shall require the following information for registration of a drone:
   a. The name of the public body that owns or operates the drone.
   b. The name and contact information of the individuals who operate the drone.
   c. Identifying information for the drone as required by the department by rule.
6. A public body that registers one or more drones under this section shall provide an annual report to the Oregon Department of Aviation that summarizes:
   a. The frequency of use of the drones by the public body during the preceding calendar year; and
   b. The purposes for which the drones have been used by the public body during the preceding calendar year.
7. The State Aviation Board may adopt all rules necessary for the registration of drones in Oregon that are consistent with federal laws and regulations.

SECTION 9.

1. Except as provided in subsection (2) of this section, section 8 of this 2013 Act becomes operative January 2, 2016.
2. The Oregon Department of Aviation and the State Aviation Board may take any action before January 2, 2016, including the adoption of rules, that is necessary to allow implementation of section 8 of this 2013 Act on January 2, 2016.

PROHIBITION ON USE OF WEAPONIZED DRONES BY PUBLIC BODIES

SECTION 10.

A public body may not operate a drone that is capable of firing a bullet or other projectile, directing a laser or otherwise being used as a weapon.

USE OF INFORMATION ACQUIRED BY PUBLIC BODY DRONES

SECTION 11.

Any image or other information that is acquired by a public body through the use of a drone that has not been approved by the Federal Aviation Administration, and any evidence derived from that image or information:

1. Is not admissible in, and may not be disclosed in, a judicial proceeding, administrative proceeding, arbitration proceeding or other adjudicatory proceeding; and
2. May not be used to establish reasonable suspicion or probable cause to believe that an offense has been committed.

SECTION 12.
Section 11 of this 2013 Act is repealed January 2, 2016.

CRIMES INVOLVING DRONES

SECTION 13.

1. A person commits a Class A felony if the person possesses or controls a drone and intentionally causes, or attempts to cause, the drone to:
   a. Fire a bullet or other projectile at an aircraft while the aircraft is in the air;
   b. Direct a laser at an aircraft while the aircraft is in the air; or
   c. Crash into an aircraft while the aircraft is in the air.
2. A person who intentionally interferes with, or gains unauthorized control over, a drone licensed by the Federal Aviation Administration, or operated by the Armed Forces of the United States as defined in ORS 351.642, an agency of the United States or a federal, state or local law enforcement agency, commits a Class C felony.

CIVIL REMEDIES

SECTION 14.
In addition to any other remedies allowed by law, a person who intentionally interferes with, or gains unauthorized control over, a drone licensed by the Federal Aviation Administration, or operated by the Armed Forces of the United States as defined in ORS 351.642, an agency of the United States or a federal, state or local law enforcement agency, is liable to the owner of the drone in an amount of not less than $5,000. The court shall award reasonable attorney fees to a prevailing plaintiff in an action under this section.

SECTION 15.

1. Except as provided in subsection (2) of this section, a person who owns or lawfully occupies real property in this state may bring an action against any person or public body that operates a drone that is flown at a height of less than 400 feet over the property if:
   a. The operator of the drone has flown the drone over the property at a height of less than 400 feet on at least one previous occasion; and
   b. The person notified the owner or operator of the drone that the person did not want the drone flown over the property at a height of less than 400 feet.
2. A person may not bring an action under this section if:

   a. The drone is lawfully in the flight path for landing at an airport, airfield or runway; and
   b. The drone is in the process of taking off or landing.
3. A prevailing plaintiff may recover treble damages for any injury to the person or the property by reason of a trespass by a drone as described in this section, and may be awarded injunctive relief in the action.
4. A prevailing plaintiff may recover attorney fees under ORS 20.080 if the amount pleaded in an action under this section is $10,000 or less.
5. The Attorney General, on behalf of the State of Oregon, may bring an action or claim for relief alleging nuisance or trespass arising from the operation of a drone in the airspace over this state. A court shall award reasonable attorney fees to the Attorney General if the Attorney General prevails in an action under this section.

APPLICABILITY TO ARMED FORCES

SECTION 16.

Sections 1 to 17 of this 2013 Act do not apply to the Armed Forces of the United States as defined in ORS 351.642.

PREEMPTION OF LOCAL LAWS REGULATING DRONES

SECTION 17.

Except as expressly authorized by state statute, the authority to regulate the ownership or operation of drones is vested solely in the Legislative Assembly. Except as expressly authorized by state statute, a local government, as defined ORS 174.116, may not enact an ordinance or resolution that regulates the ownership or operation of drones or otherwise engage in the regulation of the ownership or operation of drones.

REPORT TO LEGISLATURE

SECTION 18.

On or before November 1, 2014, the Oregon Department of Aviation shall report to a joint interim committee of the Legislative Assembly related to the judiciary, or other appropriate interim committees, on:

1. The status of federal regulations relating to unmanned aerial vehicles; and
2. Whether unmanned aerial vehicles operated by private parties should be registered in Oregon in a manner similar to that required for other aircraft.

CAPTIONS

SECTION 19.

The unit captions used in this 2013 Act are provided only for the convenience of the reader and do not become part of the statutory law of this state or express any legislative intent in the enactment of this 2013 Act.

EMERGENCY CLAUSE

SECTION 20.

This 2013 Act being necessary for the immediate preservation of the public peace, health and safety, an emergency is declared to exist, and this 2013 Act takes effect on its passage.[358]

House Bill (HB) 2534: On May 18, 2015, Oregon Legislative assembly passed HB 2534 into law in Chapter 61, which went into effect January 1, 2016. This is an act relating to the regulation of drones by the State Fish and Wildlife Commission.

Whereas it is the intent of the Legislative Assembly that the wildlife laws embody the sporting ethic of fair chase hunting as it relates to the use of drones in the pursuit of wildlife; now, therefore,

Be it enacted by the People of the State of Oregon:

SECTION 1.

1. The State Fish and Wildlife Commission shall adopt rules prohibiting the use of drones for the following purposes related to the pursuit of wildlife:
   a. Angling;
   b. Hunting;
   c. Trapping;
   d. Aiding angling, hunting or trapping through the use of drones to harass, track, locate or scout wildlife; and
   e. Interfering in the acts of a person who is lawfully angling, hunting or trapping.
2. Rules adopted to carry out the prohibitions provided for in this section may include exemptions for:
   a. Subject to ORS 837.360 and 837.365, the State Department of Fish and Wildlife and the department's agents and contractors for the use of drones in carrying out the duties of the department; or
   b. The use of drones in a manner otherwise prohibited under this section if the purpose of the use is to benefit wildlife management or habitat or for the protection of property.

---

[358]HB 2710, Oregon Legislative Assembly, https://olis.leg.state.or.us/liz/2013R1/Downloads/MeasureDocument/HB2710/Enrolled.

3. Nothing in this section is meant to limit the use of drones by a person who is lawfully engaging in activities authorized under the commercial fishing laws.
4. As used in this section, "drone" means:
   a. An unmanned flying machine;
   b. An unmanned water-based vehicle; or
   c. Any other vehicle that is able to operate in the air, in or under the water or on land, either remotely or autonomously, and without a human occupant.[359]

House Bill (HB) 2354: On June 15, 2015, Oregon Legislative assembly passed HB 2354 into law in Chapter 315, which went into effect January 1, 2016. This is an act relating to unmanned aerial systems; amending ORS 837.300, 837.310, 837.320, 837.330, 837.335, 837.340, 837.345, 837.360, 837.365, 837.375, 837.380, 837.385 and 837.995.
Be it enacted by the People of the State of Oregon:

SECTION 1. ORS 837.300 is amended to read:
837.300. As used in ORS 837.300 to 837.390 and 837.995 [*and section 11, chapter 686, Oregon Laws 2013*]:

1. [*"Drone"*] "Unmanned aircraft system" means an unmanned flying machine, commonly known as a drone. [*"Drone"*] "Unmanned aircraft system" does not include a model aircraft as defined in section 336 of the FAA Modernization and Reform Act of 2012 (P.L. 112-95) as in effect on July 29, 2013.
2. "Law enforcement agency" means an agency that employs police officers, as defined in ORS 133.525, or that prosecutes offenses.
3. "Public body" has the meaning given that term in ORS 174.109.
4. "Warrant" means a warrant issued under ORS 133.525 to 133.703.

SECTION 2. ORS 837.310 is amended to read:
837.310.

1. Except as otherwise provided in ORS 837.310 to 837.345, a law enforcement agency may not operate [*a drone*] an unmanned aircraft system, acquire information through the operation of [*a drone*] an unmanned aircraft system or disclose information acquired through the operation of [*a drone*] an unmanned aircraft system.
2. Any image or other information that is acquired through the use of [*a drone*] an unmanned aircraft system by a law enforcement agency in violation of ORS 837.310 to 837.345, and any evidence derived from that image or information:

[359]HB 2534, Oregon Legislative Assembly, https://olis.leg.state.or.us/liz/2015R1/Downloads/MeasureDocument/HB2534.

a. Is not admissible in, and may not be disclosed in, a judicial proceeding, administrative proceeding, arbitration proceeding or other adjudicatory proceeding; and
b. May not be used to establish reasonable suspicion or probable cause to believe that an offense has been committed.

SECTION 3. ORS 837.320 is amended to read:

1. 837.320. (1) A law enforcement agency may operate [*a drone*] an unmanned aircraft system, acquire information through the operation of [*a drone*] an unmanned aircraft system, or disclose information acquired through the operation of [*a drone*] an unmanned aircraft system, if:
    a. A warrant is issued authorizing use of [*a drone*] an unmanned aircraft system; or
    b. The law enforcement agency has probable cause to believe that a person has committed a crime, is committing a crime or is about to commit a crime, and exigent circumstances exist that make it unreasonable for the law enforcement agency to obtain a warrant authorizing use of [*a drone*] an unmanned aircraft system.
2. A warrant authorizing the use of [*a drone*] an unmanned aircraft system must specify the period for which operation of the [*drone*] unmanned aircraft system is authorized. In no event may a warrant provide for the operation of [*a drone*] an unmanned aircraft system for a period of more than 30 days. Upon motion and good cause shown, a court may renew a warrant after the expiration of the 30-day period.

SECTION 4. ORS 837.330 is amended to read:
837.330. A law enforcement agency may operate [*a drone*] an unmanned aircraft system for the purpose of acquiring information about an individual, or about the individual's property, if the individual has given written consent to the use of [*a drone*] an unmanned aircraft system for those purposes.

SECTION 5. ORS 837.335 is amended to read:

1. 837.335. (1) A law enforcement agency may operate [*a drone*] an unmanned aircraft system, acquire information through the operation of [*a drone*] an unmanned aircraft system, or disclose information acquired through the operation of [*a drone*] an unmanned aircraft system, for the purpose of search and rescue activities, as defined in ORS 404.200. (2) A law enforcement agency may operate [*a drone*] an unmanned aircraft system, acquire information through the operation of [*a drone*] an unmanned aircraft system, or disclose information acquired through the operation of [*a drone*] an unmanned aircraft system, for the purpose of assisting an individual in an emergency if:

   a. The law enforcement agency reasonably believes that there is an imminent threat to the life or safety of the individual, and documents the factual basis for that belief; and
   b. Not more than 48 hours after the emergency operation begins, an official of the law enforcement agency files a sworn statement with the circuit court that describes the nature of the emergency and the need for use of [*a drone*] an unmanned aircraft system.
2. A law enforcement agency may operate [*a drone*] an unmanned aircraft system, acquire information through the operation of [*a drone*] an unmanned aircraft system, or disclose information acquired through the operation of [*a drone*] an unmanned aircraft system, during a state of emergency that is declared by the Governor under ORS chapter 401 if:
   a. The [*drone*] unmanned aircraft system is used only for the purposes of preserving public safety, protecting property or conducting surveillance for the assessment and evaluation of environmental or weather related damage, erosion or contamination; and
   b. The [*drone*] unmanned aircraft system is operated only in the geographical area specified in a proclamation pursuant to ORS 401.165 (5).

SECTION 6. ORS 837.340 is amended to read:

1. 837.340. (1) A law enforcement agency may operate [*a drone*] an unmanned aircraft system, acquire information through the operation of [*a drone*] an unmanned aircraft system, or disclose information acquired through the operation of [*a drone*] an unmanned aircraft system, for the purpose of reconstruction of a specific crime scene, or similar physical assessment, related to a specific criminal investigation.
2. The period that a law enforcement agency may operate [*a drone*] an unmanned aircraft system under this section may not exceed five days for the purpose of reconstruction of a specific crime scene, or similar physical assessment, related to a specific criminal investigation.

SECTION 7. ORS 837.345 is amended to read:

1. 837.345. (1) A law enforcement agency may operate [*a drone*] an unmanned aircraft system for the purpose of training in:
   a. The use of [*drones*] unmanned aircraft systems; and
   b. The acquisition of information through the operation of [*a drone*] an unmanned aircraft system.

2. Any image or other information that is acquired through the use of [*a drone*] an unmanned aircraft system by a law enforcement agency under this section, and any evidence derived from that image or information:
    a. Is not admissible in, and may not be disclosed in, a judicial proceeding, administrative proceeding, arbitration proceeding or other adjudicatory proceeding; and
    b. May not be used to establish reasonable suspicion or probable cause to believe that an offense has been committed.

SECTION 8. ORS 837.360 is amended to read:

1. 837.360. (1) A public body may not operate [*a drone*] an unmanned aircraft system in the airspace over this state without registering the [*drone*] unmanned aircraft system with the Oregon Department of Aviation.
2. The Oregon Department of Aviation may impose a civil penalty of up to $10,000 against a public body that violates subsection (1) of this section.
3. Evidence obtained by a public body through the use of [*a drone*] an unmanned aircraft system in violation of subsection (1) of this section is not admissible in any judicial or administrative proceeding and may not be used to establish reasonable suspicion or probable cause to believe that an offense has been committed.
4. The Oregon Department of Aviation shall establish a registry of [*drones*] unmanned aircraft systems operated by public bodies and may charge a fee sufficient to reimburse the department for the maintenance of the registry.
5. The Oregon Department of Aviation shall require the following information for registration of [*a drone*] an unmanned aircraft system:
    a. The name of the public body that owns or operates the [*drone*] unmanned aircraft system.
    b. The name and contact information of the individuals who operate the [*drone*] unmanned aircraft system.
    c. Identifying information for the [*drone*] unmanned aircraft system as required by the department by rule.
6. A public body that registers one or more [*drones*] unmanned aircraft systems under this section shall provide an annual report to the Oregon Department of Aviation that summarizes:
    a. The frequency of use of the [*drones*] unmanned aircraft systems by the public body during the preceding calendar year; and
    b. The purposes for which the [*drones*] unmanned aircraft systems have been used by the public body during the preceding calendar year.

7. The State Aviation Board may adopt all rules necessary for the registration of [*drones*] unmanned aircraft systems in Oregon that are consistent with federal laws and regulations.

SECTION 9. ORS 837.365 is amended to read:
837.365. A public body may not operate [*a drone*] an unmanned aircraft system that is capable of firing a bullet or other projectile, directing a laser or otherwise being used as a weapon.

SECTION 10. ORS 837.375 is amended to read:
837.375. In addition to any other remedies allowed by law, a person who intentionally interferes with, or gains unauthorized control over, [*a drone*] an unmanned aircraft system licensed by the Federal Aviation Administration, or operated by the Armed Forces of the United States as defined in ORS 351.642, an agency of the United States or a federal, state or local law enforcement agency, is liable to the owner of the [*drone*] unmanned aircraft system in an amount of not less than $5,000. The court shall award reasonable attorney fees to a prevailing plaintiff in an action under this section.

SECTION 11. ORS 837.380 is amended to read:

1. 837.380. (1) Except as provided in subsection (2) of this section, a person who owns or lawfully occupies real property in this state may bring an action against any person or public body that operates [*a drone*] an unmanned aircraft system that is flown [*at a height of less than 400 feet*] over the property if:
    a. The operator of the [*drone*] unmanned aircraft system has flown the [*drone*] unmanned aircraft system over the property [*at a height of less than 400 feet*] on at least one previous occasion; and
    b. The person notified the owner or operator of the [*drone*] unmanned aircraft system that the person did not want the [*drone*] unmanned aircraft system flown over the property [*at a height of less than 400 feet*].
2. A person may not bring an action under this section if:
    a. The [*drone*] unmanned aircraft system is lawfully in the flight path for landing at an airport, airfield or runway; and
    b. The [*drone*] unmanned aircraft system is in the process of taking off or landing.
3. A prevailing plaintiff may recover treble damages for any injury to the person or the property by reason of a trespass by [*a drone*] an unmanned aircraft system as described in this section, and may be awarded injunctive relief in the action.
4. A prevailing plaintiff may recover attorney fees under ORS 20.080 if the amount pleaded in an action under this section is $10,000 or less.

5. The Attorney General, on behalf of the State of Oregon, may bring an action or claim for relief alleging nuisance or trespass arising from the operation of [*a drone*] an unmanned aircraft system in the airspace over this state. A court shall award reasonable attorney fees to the Attorney General if the Attorney General prevails in an action under this section.

SECTION 12. ORS 837.385 is amended to read:
837.385. Except as expressly authorized by state statute, the authority to regulate the ownership or operation of [*drones*] unmanned aircraft systems is vested solely in the Legislative Assembly. Except as expressly authorized by state statute, a local government, as defined ORS 174.116, may not enact an ordinance or resolution that regulates the ownership or operation of [*drones*] unmanned aircraft systems or otherwise engage in the regulation of the ownership or operation of [*drones*] unmanned aircraft systems.

SECTION 13. ORS 837.995 is amended to read:

1. 837.995. (1) A person commits a Class A felony if the person possesses or controls [*a drone*] an unmanned aircraft system and intentionally causes, or attempts to cause, the [*drone*] unmanned aircraft system to:
    a. Fire a bullet or other projectile at an aircraft while the aircraft is in the air;
    b. Direct a laser at an aircraft while the aircraft is in the air; or
    c. Crash into an aircraft while the aircraft is in the air.
2. A person who intentionally interferes with, or gains unauthorized control over, [*a drone*] an unmanned aircraft system licensed by the Federal Aviation Administration, or operated by the Armed Forces of the United States as defined in ORS 351.642, an agency of the United States or a federal, state or local law enforcement agency, commits a Class C felony.[360]

PENNSYLVANIA

On March 21, 2013, the Pennsylvania House adopted Resolution (HR) 172 beseeching the US DoD to reconsider the order of precedence for the newly created Distinguished Warfare Medal intended to recognize military achievement in cyberwarfare or combat drone operations for engagements that do not include valor in live combat.[361]

Additionally, Pennsylvania has introduced HB1684, an act amending Title 18 of the Pennsylvania Consolidated Statutes, in sports and amusements, prohibiting the use of

[360]HB 2354, Oregon Legislative Assembly, https://olis.leg.state.or.us/liz/2015R1/Downloads/MeasureDocument/HB2354/Enrolled.
[361]HR 172, Pennsylvania General Assembly, https://legiscan.com/PA/rollcall/HR172/id/252258.

drones near stadiums;[362] HB 409, an act amending Title 18 of the Pennsylvania Consolidated Statutes, in wiretapping and electronic surveillance, providing for unmanned aircraft surveillance;[363] HR 295, a resolution urging the FAA to promulgate regulations and rules to protect Americans from the use of drones capable of inflicting harm or invading privacy;[364] SB 971, an act regulating unmanned aircraft in Pennsylvania;[365] and SR 238, a resolution directing the Joint Government Commission to conduct a study regarding the use of unmanned drones by state and local agencies, including law enforcement agencies.[366]

Furthermore, Pittsburgh adopted Ordinance No. 2053, amending and supplementing the City of Pittsburgh Code, Title 4, Article XI, by amending Chapter 473 with respect to the operation of drones and other activities involving airborne objects or persons located within, above or in the vicinity of City parks and playgrounds,[367] and East Goshen Township adopted Ordinance 129-B-2015 prohibiting the operation of a model aircraft, amateur rocket, or other flying objects of a similar nature at an elevation of less than 200 feet over property not owned by the operator without the permission of the property owner, and other rules.[368]

House Resolution (HR) 172 reads as follows:

Beseeching the United States Department of Defense to reconsider the order of precedence for the newly created Distinguished Warfare Medal.

WHEREAS, In February of 2013, the United States Department of Defense introduced a new medal to be known as the Distinguished Warfare Medal; and

WHEREAS, The medal is intended to recognize military achievement in cyberwarfare or combat drone operations for engagements that do not include valor in live combat; and

WHEREAS, The new medal is the first combat-related award to be created since the Bronze Star in 1944; and

WHEREAS, The United States Department of Defense has indicated that the new medal is to be ranked higher than the Bronze Star, the Purple Heart and other medals awarded to those who have participated in active combat; and

WHEREAS, The Bronze Star is the fourth highest combat decoration and rewards meritorious service in battle; and

WHEREAS, The Pentagon has indicated that the new medal gives the military a way to recognize a single act that directly affects a combat operation, does not involve an act of valor and warrants an award higher than the Bronze Star; and

---

[362] HB 1684, Pennsylvania General Assembly, http://www.legis.state.pa.us/cfdocs/billInfo/billInfo.cfm?sYear=2015&sInd=0&body=H&type=B&bn=1684.

[363] HB 409, Pennsylvania General Assembly, http://openstates.org/pa/bills/2015-2016/HB409/.

[364] HR 295, Pennsylvania General Assembly, http://openstates.org/pa/bills/2015-2016/HR295/.

[365] SB 971, Pennsylvania General Assembly, http://www.legis.state.pa.us/CFDOCS/billInfo/billInfo.cfm?syear=2015&sInd=0&body=S&type=B&bn=971.

[366] SR 238, Pennsylvania General Assembly, http://www.legis.state.pa.us/cfdocs/billInfo/billInfo.cfm?sYear=2015&sInd=0&body=s&type=r&bn=238.

[367] Ordinance No. 2053, City of Pittsburgh, https://pittsburgh.legistar.com/LegislationDetail.aspx?ID=2471542&GUID=CE412BC6-C074-4249-BEEC-EAB2EAC508F2&Options=ID%7cText%7c&Search=drone.

[368] Ordinance No. 129-B-2015, East Goshen Township, http://eastgoshen.org/wp-content/uploads/2014/08/Drone-Ordinance1.pdf.

WHEREAS, The Pentagon has also indicated that the newly created medal will reflect the changing nature of warfare in the 21st Century; and
WHEREAS, The decision to rank the Distinguished Warfare Medal higher than the Bronze Star, including a Bronze Star with "V" Device, which honors heroic conduct in combat, has been condemned by the Veterans of Foreign Wars and the Military Order of the Purple Heart; and
WHEREAS, Veterans and lawmakers across the country have complained that troops can get the new medal for actions taken far removed from the combat theater and that the ranking of the new medal above those medals awarded to individuals who served on the front line in harm's way, is not justifiable; and
WHEREAS, Because of the growing backlash that has ensued relating to the new medal, the Pentagon has announced that production of the new medal for remote warfare troops has been stopped while the Pentagon considers complaints from veterans and lawmakers, including United States Congressmen from across the country over the ranking of the new medal; and
WHEREAS, The Secretary of Defense has ordered the Chairman of the Joint Chiefs of Staff to review how the medal is ranked among the others and report back to the Secretary of Defense in 30 days; therefore be it
RESOLVED, That the House of Representatives beseech the United States Department of Defense to reconsider the order of precedence of the newly created Distinguished Warfare Medal; and be it further
RESOLVED, That a copy of this resolution be transmitted to the President of the United States, the Secretary of Defense, the presiding officers of each House of Congress and to each member of Congress from Pennsylvania.[369]

RHODE ISLAND

Rhode Island has enacted HB 7511.[370] Additionally, Rhode Island has introduced HR 7022 to reinstate and extend the Commission life and amend the membership of the special legislative commission to study and review regulation of drones and unmanned aerial vehicles,[371] and HR 5293 creating a special legislative commission to study and review regulation of drones and unmanned aerial vehicles.[372]

House Bill (HB) 7511: On July 1, 2016, HB 7511 went into effect, without the governor's signature. This law grants exclusive authority subject to federal law, to the state of Rhode

[369]HR 172, Pennsylvania General Assembly, https://legiscan.com/PA/rollcall/HR172/id/252258.
[370]HB 7511, Rhode Island General Assembly, http://webserver.rilin.state.ri.us/BillText/BillText16/HouseText16/H7511B.pdf.
[371]HR 7022, Rhode Island General Assembly, http://openstates.org/ri/bills/2016/HR7022/.
[372]HR 5293, Rhode Island General Assembly, http://openstates.org/ri/bills/2015/HR5293/.

Island and the Rhode Island Airport Corporation to regulate unpiloted aerial vehicles, remotely piloted aircraft, drones, and unmanned aircraft systems.[373]

SOUTH CAROLINA

South Carolina has introduced HB 3510, a bill to amend the code of laws of South Carolina by adding section 17-13-180 so as to define necessary terms, prohibit the operation of a public unmanned aircraft system and the disclosure of personal information acquired through the operation of a public unmanned aircraft system, to provide penalties for violations, and to provide exceptions;[374] HB 4421 to provide that it is unlawful to operate a UAV, an unpiloted aerial vehicle, or a remotely piloted aerial vehicle in an illegal airspace, to provide that the owner of these vehicles must register them with the Aeronautics Commission, and to provide penalties;[375] HB 4425 to provide that it is unlawful to operate an unpiloted aerial vehicle that is armed with a weapon, and to provide a penalty;[376] HCR 4426 to create the study committee on the use of drones as a law enforcement crime fighting tool, to provide for the membership and staffing of the study committee, and to provide for the study committee's termination;[377] and SB 498 relating to trespasses and unlawful use of land of others by willfully causing or allowing certain aerial vehicles, commonly referred to as drones, to enter upon or above the land of another, to provide for penalties for violations, and exemptions, among other rules.[378]

SOUTH DAKOTA

Aberdeen passed Ordinance No 15-12-04 regulating the use of unmanned aircraft within city limits, effective March 30, 2016.[379]

TENNESSEE

Tennessee has passed six bills, namely, HB 2376,[380] SB 2106,[381] SB 796,[382] SB 1777,[383] SB 1892,[384] and HB 153.[385] Additionally, Tennessee has introduced HB 1191 to prohibit

---

[373]HB 7511, Rhode Island General Assembly, http://webserver.rilin.state.ri.us/BillText/BillText16/HouseText16/H7511B.pdf.

[374]HB 3510, South Carolina Legislature, https://legiscan.com/SC/research/H3510/2015.

[375]HB 4421, South Carolina Legislature, https://legiscan.com/SC/bill/H4421/2015.

[376]HB 4425, South Carolina Legislature, https://legiscan.com/SC/bill/H4425/2015.

[377]HCR 4426, South Carolina Legislature, https://legiscan.com/SC/bill/H4426/2015.

[378]SB 498, South Carolina Legislature, https://legiscan.com/SC/bill/S0498/2015.

[379]City of Aberdeen, Ordinance No. 15-12-04, http://www.aberdeen.sd.us/documentcenter/view/4249.

[380]HB 2376, Tennessee General Assembly, https://legiscan.com/TN/bill/HB2376/2015.

[381]SB 2106, Tennessee General Assembly, https://legiscan.com/TN/bill/SB2106/2015.

[382]SB 796, Tennessee General Assembly, http://www.capitol.tn.gov/Bills/108/Bill/SB0796.pdf.

[383]SB 1777, Tennessee General Assembly, http://tls.mleesmith.com/downloads/2013_2014/bills/SB1777.pdf

[384]SB 1892, Tennessee General Assembly, https://legiscan.com/TN/text/SB1892/id/931596/Tennessee-2013-SB1892-Draft.pdf.

[385]HB 153, Tennessee General Assembly, https://legiscan.com/TN/text/HB0153/id/1093787/Tennessee-2015-HB0153-Draft.pdf.

state and local law enforcement agencies from owning or using certain military vehicles, aircraft, and weaponry, and requiring the divestiture or destruction of such vehicles, aircraft, or weaponry;[386] HB 1456 creating Class E felony of attaching a weapon to an unmanned aircraft;[387] HB 1564 to redefine "autonomous technology" for purposes of provision whereby no political subdivision may by ordinance, resolution, or any other means prohibit within the jurisdictional boundaries of the political subdivision the use of a motor vehicle equipped with autonomous technology if the motor vehicle otherwise complies with all safety regulations of the political subdivision;[388] HB 1811 to create the offense of using an unmanned aircraft to fly within 250 feet of a critical infrastructure facility for the purpose of conducting surveillance of, gathering evidence or collecting information about, or photographically or electronically recording, critical infrastructure data;[389] HB 759 to specify that only a duly registered land surveyor may lawfully use an unmanned aircraft for land surveying;[390] SB 1561 redefining "autonomous technology" among other rules;[391] SB 2470 adding certain situations in which the use of unmanned aircraft to capture an image is allowed;[392] SB 2634 to direct the commissioner of transportation, in consultation with the commissioner of safety, to petition the FAA for a certificate of authority to operate UAS within the navigable airspace of each of the grand divisions of the state;[393] and SB 709 to specify that only a duly registered land surveyor may lawfully use an unmanned aircraft for land surveying.[394]

House Bill (HB) 2376: This bill, passed on May 2, 2016, amends Tennessee Code Annotated (TCA) Title 39, Chapter 13. The law adds certain situations in which the use of an unmanned aircraft to capture an image is allowed.[395]

Senate Bill (SB) 2106: This bill, passed April 26, 2016, amends TCA Title 39 and Title 40. This law creates the offense of using an unmanned aircraft to fly within 250 feet of a critical infrastructure facility for the purpose of conducting surveillance of, gathering evidence or collecting information about, or photographically or electronically recording, critical infrastructure data.[396]

Senate Bill (SB) 796: On May 20, 2013, the Tennessee General Assembly passed into law SB 796, which was an act to amend Tennessee Code Annotated, Title 29 and Title 39, relative to surveillance.

Be it enacted by the General Assembly of the State of Tennessee:

---

[386]HB 1191, Tennessee General Assembly, https://legiscan.com/TN/research/HB1191/2015.
[387]HB 1456, Tennessee General Assembly, https://legiscan.com/TN/bill/HB1456/2015.
[388]HB 1564, Tennessee General Assembly, https://legiscan.com/TN/bill/HB1564/2015.
[389]HB 1811, Tennessee General Assembly, https://legiscan.com/TN/bill/HB1811/2015.
[390]HB 759, Tennessee General Assembly, https://legiscan.com/TN/bill/HB0759/2015.
[391]SB 1561, Tennessee General Assembly, https://legiscan.com/TN/bill/SB1561/2015.
[392]SB 2470, Tennessee General Assembly, https://legiscan.com/TN/bill/SB2470/2015.
[393]SB 2634, Tennessee General Assembly, https://legiscan.com/TN/bill/SB2634/2015.
[394]SB 709, Tennessee General Assembly, https://legiscan.com/TN/bill/SB0709/2015.
[395]HB 2376, Tennessee General Assembly, https://legiscan.com/TN/bill/HB2376/2015.
[396]SB 2106, Tennessee General Assembly, https://legiscan.com/TN/bill/SB2106/2015.

SECTION 1. Tennessee Code Annotated, Title 39, Chapter 13, Part 6, is amended by adding the following language as a new section:

a. This section shall be known and may be cited as the "Freedom from Unwarranted Surveillance Act".
b. As used in this section:
    1. "Drone" means a powered, aerial vehicle that:
        A. Does not carry a human operator;
        B. Uses aerodynamic forces to provide vehicle lift;
        C. Can fly autonomously or be piloted remotely;
        D. Can be expendable or recoverable; and
        E. Can carry a lethal or nonlethal payload; and
    2. "Law enforcement agency" means a lawfully established state or local public agency that is responsible for the prevention and detection of crime, local government code enforcement, and the enforcement of penal, traffic, regulatory, game, or controlled substance laws.
c. Notwithstanding any law to the contrary, no law enforcement agency shall use a drone to gather evidence or other information.
d. This section shall not prohibit the use of a drone:
    1. To counter a high risk of a terrorist attack by a specific individual or organization if the United States secretary of homeland security determines that credible intelligence indicates that there is such a risk;
    2. If the law enforcement agency first obtains a search warrant signed by a judge authorizing the use of a drone; or
    3. If the law enforcement agency possesses reasonable suspicion that, under particular circumstances, swift action is needed to prevent imminent danger to life.
e. An aggrieved party may initiate a civil action against a law enforcement agency to obtain all appropriate relief, as determined by the court, in order to prevent or remedy a violation of this section.
f. Evidence obtained or collected in violation of this section shall not be admissible as evidence in a criminal prosecution in any court of law in this state. SECTION 2. This act shall take effect July 1, 2013, the public welfare requiring it.[397]

Senate Bill (SB) 1777: On April 4, 2014, the Governor of Tennessee signed into law SB 1777. This is an act to amend Tennessee Code Annotated, Title 70, Chapter 4, Part 3, relative to hunter protection. The act reads as follows:

---

[397]SB 796, TENNESSEE GENERAL ASSEMBLY, http://www.capitol.tn.gov/Bills/108/Bill/SB0796.pdf.

Be it enacted by the General Assembly of the State of Tennessee:
SECTION 1. Tennessee Code Annotated, Section 70-4-301, is amended by adding the following language as a new, appropriately designated subdivision:
"Drone" means a drone as defined in § 39-13-609;
SECTION 2. Tennessee Code Annotated, Section 70-4-302, is amended by deleting the section in its entirety and by substituting instead the follow:
70-4-302.

a. Any person who performs any of the following commits a Class C misdemeanor:
    1. Interferes with the lawful taking of a wild animal by another with intent to prevent the taking;
    2. Disturbs or engages in an activity that will tend to disturb wild animals, with intent to prevent their lawful taking;
    3. Disturbs another person who is engaged in the lawful taking of a wild animal or who is engaged in the process of taking, with intent to dissuade or otherwise prevent the taking;
    4. Enters or remains upon public lands, or upon private lands without permission of the owner or the owner's agent, with intent to violate this section;
    5. Fails to obey the order of a peace officer to desist from conduct in violation of this section if the officer observes such conduct, or has reasonable grounds to believe that the person has engaged in such conduct that day or that the person plans or intends to engage in such conduct that day on a specific premises; or
    6. Uses a drone to conduct video surveillance of private citizens who are lawfully hunting or fishing.
b. As used in subsection (a), "any person" means any individual, firm, association, company, partnership, corporation, public or private organization, institution or similar entity.

SECTION 3. This act shall take effect July 1, 2014, the public welfare requiring it.[398]
Senate Bill (SB) 1892: On May 16, 2014, SB 1892 was signed into Public Chapter 876. This is an act to amend Tennessee Code Annotated, Title 29 and Title 39, relative to the protection of privacy. The act reads as follows:
Be it enacted by the General Assembly of the State of Tennessee:
SECTION 1. Tennessee Code Annotated, Title 39, Chapter 13, is amended by adding Sections 2 through 8 below as a new part. SECTION 2. As used in this part.

1. "Image" means any capturing of sound waves, thermal, infrared, ultraviolet, visible light, or other electromagnetic waves, odor, or other conditions existing on or about real property in this state or an individual located on that property; and

---

[398]SB 1777, Tennessee General Assembly, http://tls.mleesmith.com/downloads/2013_2014/bills/SB1777.pdf.

2. "Unmanned aircraft" means an airborne device that is operated without an individual in or on the device.

SECTION 3.

a. Notwithstanding Section 4, it is lawful to capture an image using an unmanned aircraft in this state:
   1. For purposes of professional or scholarly research and development by a person acting on behalf of an institution of higher education, as defined by § 49-7-802, including a person who:
      A. Is a professor, employee, or student of the institution; or
      B. Is under contract with or otherwise acting under the direction or on behalf of the institution;
   2. In airspace designated as a test site or range authorized by the federal aviation administration for the purpose of integrating unmanned aircraft systems into the national airspace;
   3. As part of an operation, exercise, or mission of any branch of the United States military, consistent with the Constitution of the United States;
   4. If the image is captured by a satellite for the purposes of mapping; provided, the image of any person or thing on private property captured in the course of mapping shall be subject to section 6 as an image captured incidental to the lawful capturing of an image.
   5. If the image is captured by or for an electric or natural gas utility:
      A. For operations and maintenance of utility facilities for the purpose of maintaining utility system reliability and integrity;
      B. For inspecting utility facilities to determine repair, maintenance, or replacement needs during and after construction of such facilities;
      C. For assessing vegetation growth for the purpose of maintaining clearances on utility easements; or
      D. For utility facility routing and siting for the purpose of providing utility service;
   6. With the consent of the individual who owns or lawfully occupies the real property captured in the image;
   7. For law enforcement purposes, as permitted by § 39-13-609;
   8. If the image is captured by state or local law enforcement authorities, or a person who is under contract with or otherwise acting under the direction or on behalf of state authorities, for the purpose of:
      A. Surveying the scene of a catastrophe or other damage to determine whether a state of emergency should be declared;
      B. Preserving public safety, protecting property, or surveying damage or contamination during a lawfully declared state of emergency; or;

C. Conducting routine air quality sampling and monitoring, as provided by state or local law;

9. At the scene of a spill, or a suspected spill, of hazardous materials;
10. For the purpose of fire suppression;
11. For the purpose of rescuing a person whose life or well-being is in imminent danger;
12. If the image is captured by a Tennessee licensed real estate broker in connection with the marketing, sale, or financing of real property, provided that no individual is identifiable in the image;
13. Of public real property or a person on that property;
14. If the image is captured by the owner or operator of an oil, gas, water, or other pipeline for the purpose of inspecting, maintaining, or repairing pipelines or other related facilities, and is captured without the intent to conduct surveillance on an individual or real property located in this state;
15. In connection with oil and gas pipeline and well safety and protection; or
16. In connection with port authority surveillance and security.

b. An image captured by a state or local government agency, or by a person who is under contract with or otherwise acting under the direction or on behalf of such agency, shall be handled in accordance with § 39-13-609 and shall not be used for any purpose other than the lawful purpose for which the image was captured as permitted by this section.

SECTION 4.

a. Subject to the exceptions set forth in Section 3(a), a person commits an offense if the person:
   1. Uses an unmanned aircraft to capture an image of an individual or privately owned real property in this state with the intent to conduct surveillance on the individual or property captured in the image; or
   2. Knowingly uses an image in a manner prohibited by Section 3(b).

b. An offense under this section is a Class C misdemeanor.

c. It is a defense to prosecution under this section that the person destroyed the image.

SECTION 5.

a. A person commits an offense if the person:
   1. Captures an image in violation of Section 4; and (2)

   A. Possesses that image; or
   B. Discloses, displays, distributes, or otherwise uses that image.

b. An offense under subdivision (a)(2)(A) is a Class C misdemeanor. An offense under subdivision (a)(2)(B) is a Class B misdemeanor.
c. Each image a person possesses, discloses, displays, distributes, or otherwise uses in violation of this section is a separate offense.
d. It is a defense to prosecution under this section for the possession of an image that the person destroyed the image as soon as the person had knowledge that the image was captured in violation of Section 4.
e. It is a defense to prosecution under this section for the disclosure, display, distribution, or other use of an image that the person stopped disclosing, displaying, distributing, or otherwise using the image as soon as the person had knowledge that the image was captured in violation of Section 4.
   1. As soon as the person had knowledge that the image was captured in violation of this section; and
   2. Without disclosing, displaying, or distributing the image to a third party.

SECTION 6.

a. Except as otherwise provided by subsection (b), an image captured in violation of Section 4, or an image captured by an unmanned aircraft that was incidental to the lawful capturing of an image:
   1. May not be used as evidence in any criminal or juvenile proceeding, civil action, or administrative proceeding;
   2. Is not subject to disclosure, inspection, or copying under title 10, chapter 7; and
   3. Is not subject to discovery, subpoena, or other means of legal compulsion for its release. (b) An image described by subsection (a) may be disclosed and used as evidence to prove a violation of this part and is subject to discovery, subpoena, or other means of legal compulsion for that purpose.

SECTION 7.

This part shall not apply to the manufacture, assembly, distribution, or sale of an unmanned aircraft.

SECTION 8.

Nothing in this part shall be construed as permitting any act prohibited by other law.

SECTION 9.

Tennessee Code Annotated, Section 39-14-405, is amended by deleting the period at the end of subsection (d) and adding the following:

or when a person causes an unmanned aircraft to enter that portion of the airspace above the owner's land not regulated as navigable airspace by the federal aviation administration.

SECTION 10.

Tennessee Code Annotated, Title 39, Chapter 13, Part 6, is amended by adding a new, appropriately designated section as follows:

a. This section establishes the offenses of surreptitious commercial surveillance and aggravated surreptitious commercial surveillance.
b. As used in this section:
    1. "Commercial surveillance" means the taking of electronic or photographic images or audio recording of the premises or operations of an employer in the state; and
    2. "Employer" means a person who operates a business or commercial enterprise of any type or nature;
    3. "Surreptitious" means undertaken in a manner to avoid detection.
c. The offense of surreptitious commercial surveillance is committed when an employee of an employer or an individual on the premises of an employer for the purpose of selling or delivering goods or services to the employer, engages in surreptitious commercial surveillance without the consent of the employer.
d. The offense of aggravated surreptitious commercial surveillance is committed when an individual who has committed surreptitious commercial surveillance delivers, transmits or distributes to another person electronic or photographic images or audio recordings obtained through commercial surveillance under false pretenses.
e. 1. A first violation of subsection (c) is a Class B misdemeanor.
    2. The second and subsequent violation of subsection (c) is a Class A misdemeanor.
    3. A violation of subsection (d) is a Class A misdemeanor.

SECTION 11.

If any provision of this act or the application of any provision of this act to any person or circumstance is held invalid, the invalidity shall not affect other provisions or applications of the act that can be given effect without the invalid provision or application, and to that end, the provisions of this act are declared to be severable.

SECTION 12.

This act shall take effect July 1, 2014, the public welfare requiring it.[399]

House Bill (HB) 153: On April 30, 2015, the Tennessee General Assembly passed *House Bill (HB) 153* into law. This is an act to amend Tennessee Code Annotated, Title 39, relative to criminal offenses. It reads as follows:

Be it enacted by the General Assembly of the State of Tennessee:

SECTION 1. Tennessee Code Annotated, Section 39-13-903(a), is amended by adding a new, appropriately designated subdivision:

Without the venue owner or operator's consent, uses an unmanned aircraft to capture an image of an individual or event at an open-air event venue wherein more than one hundred (100) individuals could gather for an event.

SECTION 2. This act shall take effect July 1, 2015, the public welfare requiring it.[400]

TEXAS

There are seven bills or resolutions in Texas, namely HR 3035,[401] HB 912,[402] SR 1084,[403] HCR 217,[404] HB 3628,[405] HB 2167,[406] and HB 1481.[407]

House Resolution (HR) 3035: On May 29, 2013, the Texas House of Representatives passed HR 3035, which reads as follows:

Suspending limitations on conference committee jurisdiction, HB 912

The committee to omit text not in disagreement in proposed SECTION 2 of the bill, in added Section 423.002(a), Government Code, by omitting proposed Subdivisions (5) and (6), which read as follows:

1. by a manufacturer or distributor of the unmanned vehicle or unmanned aircraft in connection with the development, manufacture, testing, or research of the vehicle or the aircraft;
2. by a manufacturer or distributor incidental to the activities authorized under Subdivision (5) and the manufacturer or distributor discloses the image voluntarily

[399] SB 1892, Tennessee General Assembly, https://legiscan.com/TN/text/SB1892/id/931596/Tennessee-2013-SB1892-Draft.pdf.

[400] HB 153, Tennessee General Assembly, https://legiscan.com/TN/text/HB0153/id/1093787/Tennessee-2015-HB0153-Draft.pdf.

[401] HR 3035, Texas House of Representatives, http://www.legis.state.tx.us/tlodocs/83R/billtext/pdf/HR03035I.pdf.

[402] HB 912, Texas Legislature, http://www.statutes.legis.state.tx.us/Docs/GV/htm/GV.423.htm.

[403] SR 1084, Texas Legislature, http://www.legis.state.tx.us/tlodocs/83R/billtext/pdf/SR01084I.pdf.

[404] HCR 217, Texas Legislature, http://www.legis.state.tx.us/tlodocs/83R/billtext/pdf/HC00217F.pdf.

[405] HB 3628, Texas Legislature, http://www.capitol.state.tx.us/tlodocs/84R/billtext/pdf/HB03628F.pdf.

[406] HB 2167, Texas Legislature, http://www.capitol.state.tx.us/tlodocs/84R/billtext/pdf/HB02167F.pdf.

[407] HB 1481, Texas Legislature, http://www.legis.state.tx.us/tlodocs/84R/billtext/pdf/HB01481F.pdf.

to law enforcement because the manufacturer or distributor reasonably believes the image shows the commission of an offense;

Explanation: The omission of the text is necessary to remove exceptions to the application of the chapter.

1. House Rule 13, Section 9(a)(2), is suspended to permit the committee to omit text not in disagreement in proposed SECTION 2 of the bill, in added Section 423.003(a), Government Code, by omitting proposed Subdivision (2), which reads as follows:
2. real property in this state, on which a primary or secondary school or a licensed childcare facility is operated or an individual located on that property, with the intent to conduct surveillance

Explanation: This change is necessary to remove an offense.

1. House Rule 13, Section 9(a)(4), is suspended to permit the committee to add text on a matter which is not included in either the house or the senate version of the bill by adding the following text to SECTION 2 of the bill:

Sec. 423.008. REPORTING BY LAW ENFORCEMENT AGENCY.

a. No earlier than January 1 and no later than January 15 of each odd-numbered year, a municipal or county law enforcement agency located in a city or county with a population greater than 150,000, or a state law enforcement agency, that used or operated an unmanned aircraft during the preceding 24 months shall issue a written report to each member of the Texas Legislature, including the Governor and Lieutenant Governor, retain the report for public viewing and post the report on the law enforcement agency's publicly accessible website, if one exists. The report shall contain:
b. the number of times an unmanned aircraft was operated, organized by date, time, location, types of incidents, and the types of justification for operation;
c. the number of criminal investigations aided by the operation of an unmanned aircraft and a description of how the unmanned aircraft aided each investigation;
d. the date, location, and number of times an unmanned aircraft was operated for reasons other than a criminal investigation and a description of how the unmanned aircraft aided each operation;
e. the frequency and type of information collected on an individual, residence, property or area that was not the subject of an operation; and
f. the total cost of acquisition, maintenance, repair, and operation the unmanned aircraft or unmanned aircrafts for the preceding 24 months.

Explanation: This addition is necessary to provide for a publicly available report regarding law enforcement use of unmanned aircraft in this state.[408]

House Bill (HB) 912: On September 1, 2013, HB 912 went into effect, which reads as follows:

TITLE 4. EXECUTIVE BRANCH

SUBTITLE B. LAW ENFORCEMENT AND PUBLIC PROTECTION

CHAPTER 423. USE OF UNMANNED AIRCRAFT

Sec. 423.001. DEFINITION.

In this chapter, "image" means any capturing of sound waves, thermal, infrared, ultraviolet, visible light, or other electromagnetic waves, odor, or other conditions existing on or about real property in this state or an individual located on that property.

Added by Acts 2013, 83rd Leg., R.S., Ch. 1390 (H.B. 912), Sec. 2, eff. September 1, 2013.

This section was amended by the 84th Legislature. Pending publication of the current statutes, see H.B. 2167, 84th Legislature, Regular Session, for amendments affecting this section.

Sec. 423.002. NONAPPLICABILITY.

a. It is lawful to capture an image using an unmanned aircraft in this state:
    1. for purposes of professional or scholarly research and development by a person acting on behalf of an institution of higher education, as defined by Section 61.003, Education Code, including a person who:
        A. is a professor, employee, or student of the institution; or
        B. is under contract with or otherwise acting under the direction or on behalf of the institution;
    2. in airspace designated as a test site or range authorized by the Federal Aviation Administration for the purpose of integrating unmanned aircraft systems into the national airspace;
    3. as part of an operation, exercise, or mission of any branch of the United States military;
    4. if the image is captured by a satellite for the purposes of mapping;
    5. if the image is captured by or for an electric or natural gas utility:
        A. for operations and maintenance of utility facilities for the purpose of maintaining utility system reliability and integrity;

[408]HR 3035, Texas House of Representatives, http://www.legis.state.tx.us/tlodocs/83R/billtext/pdf/HR03035I.pdf.

  - B. for inspecting utility facilities to determine repair, maintenance, or replacement needs during and after construction of such facilities;
  - C. for assessing vegetation growth for the purpose of maintaining clearances on utility easements; and
  - D. for utility facility routing and siting for the purpose of providing utility service;
6. with the consent of the individual who owns or lawfully occupies the real property captured in the image;
7. pursuant to a valid search or arrest warrant;
8. if the image is captured by a law enforcement authority or a person who is under contract with or otherwise acting under the direction or on behalf of a law enforcement authority:
   - A. in immediate pursuit of a person law enforcement officers have reasonable suspicion or probable cause to suspect has committed an offense, not including misdemeanors or offenses punishable by a fine only;
   - B. for the purpose of documenting a crime scene where an offense, not including misdemeanors or offenses punishable by a fine only, has been committed;
   - C. for the purpose of investigating the scene of:
     - i. a human fatality;
     - ii. a motor vehicle accident causing death or serious bodily injury to a person; or
     - iii. any motor vehicle accident on a state highway or federal interstate or highway;
   - D. in connection with the search for a missing person;
   - E. for the purpose of conducting a high-risk tactical operation that poses a threat to human life; or
   - F. of private property that is generally open to the public where the property owner consents to law enforcement public safety responsibilities;
9. if the image is captured by state or local law enforcement authorities, or a person who is under contract with or otherwise acting under the direction or on behalf of state authorities, for the purpose of:
   - A. surveying the scene of a catastrophe or other damage to determine whether a state of emergency should be declared;
   - B. preserving public safety, protecting property, or surveying damage or contamination during a lawfully declared state of emergency; or
   - C. conducting routine air quality sampling and monitoring, as provided by state or local law;

10. at the scene of a spill, or a suspected spill, of hazardous materials;
11. for the purpose of fire suppression;
12. for the purpose of rescuing a person whose life or well-being is in imminent danger;
13. if the image is captured by a Texas licensed real estate broker in connection with the marketing, sale, or financing of real property, provided that no individual is identifiable in the image;
14. of real property or a person on real property that is within 25 miles of the United States border;
15. from a height no more than eight feet above ground level in a public place, if the image was captured without using any electronic, mechanical, or other means to amplify the image beyond normal human perception;
16. of public real property or a person on that property;
17. if the image is captured by the owner or operator of an oil, gas, water, or other pipeline for the purpose of inspecting, maintaining, or repairing pipelines or other related facilities, and is captured without the intent to conduct surveillance on an individual or real property located in this state;
18. in connection with oil pipeline safety and rig protection; or
19. in connection with port authority surveillance and security.

b. This chapter does not apply to the manufacture, assembly, distribution, or sale of an unmanned aircraft.

Added by Acts 2013, 83rd Leg., R.S., Ch. 1390 (H.B. 912), Sec. 2, eff. September 1, 2013.

Sec. 423.003. OFFENSE: ILLEGAL USE OF UNMANNED AIRCRAFT TO CAPTURE IMAGE.

a. A person commits an offense if the person uses an unmanned aircraft to capture an image of an individual or privately owned real property in this state with the intent to conduct surveillance on the individual or property captured in the image.

b. An offense under this section is a Class C misdemeanor.

c. It is a defense to prosecution under this section that the person destroyed the image:
   1. as soon as the person had knowledge that the image was captured in violation of this section; and
   2. without disclosing, displaying, or distributing the image to a third party.

d. In this section, "intent" has the meaning assigned by Section 6.03, Penal Code.

Added by Acts 2013, 83rd Leg., R.S., Ch. 1390 (H.B. 912), Sec. 2, eff. September 1, 2013.

Sec. 423.004. OFFENSE: POSSESSION, DISCLOSURE, DISPLAY, DISTRIBUTION, OR USE OF IMAGE.

a. A person commits an offense if the person:
   1. captures an image in violation of Section 423.003; and
   2. possesses, discloses, displays, distributes, or otherwise uses that image.

b. An offense under this section for the possession of an image is a Class C misdemeanor. An offense under this section for the disclosure, display, distribution, or other use of an image is a Class B misdemeanor.

c. Each image a person possesses, discloses, displays, distributes, or otherwise uses in violation of this section is a separate offense.

d. It is a defense to prosecution under this section for the possession of an image that the person destroyed the image as soon as the person had knowledge that the image was captured in violation of Section 423.003.

e. It is a defense to prosecution under this section for the disclosure, display, distribution, or other use of an image that the person stopped disclosing, displaying, distributing, or otherwise using the image as soon as the person had knowledge that the image was captured in violation of Section 423.003.

Added by Acts 2013, 83rd Leg., R.S., Ch. 1390 (H.B. 912), Sec. 2, eff. September 1, 2013.

Sec. 423.005. ILLEGALLY OR INCIDENTALLY CAPTURED IMAGES NOT SUBJECT TO DISCLOSURE.

a. Except as otherwise provided by Subsection (b), an image captured in violation of Section 423.003, or an image captured by an unmanned aircraft that was incidental to the lawful capturing of an image:
   1. may not be used as evidence in any criminal or juvenile proceeding, civil action, or administrative proceeding;
   2. is not subject to disclosure, inspection, or copying under Chapter 552; and
   3. is not subject to discovery, subpoena, or other means of legal compulsion for its release.

b. An image described by Subsection (a) may be disclosed and used as evidence to prove a violation of this chapter and is subject to discovery, subpoena, or other means of legal compulsion for that purpose.

Added by Acts 2013, 83rd Leg., R.S., Ch. 1390 (H.B. 912), Sec. 2, eff. September 1, 2013.

Sec. 423.006. CIVIL ACTION.

a. An owner or tenant of privately owned real property located in this state may bring against a person who, in violation of Section 423.003, captured an image of the property or the owner or tenant while on the property an action to:

1. enjoin a violation or imminent violation of Section 423.003 or 423.004;
2. recover a civil penalty of:
   A. $5,000 for all images captured in a single episode in violation of Section 423.003; or
   B. $10,000 for disclosure, display, distribution, or other use of any images captured in a single episode in violation of Section 423.004; or
3. recover actual damages if the person who captured the image in violation of Section 423.003 discloses, displays, or distributes the image with malice.

b. For purposes of recovering the civil penalty or actual damages under Subsection all owners of a parcel of real property are considered to be a single owner and all tenants of a parcel of real property are considered to be a single tenant.

c. In this section, "malice" has the meaning assigned by Section 41.001, Civil Practice and Remedies Code.

d. In addition to any civil penalties authorized under this section, the court shall award court costs and reasonable attorney's fees to the prevailing party.

e. Venue for an action under this section is governed by Chapter 15, Civil Practice and Remedies Code.

f. An action brought under this section must be commenced within two years from the date the image was:
   1. captured in violation of Section 423.003; or
   2. initially disclosed, displayed, distributed, or otherwise used in violation of Section 423.004.

Added by Acts 2013, 83rd Leg., R.S., Ch. 1390 (H.B. 912), Sec. 2, eff. September 1, 2013.

Sec. 423.007. RULES FOR USE BY LAW ENFORCEMENT.

The Department of Public Safety shall adopt rules and guidelines for use of an unmanned aircraft by a law enforcement authority in this state.

Added by Acts 2013, 83rd Leg., R.S., Ch. 1390 (H.B. 912), Sec. 2, eff. September 1, 2013.

Sec. 423.008. REPORTING BY LAW ENFORCEMENT AGENCY.

a. Not earlier than January 1 and not later than January 15 of each odd-numbered year, each state law enforcement agency and each county or municipal law enforcement agency located in a county or municipality, as applicable, with a population greater than 150,000, that used or operated an unmanned aircraft during the preceding 24 months shall issue a written report to the governor, the lieutenant governor, and each member of the legislature and shall:
   1. retain the report for public viewing; and

   2. post the report on the law enforcement agency's publicly accessible website, if one exists.
b. The report must include:
   1. the number of times an unmanned aircraft was used, organized by date, time, location, and the types of incidents and types of justification for the use;
   2. the number of criminal investigations aided by the use of an unmanned aircraft and a description of how the unmanned aircraft aided each investigation;
   3. the number of times an unmanned aircraft was used for a law enforcement operation other than a criminal investigation, the dates and locations of those operations, and a description of how the unmanned aircraft aided each operation;
   4. the type of information collected on an individual, residence, property, or area that was not the subject of a law enforcement operation and the frequency of the collection of this information; and
   5. he total cost of acquiring, maintaining, repairing, and operating or otherwise using each unmanned aircraft for the preceding 24 months.

Added by Acts 2013, 83rd Leg., R.S., Ch. 1390 (H.B. 912), Sec. 2, eff. September 1, 2013.[409]

Senate Resolution (SR) 1084: On May 26, 2013, the Texas Senate passed SR 1084, which reads as follows:

S.R. No. 1084 committee to omit text not in disagreement in proposed SECTION 2 of the bill, in added Section 423.002(a), Government Code, by omitting proposed Subdivisions (5) and (6), which read as follows:

1. by a manufacturer or distributor of the unmanned vehicle or unmanned aircraft in connection with the development, manufacture, testing, or research of the vehicle or the aircraft;
2. by a manufacturer or distributor incidental to the activities authorized under Subdivision (5) and the manufacturer or distributor discloses the image voluntarily to law enforcement because the manufacturer or distributor reasonably believes the image shows the commission of an offense;

Explanation: The omission of the text is necessary to remove exceptions to the application of the chapter.

1. Senate Rule 12.03(2) is suspended to permit the committee to omit text not in disagreement in proposed SECTION 2 of the bill, in added Section 423.003(a), Government Code, by omitting proposed Subdivision (2), which reads as follows:

---

[409]HB 912, Texas Legislature, http://www.statutes.legis.state.tx.us/Docs/GV/htm/GV.423.htm.

2. real property in this state, on which a primary or secondary school or a licensed childcare facility is operated or an individual located on that property, with the intent to conduct surveillance

Explanation: This change is necessary to remove an offense.

1. Senate Rule 12.03(4) is suspended to permit the committee to add text on a matter which is not included in either the house or the senate version of the bill by adding the following text to SECTION 2 of the bill:

Sec. 423.008. REPORTING BY LAW ENFORCEMENT AGENCY.

a. No earlier than January 1 and no later than January 15 of each odd-numbered year, a municipal or county law enforcement agency located in a city or county with a population greater than 150,000, or a state law enforcement agency, that used or operated an unmanned aircraft during the preceding 24 months shall issue a written report to each member of the Texas Legislature, including the Governor and Lieutenant Governor, retain the report for public viewing and post the report on the law enforcement agency's publicly accessible website, if one exists. The report shall contain:
    a. the number of times an unmanned aircraft was operated, organized by date, time, location, types of incidents, and the types of justification for operation;
    b. the number of criminal investigations aided by the operation of an unmanned aircraft and a description of how the unmanned aircraft aided each investigation;
    c. the date, location, and number of times an unmanned aircraft was operated for reasons other than a criminal investigation and a description of how the unmanned aircraft aided each operation;
    d. the frequency and type of information collected on an individual, residence, property or area that was not the subject of an operation; and
    e. the total cost of acquisition, maintenance, repair, and operation the unmanned aircraft or unmanned aircrafts for the preceding 24 months.

Explanation: This addition is necessary to provide for a publicly available report regarding law enforcement use of unmanned aircraft in this state.[410]

House Concurrent Resolution (HCR) 217: On June 14, 2013, the Texas House of Representatives passed HCR 217, which reads as follows:

WHEREAS, House Bill No. 912 has been adopted by the house of representatives and the senate and is being prepared for enrollment; and

[410]SR 1084, Texas Legislature, http://www.legis.state.tx.us/tlodocs/83R/billtext/pdf/SR01084Lpdf.

WHEREAS, The bill contains technical errors that should be corrected; now, therefore, be it RESOLVED by the 83rd Legislature of the State of Texas, That the enrolling clerk of the house of representatives be instructed to correct House Bill No. 912, in SECTION 2 of the bill, as follows:

1. In added Section 423.002(a), Government Code, strike "It is lawful to capture an image by an unmanned aircraft" and substitute "It is lawful to capture an image using an unmanned aircraft".
2. In added Section 423.002(a)(4), Government Code, between "(4)" and "by", insert "if the image is captured".
3. In added Section 423.002(a)(5), Government Code, between "(5)" and "by", insert "if the image is captured".
4. In added Section 423.002(a)(8), Government Code, between "(8)" and "by", insert "if the image is captured".
5. In added Section 423.002(a)(9), Government Code, between "(9)" and "by", insert "if the image is captured".
6. In added Section 423.002(a)(13), Government Code, between "(13)" and "by", insert "if the image is captured".
7. In added Section 423.002(a)(17), Government Code, between "(17)" and "by", insert "if the image is captured".
8. In added Section 423.002(a)(17), Government Code, immediately following "related facilities," strike "if the image" and substitute "and".
9. In added Section 423.002(a)(18), Government Code, following the underlined semicolon, insert "or".
10. In added Section 423.002(a)(19), Government Code, strike the underlined semicolon and substitute an underlined period.
11. Strike added Section 423.008, Government Code, and substitute the following:

Sec.A423.008.AAREPORTING BY LAW ENFORCEMENT AGENCY.

a. Not earlier than January 1 and not later than January 15 of each odd-numbered year, each state law enforcement agency and each county or municipal law enforcement agency located in a county or municipality, as applicable, with a population greater than 150,000, that used or operated an unmanned aircraft during the preceding 24 months shall issue a written report to the governor, the lieutenant governor, and each member of the legislature and shall:
    1. retain the report for public viewing; and
    2. post the report on the law enforcement agency's publicly accessible website, if one exists.

b. The report must include:
   1. the number of times an unmanned aircraft was used, organized by date, time, location, and the types of incidents and types of justification for the use;
   2. the number of criminal investigations aided by the use of an unmanned aircraft and a description of how the unmanned aircraft aided each investigation;
   3. the number of times an unmanned aircraft was used for a law enforcement operation other than a criminal investigation, the dates and locations of those operations, and a description of how the unmanned aircraft aided each operation;
   4. the type of information collected on an individual, residence, property, or area that was not the subject of a law enforcement operation and the frequency of the collection of this information; and
   5. the total cost of acquiring, maintaining, repairing, and operating or otherwise using each unmanned aircraft for the preceding 24 months.[411]

House Bill (HB) 3628: On May 28, 2015, the Texas House of Representatives passed HB 3628, which reads as follows:

An act relating to the adoption by the Department of Public Safety of rules governing the use of unmanned aircraft in the Capitol Complex; creating a criminal offense.
Be it enacted by the Legislature of the State of Texas:
SECTION 1. Section 411.062, Government Code, is amended by adding Subsection (d-1) to read as follows:

(d-1) The director shall adopt rules governing the use of unmanned aircraft in the Capitol Complex. The rules adopted under this subsection may:

1. prohibit the use of unmanned aircraft in the Capitol Complex; or
2. authorize limited use of unmanned aircraft in the Capitol Complex.

SECTION 2. Section 411.065(b), Government Code, is amended to read as follows:

a. An offense under this section is a Class C misdemeanor, except that an offense is a Class B misdemeanor if the person violates a rule adopted under Section 411.062 (d-1).

SECTION 3.AA

a. The director of the Department of Public Safety shall adopt the rules required by Section 411.062 (d-1), Government Code, as added by this Act, not later than December 1, 2015.

---

[411]HCR 217, Texas Legislature, http://www.legis.state.tx.us/tlodocs/83R/billtext/pdf/HC00217F.pdf.

b. The change in law made by this Act to Section 411.065, Government Code, applies only to an offense committed on or after December 1, 2015. An offense committed before that date is governed by the law in effect immediately before the effective date of this Act, and the former law is continued in effect for that purpose. For purposes of this section, an offense was committed before December 1, 2015, if any element of the offense occurred before that date.

SECTION 4. This Act takes effect September 1, 2015.[412]

House Bill (HB) 2167: On June 9, 2015, HB 2167 became law. It reads as follows:

An act relating to certain images captured by an unmanned aircraft.

Be it enacted by the Legislature of the State of Texas:

SECTION 1. Section 423.002 (a), Government Code, is amended to read as follows:

a. It is lawful to capture an image using an unmanned aircraft in this state:
   1. for the purpose of professional or scholarly research and development or for another academic purpose by a person acting on behalf of an institution of higher education or a private or independent institution of higher education, as those terms are defined by Section 61.003, Education Code, including a person who:
      A. is a professor, employee, or student of the institution; or
      B. is under contract with or otherwise acting under the direction or on behalf of the institution;
   2. in airspace designated as a test site or range authorized by the Federal Aviation Administration for the purpose of integrating unmanned aircraft systems into the national airspace;
   3. as part of an operation, exercise, or mission of any branch of the United States military;
   4. if the image is captured by a satellite for the purposes of mapping;
   5. if the image is captured by or for an electric or natural gas utility:
      A. for operations and maintenance of utility facilities for the purpose of maintaining utility system reliability and integrity;
      B. for inspecting utility facilities to determine repair, maintenance, or replacement needs during and after construction of such facilities;
      C. for assessing vegetation growth for the purpose of maintaining clearances on utility easements; and

[412] HB 3628, Texas Legislature, http://www.capitol.state.tx.us/tlodocs/84R/billtext/pdf/HB03628F.pdf.

   D. for utility facility routing and siting for the purpose of providing utility service;
6. with the consent of the individual who owns or lawfully occupies the real property captured in the image;
7. pursuant to a valid search or arrest warrant;
8. if the image is captured by a law enforcement authority or a person who is under contract with or otherwise acting under the direction or on behalf of a law enforcement authority:
   A. in immediate pursuit of a person law enforcement officers have reasonable suspicion or probable cause to suspect has committed an offense, not including misdemeanors or offenses punishable by a fine only;
   B. for the purpose of documenting a crime scene where an offense, not including misdemeanors or offenses punishable by a fine only, has been committed;
   C. for the purpose of investigating the scene of:
      i. a human fatality;
      ii. a motor vehicle accident causing death or serious bodily injury to a person; or
      iii. any motor vehicle accident on a state highway or federal interstate or highway;
   D. in connection with the search for a missing person;
   E. for the purpose of conducting a high-risk tactical operation that poses a threat to human life; or
   F. of private property that is generally open to the public where the property owner consents to law enforcement public safety responsibilities;
9. if the image is captured by state or local law enforcement authorities, or a person who is under contract with or otherwise acting under the direction or on behalf of state authorities, for the purpose of:
   A. surveying the scene of a catastrophe or other damage to determine whether a state of emergency should be declared;
   B. preserving public safety, protecting property, or surveying damage or contamination during a lawfully declared state of emergency; or
   C. conducting routine air quality sampling and monitoring, as provided by state or local law;
10. at the scene of a spill, or a suspected spill, of hazardous materials;
11. for the purpose of fire suppression;
12. for the purpose of rescuing a person whose life or well-being is in imminent danger;

13. if the image is captured by a Texas licensed real estate broker in connection with the marketing, sale, or financing of real property, provided that no individual is identifiable in the image;
14. of real property or a person on real property that is within 25 miles of the United States border;
15. from a height no more than eight feet above ground level in a public place, if the image was captured without using any electronic, mechanical, or other means to amplify the image beyond normal human perception;
16. of public real property or a person on that property;
17. if the image is captured by the owner or operator of an oil, gas, water, or other pipeline for the purpose of inspecting, maintaining, or repairing pipelines or other related facilities, and is captured without the intent to conduct surveillance on an individual or real property located in this state;
18. in connection with oil pipeline safety and rig protection;
19. in connection with port authority surveillance and security;
20. if the image is captured by a registered professional land surveyor in connection with the practice of professional surveying, as those terms are defined by Section 1071.002, Occupations Code, provided that no individual is identifiable in the image; or
21. if the image is captured by a professional engineer licensed under Subchapter G, Chapter 1001, Occupations Code, in connection with the practice of engineering, as defined by Section 1001.003, Occupations Code, provided that no individual is identifiable in the image.

SECTION 2.

The change in law made by this Act applies only to an offense committed on or after the effective date of this Act. An offense committed before the effective date of this Act is governed by the law in effect on the date the offense was committed, and the former law is continued in effect for that purpose. For purposes of this section, an offense was committed before the effective date of this Act if any element of the offense occurred before that date. SECTION 3.AA

This Act takes effect September 1, 2015.[413]

House Bill (HB) 1481: On June 19, 2015, the Texas House of Representatives passed HB 1481, which becomes effective September 1, 2015, and reads as follows:

An act relating to prohibiting the operation of an unmanned aircraft over certain facilities; creating a criminal offense.

Be it enacted by the Legislature of the State of Texas:

---

[413]HB 2167, Texas Legislature, http://www.capitol.state.tx.us/tlodocs/84R/billtext/pdf/HB02167F.pdf.

SECTION 1. Chapter 423, Government Code, is amended by adding Section 423.0045 to read as follows:

Sec. 423.0045. OFFENSE: OPERATION OF UNMANNED AIRCRAFT OVER CRITICAL INFRASTRUCTURE FACILITY.

a. In this section:
    1. "Critical infrastructure facility" means:
        A. one of the following, if completely enclosed by a fence or other physical barrier that is obviously designed to exclude intruders, or if clearly marked with a sign or signs that are posted on the property, are reasonably likely to come to the attention of intruders, and indicate that entry is forbidden:
            i. a petroleum or alumina refinery;
            ii. an electrical power generating facility, substation, switching station, or electrical control center;
            iii. a chemical, polymer, or rubber manufacturing facility;
            iv. a water intake structure, water treatment facility, waste water treatment plant, or pump station;
            v. a natural gas compressor station;
            vi. a liquid natural gas terminal or storage facility; switching office;
            vii. a telecommunications central
            viii. a port, railroad switching yard, trucking terminal, or other freight transportation facility;
            ix. a gas processing plant, including a plant used in the processing, treatment, or fractionation of natural gas;
            x. a transmission facility used by a federally licensed radio or television station;
            xi. a steelmaking facility that uses an electric arc furnace to make steel; or
            xii. a dam that is classified as a high hazard by the Texas Commission on Environmental Quality; or
        B. any portion of an aboveground oil, gas, or chemical pipeline that is enclosed by a fence or other physical barrier that is obviously designed to exclude intruders.
    2. "Dam" means any barrier, including any appurtenant structures, that is constructed for the purpose of permanently or temporarily impounding water.
b. A person commits an offense if the person intentionally or knowingly:
    1. operates an unmanned aircraft over a critical infrastructure facility and the unmanned aircraft is not higher than 400 feet above ground level;

2. allows an unmanned aircraft to make contact with a critical infrastructure facility, including any person or object on the premises of or within the facility; or
3. allows an unmanned aircraft to come within a distance of a critical infrastructure facility that is close enough to interfere with the operations of or cause a disturbance to the facility.

c. This section does not apply to conduct described by Subsection (b) that is committed by:
1. the federal government, the state, or a governmental entity;
2. a person under contract with or otherwise acting under the direction or on behalf of the federal government, the state, or a governmental entity;
3. a law enforcement agency;
4. a person under contract with or otherwise acting under the direction or on behalf of a law enforcement agency;
5. an owner or operator of the critical infrastructure facility;
6. a person under contract with or otherwise acting under the direction or on behalf of an owner or operator of the critical infrastructure facility;
7. a person who has the prior written consent of the owner or operator of the critical infrastructure facility;
8. the owner or occupant of the property on which the critical infrastructure facility is located or a person who has the prior written consent of the owner or occupant of that property; or
9. an operator of an unmanned aircraft that is being used for a commercial purpose, if the operator is authorized by the Federal Aviation Administration to conduct operations over that airspace.

d. An offense under this section is a Class B misdemeanor, except that the offense is a Class A misdemeanor if the actor has previously been convicted under this section. SECTION 2.AA

This Act takes effect September 1, 2015.[414]

UTAH

Utah has passed HB 126,[415] HB 3003,[416] SB 167,[417] and HB 296.[418] Additionally Ski Utah has made its own drone policies, depending on whether one of three categories applies: (1) No Fly "NO"; (2) Special Permit Required; and (3) No Limitations.[419]

---

[414]HB 1481, Texas Legislature, http://www.legis.state.tx.us/tlodocs/84R/billtext/pdf/HB01481F.pdf.
[415]HB 126, Utah Legislature, http://le.utah.gov/~2016/bills/static/HB0126.html.
[416]HB 3003, Utah Legislature, http://le.utah.gov/~2016s3/bills/static/HB3003.html.
[417]SB 167, Utah Legislature, http://le.utah.gov/~2014/bills/sbillenr/SB0167.pdf.
[418]HB 296, Utah Legislature, http://le.utah.gov/~2015/bills/hbillenr/HB0296.pdf.
[419]*Utah Resort Drone Policies,* Ski Utah, https://www.skiutah.com/blog/authors/yeti/utah-resort-drone-policies.

House Bill (HB) 126: Effective May 10, 2106, this law makes it a class B misdemeanor to operate a UAS within a certain distance of a wildfire. It becomes a class A misdemeanor if the UAS causes an aircraft fighting the wildfire to drop a payload in the wrong location or to land without dropping the payload. It is a third degree felony if the UAS crashes into a manned aircraft and a second degree felony if that causes the manned aircraft to crash.[420]

House Bill (HB) 3003: Effective July 17, 2016, this law increases the penalties for offenses related to operating within a certain distance of a wildfire and permits certain law enforcement officers to disable a drone that is flying in a prohibited area near a wildland fire.[421]

Senate Bill (SB) 167: On April 1, 2014, the governor of Utah signed SB 167, which reads as follows:

This bill establishes provisions for the appropriate use of an unmanned aerial vehicle.
Highlighted Provisions:
This bill:

- defines terms;
- enacts the "Government Use of Unmanned Aerial Vehicles Act";
- prohibits a law enforcement agency from obtaining data through an unmanned aerial vehicle unless the data was obtained:
    - pursuant to a warrant;
    - in accordance with judicially recognized exceptions to warrant requirements; or
    - under certain conditions, from a nongovernment actor;
- establishes requirements for the retention and use of data collected by an unmanned aerial vehicle;
- establishes reporting requirements for:
    - a law enforcement agency that operates an unmanned aerial vehicle; and
    - the Utah Department of Public Safety; and provides a statement of intent.

Money Appropriated in this Bill:
None
Other Special Clauses:
None
Utah Code Sections Affected:

ENACTS:
63G-18-101, Utah Code Annotated 1953
63G-18-102, Utah Code Annotated 1953
63G-18-103, Utah Code Annotated 1953

[420]HB 126, Utah Legislature, http://le.utah.gov/~2016/bills/static/HB0126.html.
[421]HB 3003, Utah Legislature, http://le.utah.gov/~2016s3/bills/static/HB3003.html.

63G-18-104, Utah Code Annotated 1953
63G-18-105, Utah Code Annotated 1953
Uncodified Material Affected:

ENACTS UNCODIFIED MATERIAL

Be it enacted by the Legislature of the state of Utah:
Section 1. Section 63G-18-101 is enacted to read:

CHAPTER 18. GOVERNMENT USE OF UNMANNED AERIAL VEHICLES ACT

63G-18-101. Title.

This chapter is known as the "Government Use of Unmanned Aerial Vehicles Act."

Section 2. Section 63G-18-102 is enacted to read:
63G-18-102. Definitions.
As used in this chapter:

1. "Law enforcement agency" means an entity of the state or an entity of a political subdivision of the state, including an entity of a state institution of higher education, that exists primarily to prevent, detect, or prosecute crime and enforce criminal statutes or ordinances.
2. "Nongovernment actor" means a person that is not:
   a. an agency, department, division, or other entity within state government;
   b. a person employed by or acting in an official capacity on behalf of the state;
   c. a political subdivision of the state; or
   d. a person employed by or acting in an official capacity on behalf of a political subdivision of the state.
3. "Target" means a person upon whom, or a structure or area upon which a person:
   a. has intentionally collected or attempted to collect information through the operation of an unmanned aerial vehicle; or
   b. plans to collect or attempt to collect information through the operation of an unmanned aerial vehicle.
4. "Unmanned aerial vehicle" means an aircraft that:
      i. is capable of sustaining flight; and
      ii. operates with no possible direct human intervention from on or within the aircraft.
   a. "Unmanned aerial vehicle" does not include an unmanned aircraft that is flown:
      i. within visual line of sight of the individual operating the aircraft; and
      ii. strictly for hobby or recreational purposes.

Section 3. Section 63G-18-103 is enacted to read:
63G-18-103. Warrant required—Exceptions.

1. A law enforcement agency may not obtain, receive, or use data acquired through an unmanned aerial vehicle unless the data is obtained:
   a. pursuant to a search warrant;
   b. in accordance with judicially recognized exceptions to warrant requirements; or
   c. subject to Subsection (2), from a person who is a nongovernment actor.
2. A nongovernment actor may only disclose data acquired through an unmanned aerial vehicle to a law enforcement agency if:
   a. the data appears to pertain to the commission of a crime; or
   b. the nongovernment actor believes, in good faith, that:
      i. the data pertains to an imminent or ongoing emergency involving danger of death or serious bodily injury to an individual; and
      ii. disclosing the data would assist in remedying the emergency.

Section 4. Section 63G-18-104 is enacted to read:
63G-18-104. Data retention.

1. Except as provided in this section, a law enforcement agency:
   a. may not use, copy, or disclose data collected by an unmanned aerial vehicle on a person, structure, or area that is not a target; and
   b. shall ensure that data described in Subsection (1)(a) is destroyed as soon as reasonably possible after the law enforcement agency collects or receives the data.
2. A law enforcement agency is not required to comply with Subsection (1) if:
   a. deleting the data would also require the deletion of data that:
      i. relates to the target of the operation; and
      ii. is requisite for the success of the operation;
   b. the law enforcement agency receives the data:
      i. through a court order that:
         A. requires a person to release the data to the law enforcement agency; or
         B. prohibits the destruction of the data; or
      ii. from a person who is a nongovernment actor;
   c. i. the data was collected inadvertently; and
      ii. the data appears to pertain to the commission of a crime;
   d. i. the law enforcement agency reasonably determines that the data pertains to an emergency situation; and
      ii. using or disclosing the data would assist in remedying the emergency; or

e. the data was collected through the operation of an unmanned aerial vehicle over public lands outside of municipal boundaries.

Section 5. Section 63G-18-105 is enacted to read:

63G-18-105. Reporting.

1. (a) Except as provided by Subsection (1)(b), before March 31 of each year, a law enforcement agency that operated an unmanned aerial vehicle in the previous calendar year shall submit to the Utah Department of Public Safety, and make public on the law enforcement agency's website, a written report containing:
   i. the number of times the law enforcement agency operated an unmanned aerial vehicle in the previous calendar year;
   ii. the number of criminal investigations aided by the use of an unmanned aerial vehicle operated by the law enforcement agency in the previous calendar year;
   iii. a description of how the unmanned aerial vehicle was helpful to each investigation described in Subsection (1)(a)(ii);
   iv. the frequency with which data was collected, and the type of data collected, by an unmanned aerial vehicle operated by the law enforcement agency on any person, structure, or area other than a target in the previous calendar year;
   v. the number of times a law enforcement agency received, from a person who is not a law enforcement agency, data collected by an unmanned aerial vehicle; and
   vi. the total cost of the unmanned aerial vehicle program operated by the law enforcement agency in the previous calendar year.

   b. i A law enforcement agency that submits a report described in Subsection (1)(a) may exclude from the report information pertaining to an ongoing investigation.
   ii. A law enforcement agency that excludes information under Subsection (1)(b)(i) from the report shall report the excluded information to the Utah Department of Public Safety on the annual report in the year following the year in which the information was excluded.
2. Before May 31 of each year, the Utah Department of Public Safety shall, for all reports received under Subsection (1) during the previous calendar year:
   a. transmit to the Government Operations Interim Committee and post on the department's website a report containing:
      i. a summary of the information reported to the department;
      ii. the total number of issued warrants authorizing the operation of an unmanned aerial vehicle; and

iii. the number of denied warrants for the operation of an unmanned aerial vehicle; and (b) post on the department's website each report the department received.

Section 6. Statement of intent.
This chapter is intended to govern the use of an unmanned aerial vehicle by a law enforcement agency. Nothing herein is intended to prohibit or impede the public and private research, development, or manufacture of unmanned aerial vehicles. Unmanned aerial vehicles will provide promising technological advances, which, if properly developed, will prove beneficial to the health, safety, and welfare of the citizens of this state and greater society.[422]

House Bill (HB) 296: On March 27, 2015, the governor of Utah signed HB 296 into law. The bill reads as follows:

GOVERNMENT USE OF UNMANNED AERIAL VEHICLES—AMENDMENTS

This bill amends the provisions of Title 63G, Chapter 18, Government Use of Unmanned Aerial Vehicles Act.
Highlighted Provisions:
This bill:

- allows a law enforcement agency to use an unmanned aircraft system to collect certain types of data;
- institutes testing requirements for a law enforcement agency's use of an unmanned aircraft system;
- amends the reporting requirements for a law enforcement agency that operates an unmanned aircraft system; and
- makes technical changes.

Money Appropriated in this Bill:
None
Other Special Clauses:
This bill provides a special effective date.
Utah Code Sections Affected:

AMENDS:
63G-18-101, as enacted by Laws of Utah 2014, Chapter 399
63G-18-102, as enacted by Laws of Utah 2014, Chapter 399
63G-18-103, as enacted by Laws of Utah 2014, Chapter 399

422SB 167, Utah Legislature, http://le.utah.gov/~2014/bills/sbillenr/SB0167.pdf.

63G-18-104, as enacted by Laws of Utah 2014, Chapter 399
63G-18-105, as enacted by Laws of Utah 2014, Chapter 399

Be it enacted by the Legislature of the state of Utah:

Section 1. Section 63G-18-101 is amended to read:

CHAPTER 18. GOVERNMENT USE OF UNMANNED AIRCRAFT SYSTEMS ACT

63G-18-101. Title.

This chapter is known as the "Government Use of Unmanned Aircraft Systems Act."

Section 2. Section 63G-18-102 is amended to read:

63G-18-102. Definitions.

As used in this chapter:

1. "Law enforcement agency" means an entity of the state or an entity of a political subdivision of the state, including an entity of a state institution of higher education, that exists primarily to prevent, detect, or prosecute crime and enforce criminal statutes or ordinances.
2. "Nongovernment actor" means a person that is not:
    a. an agency, department, division, or other entity within state government;
    b. a person employed by or otherwise acting in an official capacity on behalf of the state;
    c. a political subdivision of the state; or
    d. a person employed by or otherwise acting in an official capacity on behalf of a political subdivision of the state.
3. "Target" means a person upon whom, or a structure or area upon which, a person:
    a. has intentionally collected or attempted to collect information through the operation of an unmanned aircraft system; or
    b. plans to collect or attempt to collect information through the operation of an unmanned aircraft system.
4. "Testing site" means an area that:
    a. has boundaries that are clearly identified using GPS coordinates;
    b. a law enforcement agency identifies in writing to the Department of Public Safety, including the boundaries identified under Subsection (4)(a);
    c. is not more than three square miles; and
    d. contains no occupied structures.
5. "Unmanned aircraft system" means an aircraft that:
    a. is capable of sustaining flight; and
    b. operates with no possible direct human intervention from on or within the aircraft.

c. "Unmanned aircraft system" does not include an unmanned aircraft that is flown:
   i. within visual line of sight of the individual operating the aircraft; and
   ii. strictly for hobby or recreational purposes.

Section 3. Section 63G-18-103 is amended to read:
63G-18-103. Unmanned aircraft system use requirements—Exceptions—Testing.

1. A law enforcement agency may not obtain, receive, or use data acquired through an unmanned aircraft system unless the data is obtained:
   a. pursuant to a search warrant;
   b. in accordance with judicially recognized exceptions to warrant requirements;
   c. subject to Subsection (2), from a person who is a nongovernment actor;
   d. at a testing site; or
   e. to locate a lost or missing person in an area in which a person has no reasonable expectation of privacy.
2. A nongovernment actor may only disclose data acquired through an unmanned aircraft system to a law enforcement agency if:
   a. the data appears to pertain to the commission of a crime; or
   b. the nongovernment actor believes, in good faith, that:
      i. the data pertains to an imminent or ongoing emergency involving danger of death or serious bodily injury to an individual; and
      ii. disclosing the data would assist in remedying the emergency.
3. A law enforcement agency that obtains, receives, or uses data acquired under Subsection (1)(d) or (e) shall destroy the data as soon as reasonably possible after the law enforcement agency obtains, receives, or uses the data.
4. A law enforcement agency that operates an unmanned aircraft system under Subsection (1)(d) may not operate the unmanned aircraft system outside of the testing site.

Section 4. Section 63G-18-104 is amended to read:
63G-18-104. Data retention.

1. Except as provided in this section, a law enforcement agency:
   a. may not use, copy, or disclose data collected by an unmanned aircraft system on a person, structure, or area that is not a target; and
   b. shall ensure that data described in Subsection (1)(a) is destroyed as soon as reasonably possible after the law enforcement agency collects or receives the data.

2. A law enforcement agency is not required to comply with Subsection (1) if:
    a. deleting the data would also require the deletion of data that:
        i. relates to the target of the operation; and
        ii. is requisite for the success of the operation;
    b. the law enforcement agency receives the data:
        i. through a court order that:
            A. requires a person to release the data to the law enforcement agency; or
            B. prohibits the destruction of the data; or
        ii. from a person who is a nongovernment actor;
    c. the data was collected inadvertently; and
        i. the data appears to pertain to the commission of a crime;
    d. i. the law enforcement agency reasonably determines that the data pertains to an emergency situation; and
        ii. using or disclosing the data would assist in remedying the emergency; or
    e. the data was collected through the operation of an unmanned aircraft system over public lands outside of municipal boundaries.

Section 5. Section 63G-18-105 is amended to read:
63G-18-105. Reporting.

1. Except as provided by Subsections (2) and (3), before March 31 of each year, a law enforcement agency that operated an unmanned aircraft system in the previous calendar year shall submit to the Utah Department of Public Safety, and make public on the law enforcement agency's website, a written report containing:
    a. the number of times the law enforcement agency operated an unmanned aircraft system in the previous calendar year;
    b. the number of criminal investigations aided by the use of an unmanned aircraft system operated by the law enforcement agency in the previous calendar year;
    c. a description of how the unmanned aircraft system was helpful to each investigation described in Subsection (1)(b);
    d. the frequency with which data was collected, and the type of data collected, by an unmanned aircraft system operated by the law enforcement agency on any person, structure, or area other than a target in the previous calendar year;
    e. the number of times a law enforcement agency received, from a person who is not a law enforcement agency, data collected by an unmanned aircraft system; and

f. the total cost of the unmanned aircraft system program operated by the law enforcement agency in the previous calendar year, including the source of any funds used to operate the program.

2. a. A law enforcement agency that submits a report described in Subsection (1) may exclude from the report information pertaining to an ongoing investigation.
   b. A law enforcement agency that excludes information under Subsection (2)(a) from the report shall report the excluded information to the Utah Department of Public Safety on the annual report in the year following the year in which the investigation to which the information pertains is concluded.
3. A law enforcement agency is not required to submit, under Subsection (1), to the Department of Public Safety information pertaining to the use of an unmanned aircraft system operated at a testing site.
4. Before May 31 of each year, the Utah Department of Public Safety shall, for all reports received under Subsection (1) during the previous calendar year:
   a. transmit to the Government Operations Interim Committee and post on the department's website a report containing:
      i. a summary of the information reported to the department;
      ii. the total number of issued warrants authorizing the operation of an unmanned aircraft system; and
      iii. the number of denied warrants for the operation of an unmanned aircraft system; and (b) post on the department's website each report the department received.

Section 6. Effective date.
If approved by two-thirds of all the members elected to each house, this bill takes effect upon approval by the governor, or the day following the constitutional time limit of Utah Constitution, Article VII, Section 8, without the governor's signature, or in the case of a veto, the date of veto override.[423]

VERMONT

Vermont has introduced HB 302, relating to the use of drones by proposing to establish regulations for the use of drones and permitting a law enforcement agency to use a drone only if the agency obtains a warrant or if emergency circumstances exist.[424]

Vermont passed SB 155 on June 6, 2016. This was an act relating to privacy protection and a code of administrative rules.[425]

---

[423]HB 296, UTAH LEGISLATURE, http://le.utah.gov/~2015/bills/hbillenr/HB0296.pdf.
[424]HB 302, VERMONT GENERAL ASSEMBLY, http://legislature.vermont.gov/bill/status/2016/H.302.
[425]SB 155, VERMONT GENERAL ASSEMBLY, https://legiscan.com/VT/bill/S0155/2015.

Senate Bill (SB) 155: This law regulates the use of drones by law enforcement and requires law enforcement to annually report on the use of drones by the department. It also prohibits the weaponization of drones.[426]

VIRGINIA

There are five bills currently that have been signed into law. These are HB 412,[427] HB 2012,[428] HB 2125,[429] SB 1331,[430] and SB 1301.[431] Virginia has introduced HB 1141 dealing with invasion of privacy with respect to unmanned aircraft;[432] HB 1372 relating to autonomous vehicles and piloted vehicles;[433] HB 269 to create a civil cause of action for the physical and constructive invasion of privacy where a person enters onto the land or into the airspace above the land of another person to capture an image of private property without consent of the owner using a UAS, and allowing a plaintiff to recover the greater of actual damages or $1,000, along with reasonable attorney fees and costs, as well as allowing the court to award punitive damages when actual damages are awarded;[434] HB 413 to create the Fly Wisely Unmanned Aerial Systems Accelerator Authority for the purpose of assisting the coal-producing areas of the Commonwealth to achieve some degree of economic diversification;[435] HB 538 to provide that a locality may by ordinance prohibit the use of unmanned aircraft to trespass upon the property of others within its boundaries, and such ordinance may provide that landowners may exclude such aircraft from a column or airspace extending from the surface of their land up to 350 feet above ground level, and a person shall be guilty of a Class 4 misdemeanor if he operates such aircraft in violation of the ordinance;[436] HB 561 making the use of UAS for surveillance a Class 3 misdemeanor, and the sale or dissemination of these images a Class 2 misdemeanor;[437] HB 737 making it a Class 3 misdemeanor to use an unmanned aircraft system to capture an image of private property or an individual located on such private property after having been given written notice by the landowner or other person lawfully in charge of such private property that the individual, landowner, or other person objects to having such images taken, as well as making the sale or dissemination of such a captured image punishable as a Class 2 misdemeanor;[438] HB 878 creating a civil cause of action for the

[426] *Id.*
[427] HB 412, Virginia Legislature, https://legiscan.com/VA/bill/HB412/2016.
[428] HB 2012, Virginia Legislature, http://lis.virginia.gov/cgi-bin/legp604.exe?131+ful+CHAP0755+pdf.
[429] HB 2125, Virginia Legislature, https://lis.virginia.gov/cgi-bin/legp604.exe?151+ful+CHAP0764.
[430] SB 1331, Virginia Legislature, https://lis.virginia.gov/cgi-bin/legp604.exe?131+ful+CHAP0796+pdf,
[431] SB 1301, Virginia Legislature, https://lis.virginia.gov/cgi-bin/legp604.exe?151+ful+CHAP0774+pdf.
[432] HB 1141, Virginia Legislature, https://legiscan.com/VA/research/HB1141/2016.
[433] HB 1372, Virginia Legislature, https://legiscan.com/VA/bill/HB1372/2016.
[434] HB 269, Virginia Legislature, https://legiscan.com/VA/bill/HB269/2016.
[435] HB 413, Virginia Legislature, https://legiscan.com/VA/bill/HB413/2016.
[436] HB 538, Virginia Legislature, https://legiscan.com/VA/bill/HB538/2016.
[437] HB 561, Virginia Legislature, https://legiscan.com/VA/bill/HB561/2016.
[438] HB 737, Virginia Legislature, https://legiscan.com/VA/bill/HB737/2016.

physical and constructive invasion of privacy where a person, with the intent to coerce, intimidate, or harass enters onto the land or into the airspace above the land of another person to capture an image without consent or uses a UAS to capture the image, and allowing the plaintiff to recover the greater of actual damages or $1,000, along with reasonable attorney fees and costs, and allowing a court to award punitive damages when actual damages are awarded;[439] HB 89 to provide that a person is guilty of a Class 1 misdemeanor if he uses or attempts to use an unmanned aircraft system while committing or attempting to commit a crime or obstructing or attempting to obstruct law enforcement officers, animal control officers, or emergency medical services agency personnel in the performance of their duties;[440] SB 584 creating a civil cause of action for the physical and constructive invasion of privacy where a person uses an unmanned aircraft system to enter into the airspace above the land of another person to capture an image of private property or an individual located on the private property without consent or uses an unmanned aircraft system to capture such an image, and allowing the plaintiff to recover the greater of actual damages of $1,000, along with reasonable attorney fees and costs, and allowing a court to award punitive damages when actual damages are awarded;[441] SB 729 providing that a person is guilty of a Class 1 misdemeanor if he knowingly uses or attempts to use an unmanned aircraft system (i) to commit or attempt to commit a crime, (ii) while obstructing or attempting to obstruct the delivery of emergency medical services by emergency medical services agency personnel, or (iii) while obstructing or attempting to obstruct a law enforcement officer or animal control officer in the performance of his duties;[442] and SB 954 making it unlawful to willfully and intentionally impede the lawful hunting or trapping of wild birds or wild animals utilizing a UAV for the sole purpose of monitoring and photographing persons who are lawfully hunting on private land without the landowner's permission.[443]

Charlottesville, in 2013, was the first city in the nation to pass an anti-drone resolution making the city a no-drone zone and banning the use of drones for surveillance and other uses.[444]

House Bill (HB) 412: This law, passed on March 11, 2016, provides that no locality may regulate the use of privately owned, unmanned aircraft systems within it boundaries. The provisions of the bill expire on July 1, 2019.[445]

House Bill (HB) 2012: On April 3, 2013, the Governor of Virginia signed HB 2012 into law. It reads as follows:

---

[439]HB 878, Virginia Legislature, https://legiscan.com/VA/bill/HB878/2016.
[440]HB 89, Virginia Legislature, https://legiscan.com/VA/bill/HB89/2016.
[441]SB 584, Virginia Legislature, https://legiscan.com/VA/bill/SB584/2016.
[442]SB 729, Virginia Legislature, https://legiscan.com/VA/bill/SB729/2016.
[443]SB 954, Virginia Legislature, http://leg1.state.va.us/cgi-bin/legp504.exe?131+ful+SB954.
[444]*Charlottesville becomes first US city to pass anti-drone legislation,* NBC29, http://www.nbc29.com/story/20963560/charlottesville-city-council-passes-anti-drone-resolution.
[445]HB 412, Virginia Legislature, https://legiscan.com/VA/bill/HB412/2016.

VIRGINIA ACTS OF ASSEMBLY—2013 RECONVENED SESSION CHAPTER 755

An Act to place a moratorium on the use of unmanned aircraft systems.
Be it enacted by the General Assembly of Virginia:

1. § 1. No state or local government department, agency, or instrumentality having jurisdiction over criminal law enforcement or regulatory violations, including but not limited to the Department of State Police, and no department of law enforcement as defined in § 15.2-836 of the Code of Virginia of any county, city, or town shall utilize an unmanned aircraft system before July 1, 2015.

Notwithstanding the prohibition in this section, an unmanned aircraft system may be deployed before July 1, 2015, (i) when an Amber Alert is activated pursuant to § 52-34.3 of the Code of Virginia, (ii) when a Senior Alert is activated pursuant to § 52-34.6 of the Code of Virginia, (iii) when a Blue Alert is activated pursuant to § 52-34.9 of the Code of Virginia, (iv) for the purpose of a search or rescue operation where use of an unmanned aircraft system is determined to be necessary to alleviate an immediate danger to any person, or (v) for training exercises related to such uses. In no case may a weaponized unmanned aircraft system be deployed or its use facilitated by a state or local agency in Virginia.

The prohibitions in this section shall not apply to the Virginia National Guard while utilizing unmanned aircraft systems during training required to maintain readiness for its federal mission, when facilitating training for other United States Department of Defense units, or when such systems are utilized to support the Commonwealth for purposes other than law enforcement, including damage assessment, traffic assessment, flood stages, and wildfire assessment. Nothing herein shall prohibit use of unmanned aircraft systems solely for research and development purposes by institutions of higher education and other research organizations or institutions.

1. That the Department of Criminal Justice Services, in consultation with the Office of the Attorney General and other state agencies, shall develop model protocols for use of unmanned aircraft systems by law-enforcement agencies and shall report such findings to the Governor and the General Assembly on or before November 1, 2013.[446]

House Bill (HB) 2125: On April 30, 2015, the governor signed HB 2125 into law. This was an act to amend the Code of Virginia by adding in Chapter 5 of Title 19.2 a section

[446]HB 2012, VIRGINIA LEGISLATURE, http://lis.virginia.gov/cgi-bin/legp604.exe?131+ful+CHAP0755+pdf.

numbered 19.2-60.1, relating to use of unmanned aircraft systems by public bodies; search warrant required. It reads as follows:

Be it enacted by the General Assembly of Virginia:

2. That the Code of Virginia is amended by adding in Chapter 5 of Title 19.2 a section numbered 19.2-60.1 as follows:

   § 19.2-60.1. Use of unmanned aircraft systems by public bodies; search warrant required.

A. As used in this section, unless the context requires a different meaning:

   "Unmanned aircraft" means an aircraft that is operated without the possibility of human intervention from within or on the aircraft.

   "Unmanned aircraft system" means an unmanned aircraft and associated elements, including communication links, sensing devices, and the components that control the unmanned aircraft.

B. No state or local government department, agency, or instrumentality having jurisdiction over criminal law enforcement or regulatory violations, including but not limited to the Department of State Police, and no department of law enforcement as defined in § 15.2-836 of any county, city, or town shall utilize an unmanned aircraft system except during the execution of a search warrant issued pursuant to this chapter or an administrative or inspection warrant issued pursuant to law.

C. Notwithstanding the prohibition in this section, an unmanned aircraft system may be deployed without a warrant (i) when an Amber Alert is activated pursuant to § 52-34.3, (ii) when a Senior Alert is activated pursuant to § 52-34.6, (iii) when a Blue Alert is activated pursuant to § 52-34.9, (iv) where use of an unmanned aircraft system is determined to be necessary to alleviate an immediate danger to any person, (v) for training exercises related to such uses; or (vi) if a person with legal authority consents to the warrantless search.

D. The warrant requirements of this section shall not apply when such systems are utilized to support the Commonwealth for purposes other than law enforcement, including damage assessment, traffic assessment, flood stage assessment, and wildfire assessment. Nothing herein shall prohibit use of unmanned aircraft systems for private, commercial, or recreational use or solely for research and development purposes by institutions of higher education and other research organizations or institutions.

E. Evidence obtained through the utilization of an unmanned aircraft system in violation of this section is not admissible in any criminal or civil proceeding.

F. In no case may a weaponized unmanned aircraft system be deployed in the Commonwealth or its use facilitated in the Commonwealth by a state or local government department, agency, or instrumentality or department of law enforcement in the Commonwealth except in operations at the Space Port and Naval/Aegis facilities at Wallops Island.

G. Nothing herein shall apply to the Armed Forces of the United States or the Virginia National Guard while utilizing unmanned aircraft systems during training required to maintain readiness for its federal mission or when facilitating training for other U.S. Department of Defense units.[447]

Senate Bill (SB) 1331: On April 3, 2013, the governor of Virginia signed SB 1331 into law. It reads as follows:

VIRGINIA ACTS OF ASSEMBLY—2013 RECONVENED SESSION CHAPTER 796

An Act to place a moratorium on the use of unmanned aircraft systems.
Be it enacted by the General Assembly of Virginia:

1. § 1. No state or local government department, agency, or instrumentality having jurisdiction over criminal law enforcement or regulatory violations, including but not limited to the Department of State Police, and no department of law enforcement as defined in § 15.2-836 of the Code of Virginia of any county, city, or town shall utilize an unmanned aircraft system before July 1, 2015.

Notwithstanding the prohibition in this section, an unmanned aircraft system may be deployed before July 1, 2015, (i) when an Amber Alert is activated pursuant to § 52-34.3 of the Code of Virginia, (ii) when a Senior Alert is activated pursuant to § 52-34.6 of the Code of Virginia, (iii) when a Blue Alert is activated pursuant to § 52-34.9 of the Code of Virginia, (iv) for the purpose of a search or rescue operation where use of an unmanned aircraft system is determined to be necessary to alleviate an immediate danger to any person, or (v) for training exercises related to such uses. In no case may a weaponized unmanned aircraft system be deployed or its use facilitated by a state or local agency in Virginia.

The prohibitions in this section shall not apply to the Virginia National Guard while utilizing unmanned aircraft systems during training required to maintain readiness for its federal mission, when facilitating training for other United States Department of Defense units, or when such systems are utilized to support the Commonwealth for purposes other than law enforcement, including damage assessment, traffic assessment, flood stages, and wildfire assessment. Nothing herein shall prohibit use of unmanned aircraft systems solely

[447] HB 2125, Virginia Legislature, https://lis.virginia.gov/cgi-bin/legp604.exe?151+ful+CHAP0764.

for research and development purposes by institutions of higher education and other research organizations or institutions.

2. That the Department of Criminal Justice Services, in consultation with the Office of the Attorney General and other state agencies, shall develop model protocols for use of unmanned aircraft systems by law-enforcement agencies and shall report such findings to the Governor and the General Assembly on or before November 1, 2013.[448]

Senate Bill (SB) 1301: On April 30, 2015, the governor of Virginia signed SB 1301 into law. It reads as follows:

VIRGINIA ACTS OF ASSEMBLY—2015 RECONVENED SESSION CHAPTER 774

An Act to amend the Code of Virginia by adding in Chapter 5 of Title 19.2 a section numbered 19.2-60.1, relating to use of unmanned aircraft systems by public bodies; search warrant required.

Be it enacted by the General Assembly of Virginia:

1. That the Code of Virginia is amended by adding in Chapter 5 of Title 19.2 a section numbered 19.2-60.1 as follows:

§ 19.2-60.1. Use of unmanned aircraft systems by public bodies; search warrant required.

A. As used in this section, unless the context requires a different meaning:

"Unmanned aircraft" means an aircraft that is operated without the possibility of human intervention from within or on the aircraft.

"Unmanned aircraft system" means an unmanned aircraft and associated elements, including communication links, sensing devices, and the components that control the unmanned aircraft.

B. No state or local government department, agency, or instrumentality having jurisdiction over criminal law enforcement or regulatory violations, including but not limited to the Department of State Police, and no department of law enforcement as defined in § 15.2-836 of any county, city, or town shall utilize an unmanned aircraft system except during the execution of a search warrant issued pursuant to this chapter or an administrative or inspection warrant issued pursuant to law.

C. Notwithstanding the prohibition in this section, an unmanned aircraft system may be deployed without a warrant (i) when an Amber Alert is activated pursuant to § 52-34.3, (ii) when a Senior Alert is activated pursuant to § 52-34.6, (iii) when a Blue Alert is activated pursuant to § 52-34.9, (iv) where use of an unmanned

[448]SB 1331, VIRGINIA LEGISLATURE, https://lis.virginia.gov/cgi-bin/legp604.exe?131+ful+CHAP0796+pdf.

aircraft system is determined to be necessary to alleviate an immediate danger to any person, (v) for training exercises related to such uses, or (vi) if a person with legal authority consents to the warrantless search.

D. The warrant requirements of this section shall not apply when such systems are utilized to support the Commonwealth for purposes other than law enforcement, including damage assessment, traffic assessment, flood stage assessment, and wildfire assessment. Nothing herein shall prohibit use of unmanned aircraft systems for private, commercial, or recreational use or solely for research and development purposes by institutions of higher education and other research organizations or institutions.

E. Evidence obtained through the utilization of an unmanned aircraft system in violation of this section is not admissible in any criminal or civil proceeding.

F. In no case may a weaponized unmanned aircraft system be deployed in the Commonwealth or its use facilitated in the Commonwealth by a state or local government department, agency, or instrumentality or department of law enforcement in the Commonwealth except in operations at the Space Port and Naval/Aegis facilities at Wallops Island.

G. Nothing herein shall apply to the Armed Forces of the United States or the Virginia National Guard while utilizing unmanned aircraft systems during training required to maintain readiness for its federal mission or when facilitating training for other Department of Defense units.[449]

WASHINGTON

Washington has introduced HB 1093, an act relating to unmanned aircraft and adding a new chapter to Title 14, and prescribing penalties;[450] HB 2016, an act relating to unmanned aerial vehicles and adding new sections to chapter 9.73;[451] HB 2774 concerning the operation of unmanned aerial vehicles near certain airports;[452] SB 5499, an act relating to a special allegation of a nefarious drone enterprise, amending RCW 9.94A.533 and adding a new section to chapter 9.94A RCW, as well as prescribing penalties;[453] and SB 6437, an act relating to dangerous drone operations near correctional facilities.[454]

---

[449]SB 1301, Virginia Legislature, https://lis.virginia.gov/cgi-bin/legp604.exe?151+ful+CHAP0774+pdf.
[450]HB 1093, Washington Legislature, https://legiscan.com/WA/bill/HB1093/2015.
[451]HB 2016, Washington Legislature, https://legiscan.com/WA/bill/HB2016/2015.
[452]HB 2774, Washington Legislature, https://legiscan.com/WA/bill/HB2774/2015.
[453]SB 5499, Washington Legislature, https://legiscan.com/WA/bill/SB5499/2015.
[454]SB 6437, Washington Legislature, https://legiscan.com/WA/bill/SB6437/2015.

King County Municipal Code 18.12.265 passed an ordinance dealing with motorized models.[455] Pierce County passed Ordinance 2013-28s entitled, Freedom from Unwarranted Surveillance.[456]

King County Municipal Code 18.12.265: The code reads as follows:

It is unlawful to operate any motorized model aircraft or motorized model watercraft in any park except at places set apart by the Superintendent for such purposes or as authorized by a permit from the Superintendent. (Ord. 113436 § 17, 1987.)[457]

WEST VIRGINIA

West Virginia has enacted one law pertaining to UAS, HB 2515.[458] West Virginia introduced HB 2420 relating to unlawful methods of hunting and fishing and prohibiting the use of night vision technology and drones or other unmanned aerial vehicles in the hunting or taking of wildlife;[459] HB 2653 to prohibit the use of unmanned aircraft with exceptions, to require documentation when an unmanned aircraft is used, to provide remedies, and to prohibit the use of certain evidence and preservation of collected data;[460] HB 4170 relating to hunting and trapping and unmanned aircraft;[461] HB 4184 establishing criminal penalties for the illegal taking of deer;[462] HB 4283 to remove restrictions for hunting and trapping on private lands on Sundays;[463] SB 291 relating to law enforcement use of unmanned aircraft systems;[464] SB 331 relating to hunting or trapping on private lands;[465] and SB 278 to prohibit hunting with drone or other unmanned aircraft.[466]

---

[455]Seattle, King County Municipal Code 18.12.265, https://www.municode.com/library/wa/seattle/codes/municipal_code?searchRequest=%7B%22searchText%22:%2218.12.265%22,%22pageNum%22:1,%22resultsPerPage%22:25,%22booleanSearch%22:false,%22stemming%22:true,%22fuzzy%22:false,%22synonym%22:false,%22contentTypes%22:%5B%22CODES%22%5D,%22productIds%22:%5B%5D%7D&nodeId=TIT18PARE_CH18.12PACO_SUBCHAPTER_VIIUSRE_18.12.265MOMO#.

[456]Ordinance NO 2013-28s, Pierce County Council, https://councilonline.co.pierce.wa.us/councilonline/proposal/proposal.htm?proposal_num=2013-28.

[457]King County Municipal Code 18.12.265, Seattle, https://www.municode.com/library/wa/seattle/codes/municipal_code?searchRequest=%7B%22searchText%22:%2218.12.265%22,%22pageNum%22:1,%22resultsPerPage%22:25,%22booleanSearch%22:false,%22stemming%22:true,%22fuzzy%22:false,%22synonym%22:false,%22contentTypes%22:%5B%22CODES%22%5D,%22productIds%22:%5B%5D%7D&nodeId=TIT18PARE_CH18.12PACO_SUBCHAPTER_VIIUSRE_18.12.265MOMO#!.

[458]HB 2515, West Virginia Legislature, http://www.legis.state.wv.us/Bill_Text_HTML/2015_SESSIONS/RS/pdf_bills/HB2515%20SUB%20ENR%20PRINTED.pdf.

[459]HB 2420, West Virginia Legislature, http://www.legis.state.wv.us/Bill_Status/bills_text.cfm?billdoc=hb2420%20intr.htm&yr=2015&sesstype=RS&i=2420.

[460]HB 2653, West Virginia Legislature, http://www.legis.state.wv.us/bill_status/bills_text.cfm?billdoc=hb2653%20intr.htm&yr=2015&sesstype=RS&i=2653.

[461]HB 4170, West Virginia Legislature, http://www.legis.state.wv.us/Bill_Status/bills_text.cfm?billdoc=hb4170%20intr.htm&yr=2016&sesstype=RS&i=4170.

[462]HB 4184, West Virginia Legislature, http://openstates.org/wv/bills/2016/HB4184/.

[463]HB 4283, West Virginia Legislature, https://legiscan.com/WV/bill/HB4283/2016.

[464]SB 291, West Virginia Legislature, https://legiscan.com/WV/bill/SB291/2016.

[465]SB 331, West Virginia Legislature, https://legiscan.com/WV/bill/SB331/2016.

[466]SB 278, West Virginia Legislature, http://www.legis.state.wv.us/bill_status/bills_text.cfm?billdoc=SB278%20SUB2.htm&yr=2015&sesstype=RS&i=278.

House Bill (HB) 2515: On March 14, 2014, the governor of West Virginia passed HB 2515 into law, and it went into effect ninety days thereafter. The bill reads as follows:

An act to amend and reenact §20-2-4, §20-2-5, §20-2-5a, §20-2-5g, §20-2-22a and §20-2-42w of the Code of West Virginia, 1931, as amended; and to amend said code by adding thereto a new section, designated §20-2-5h, all relating to wildlife; disallowing elk to be possessed if struck by motor vehicle; requiring persons required to deliver wildlife to official checking station to electronically register wildlife; increasing fine for illegal taking of elk; prohibiting hunting wildlife with night vision technology, drone or other unmanned aircraft; clarifying when a person may carry certain firearms; permitting a person to carry firearm for self defense while in the woods; clarifying when a shotgun or rifle is unloaded; permitting hunting with crossbows during certain seasons and with certain limitations; establishing elk management area in Southern West Virginia; establishing elk damage fund; providing for criminal penalties for the illegal taking of elk; clarifying bear laws and Class Y permits; authorizing director to propose legislative rules; and making technical corrections. Be it enacted by the Legislature of West Virginia:

That §20-2-4, §20-2-5, §20-2-5a, §20-2-5g, §20-2-22a and §20-2- 42w of the Code of West Virginia, 1931, as amended, be amended and reenacted; and that said code be amended by adding thereto a new section, designated §20-2-5h, all to read as follows:

ARTICLE 2. WILDLIFE RESOURCES.

§20-2-4. Possession of wildlife.

a. Except for wildlife, lawfully taken, killed or obtained, no person may have in his or her possession any wildlife, or parts thereof, during closed seasons. It is unlawful to possess any wildlife, or parts thereof, which have been illegally taken, killed or obtained. Any wildlife illegally taken, killed or possessed shall be forfeited to the state and shall be counted toward the daily, seasonal, bag, creel and possession limit of the person in possession of, or responsible for, the illegal taking or killing of any wildlife.
b. Wildlife lawfully taken outside of this state is subject to the same laws and rules as wildlife taken within this state.
c. Migratory wild birds may be possessed only in accordance with the Migratory Bird Treaty Act, 16 U. S. C. §703, et seq., and its regulations.
d. The restrictions in this section do not apply to the director or duly authorized agents, who may take or maintain in captivity any wildlife for the purpose of carrying out the provisions of this chapter.
e. Wildlife, except protected birds, elk, spotted fawn, and bear cubs, killed or mortally wounded as a result of being accidentally or inadvertently struck by a motor vehicle may be lawfully possessed if the possessor of the wildlife provides

notice of the claim within twelve hours to a relevant law-enforcement agency, and obtains a nonhunting game tag within twenty-four hours of possession. The director shall propose administrative policy which addresses the means, methods and administrative procedures for implementing the provisions of this section.

f. Persons required to deliver wildlife to an official checking station shall, in accordance with rules promulgated by the director, electronically register the wildlife in lieu of the delivery to an official checking station. "Electronically register" means submission of all necessary and relevant information to the division, in the manner designated by rule, in lieu of delivery of the wildlife to an official checking station. The director may promulgate rules, pursuant to article three, chapter twenty-nine-a of this code, governing the electronic registration of wildlife.

§20-2-5. Unlawful methods of hunting and fishing and other unlawful acts.

Except as authorized by the director or by law, it is unlawful at any time for any person to:

1. Shoot at any wild bird or wild animal unless it is plainly visible;
2. Dig out, cut out, smoke out, or in any manner take or attempt to take any live wild animal or wild bird out of its den or place of refuge;
3. Use or attempt to use any artificial light or any night vision technology, including image intensification, thermal imaging or active illumination while hunting, locating, attracting, taking, trapping or killing any wild bird or wild animal: Provided, That it is lawful to hunt or take coyote, fox, raccoon, opossum or skunk by the use of artificial light or night vision technology.

   Any person violating this subdivision is guilty of a misdemeanor and, upon conviction thereof, shall for each offense be fined not less than $100 nor more than $500, and shall be confined in jail for not less than ten days nor more than one hundred days;
4. Hunt, take, kill, wound or shoot at wild animals or wild birds from an airplane or other airborne conveyance, a drone or other unmanned aircraft, an automobile or other land conveyance, or from a motor-driven water conveyance;
5. Use a drone or other unmanned aircraft to hunt, take or kill a wild bird or wild animal, or to use a drone or other unmanned aircraft to drive or herd any wild bird or wild animal for the purposes of hunting, trapping or killing;
6. Take any beaver or muskrat by any means other than a trap;
7. Catch, capture, take, hunt or kill by seine, net, bait, trap or snare or like device a bear, wild turkey, ruffed grouse, pheasant or quail;
8. Intentionally destroy or attempt to destroy the nest or eggs of any wild bird or have in his or her possession the nest or eggs;

9. Carry an uncased or loaded firearm in the woods of this state with the following permissible exceptions:
   A. A person in possession of a valid license or permit during open firearms hunting season for wild animals and nonmigratory wild birds;
   B. A person hunting or taking unprotected species of wild animals, wild birds and migratory wild birds during the open season, in the open fields, open water and open marshes of the state;
   C. A person carrying a firearm pursuant to sections six and six-a of this article; or
   D. A person carrying a firearm for self defense who is not prohibited from possessing firearms by section seven, article seven, chapter sixty-one of this code;
10. Have in his or her possession a crossbow with a nocked bolt, or a rifle or shotgun with cartridges that have not been removed or a magazine that has not been detached, in or on any vehicle or conveyance, or its attachments. For the purposes of this section, a rifle or shotgun whose magazine readily detaches is considered unloaded if the magazine is detached and no cartridges remain in the rifle or shotgun itself. Except that between five o'clock post meridian of day one and seven o'clock ante meridian, Eastern Standard Time, of the following day, any unloaded firearm or crossbow may be carried only when in a case or taken apart and securely wrapped. During the period from July 1 to September 30, inclusive, of each year, the requirements relative to carrying unloaded firearms are permissible only from eight-thirty o'clock post meridian to five o'clock ante meridian, Eastern Standard Time: Provided, That the time periods for carrying unloaded and uncased firearms are extended for one hour after the post meridian times and one hour before the ante meridian times established in this subdivision, if a person is transporting or transferring the firearms to or from a hunting site, campsite, home or other abode;
11. Hunt, catch, take, kill, trap, injure or pursue with firearms or other implement by which wildlife may be taken after the hour of five o'clock ante meridian on Sunday on private land without the written consent of the landowner any wild animals or wild birds except when a big game season opens on a Monday, the Sunday prior to that opening day will be closed for any taking of wild animals or birds after five o'clock ante meridian on that Sunday: Provided, That traps previously and legally set may be tended after the hour of five o'clock ante meridian on Sunday and the person tending the traps may carry firearms for the purpose of humanely dispatching trapped animals. Any person violating this subdivision is guilty of a misdemeanor and, upon conviction thereof, in addition to any fines that may be imposed by this or other sections of this code, is subject to a $100 fine;
12. Hunt, catch, take, kill, injure or pursue a wild animal or wild bird with the use of a ferret;

13. Buy raw furs, pelts or skins of fur-bearing animals unless licensed to do so;
14. Catch, take, kill or attempt to catch, take or kill any fish by any means other than by rod, line and hooks with natural or artificial lures: Provided, That snaring of any species of suckers, carp, fallfish and creek chubs is lawful;
15. Employ, hire, induce or persuade, with money, things of value or by any means, any person to hunt, take, catch or kill any wild animal or wild bird except those species in which there is no closed season; or to fish for, catch, take or kill any fish, amphibian or aquatic life that is protected by rule, or the sale of which is otherwise prohibited;
16. Hunt, catch, take, kill, capture, pursue, transport, possess or use any migratory game or nongame birds except as permitted by the Migratory Bird Treaty Act, 16 U. S. C. §703, et seq., and its regulations;
17. Kill, take, catch, sell, transport or have in his or her possession, living or dead, any wild bird other than a game bird including the plumage, skin or body of any protected bird, irrespective of whether the bird was captured in or out of this state, except the English or European sparrow (Passer domesticus), starling (Sturnus vulgaris) and cowbird (Molothrus ater), which may be killed at any time;
18. Use dynamite, explosives or any poison in any waters of the state for the purpose of killing or taking fish. Any person violating this subdivision is guilty of a felony and, upon conviction thereof, shall be fined not more than $500 or imprisoned for not less than six months nor more than three years, or both fined and imprisoned;
19. Have a bow and gun, or have a gun and any arrow, in the fields or woods at the same time;
20. Have a crossbow in the woods or fields, or use a crossbow to hunt for, take or attempt to take any wildlife except as otherwise provided in sections five-g and forty-two-w of this article;
21. Take or attempt to take turkey, bear, elk or deer with any arrow unless the arrow is equipped with a point having at least two sharp cutting edges measuring in excess of three fourths of an inch wide;
22. Take or attempt to take any wildlife with an arrow having an explosive head or shaft, a poisoned arrow or an arrow which would affect wildlife by any chemical action;
23. Shoot an arrow across any public highway;
24. Permit any dog owned or under his or her control to chase, pursue or follow the tracks of any wild animal or wild bird, day or night, between May 1 and August 15: Provided, That dogs may be trained on wild animals and wild birds, except deer and wild turkeys, and field trials may be held or conducted on the grounds or lands of the owner, or by his or her bona fide tenant, or upon the grounds or lands of another person with his or her written permission, or on public lands at

any time. Nonresidents may not train dogs in this state at any time except during the legal small game hunting season. A person training dogs may not have firearms or other implements in his or her possession during the closed season on wild animals and wild birds;

25. Conduct or participate in a trial, including a field trial, shoot-to-retrieve field trial, water race or wild hunt: Provided, That any person, group of persons, club or organization may hold a trial upon obtaining a permit pursuant to section fifty-six of this article. The person responsible for obtaining the permit shall prepare and keep an accurate record of the names and addresses of all persons participating in the trial and make the records readily available for inspection by any natural resources police officer upon request;
26. Hunt, catch, take, kill or attempt to hunt, catch, take or kill any wild animal, wild bird or wild fowl except during open seasons;
27. Hunting on public lands on Sunday after five o'clock ante meridian is prohibited;
28. Hunt, catch, take, kill, trap, injure or pursue with firearms or other implement which wildlife can be taken, on private lands on Sunday after the hour of five o'clock ante meridian: Provided, That the provisions of this subdivision do not apply in any county until the county commission of the county holds an election on the question of whether the provisions of this subdivision prohibiting hunting on Sunday shall apply within the county and the voters approve the allowance of hunting on Sunday in the county. The election is determined by a vote of the resident voters of the county in which the hunting on Sunday is proposed to be authorized. The county commission of the county in which Sunday hunting is proposed shall give notice to the public of the election by publication of the notice as a Class II-0 legal advertisement in compliance with the provisions of article three, chapter fifty-nine of this code and the publication area for the publication is the county in which the election is to be held. The date of the last publication of the notice shall fall on a date within the period of the fourteen consecutive days next preceding the election. On the local option election ballot shall be printed the following:

Shall hunting on Sunday be authorized on private lands only with the consent of the land owner in ________ County?

[ ]Yes

[ ]No

(Place a cross mark in the square opposite your choice.)

Any local option election to approve or disapprove of the proposed authorization of Sunday hunting within a county shall be in accordance with procedures adopted by the commission. The local option election may be held in conjunction with a primary or general election or at a special election. Approval shall be by a majority of the voters casting votes on the question of approval or disapproval of Sunday hunting at the election. If a majority votes against allowing Sunday hunting, an election on the issue may not be held for a period of one hundred four weeks. If a majority votes "yes", an election reconsidering the action may not be held for a period of five years. A local option election may thereafter be held if a written petition of qualified voters residing within the county equal to at least five percent of the number of persons who were registered to vote in the next preceding general election is received by the county commission of the county in which Sunday hunting is authorized. The petition may be in any number of counterparts. The election shall take place at the next primary or general election scheduled more than ninety days following receipt by the county commission of the petition required by this subsection: Provided, That the issue may not be placed on the ballot until all statutory notice requirements have been met. No local law or regulation providing any penalty, disability, restriction, regulation or prohibition of Sunday hunting may be enacted and the provisions of this article preempt all regulations, rules, ordinances and laws of any county or municipality in conflict with this subdivision. Amendments to this subdivision promulgated during the 2015 regular session of the Legislature shall have no effect upon the results of elections held prior to their enactment; and

29. Hunt or conduct hunts for a fee when the person is not physically present in the same location as the wildlife being hunted within West Virginia.

    §20-2-5a. Forfeiture by person causing injury or death of game or protected species of animal; additional replacement costs for antlered deer; forfeiture procedures and costs.

    a. Any person who is convicted of violating a criminal law of this state that results in the injury or death of game, as defined in section two, article one of this chapter, or a protected species of animal, in addition to any other penalty to which he or she is subject, shall forfeit the cost of replacing the game or protected species of animal to the state as follows:
        1. For each game fish or each fish of a protected species taken illegally other than by pollution kill, $10 for each pound and any fraction thereof;
        2. For each bear, $500;
        3. For each deer or raven, $200;
        4. For each wild turkey, hawk or owl, $100;
        5. For each beaver, otter or mink, $25;
        6. For each muskrat, raccoon, skunk or fox, $15;
        7. For each rabbit, squirrel, opossum, duck, quail, woodcock, grouse or pheasant, $10;
        8. For each wild boar, $200;
        9. For each bald eagle, $5,000;
        10. For each golden eagle, $5,000;

11. For each elk, $4,500; and
12. For any other game or protected species of animal, $100.

b. In addition to the replacement value for deer in subdivision (3), subsection (a) of this section, the following cost shall also be forfeited to the state by any person who is convicted of violating any criminal law of this state and the violation causes the injury or death of antlered deer:
1. For any deer in which the inside spread of the main beams of the antlers measured at the widest point equals 14 inches or greater but less than 16 inches, $1,000;
2. For any deer in which the inside spread of the main beams of the antlers measured at the widest point equals 16 inches or greater but less than 18 inches, $1,500;
3. For any deer in which the inside spread of the main beams of the antlers measured at the widest point equals 18 inches or greater but less than 20 inches, $2,000; and
4. For any deer in which the inside spread of the main beams of the antlers measured at the widest point equals 20 inches or greater, $2,500.
5. Any person convicted of a second or subsequent violation of any criminal law of this state which violation causes the injury or death of antlered deer is subject to double the authorized range of cost to be forfeited.

c. Upon conviction, the court shall order the person to forfeit to the state the amount set forth in this section for the injury or death of the game or protected species of animal. If two or more defendants are convicted for the same violation causing the injury or death of game or protected species of animal, the forfeiture shall be paid by each person in an equal amount. The forfeiture shall be paid by the person so convicted within the time prescribed by the court not to exceed sixty days. In each instance, the court shall pay the forfeiture to the Division of Natural Resources to be deposited into the License Fund-Wildlife Resources and used only for the replacement, habitat management or enforcement programs for injured or killed game or protected species of animal.

§20-2-5g. Use of a crossbow to hunt.

a. Notwithstanding any other provision of this code to the contrary, any person lawfully entitled to hunt may hunt with a crossbow during big game firearms season. A person who possesses a valid Class Y permit may also hunt with a crossbow in accordance with section forty-two-w of this article. Further, the director shall designate a separate season for crossbow hunting and identify which species of wildlife may be hunted with a crossbow.

b. Only crossbows meeting all of the following specifications may be used for hunting in West Virginia:
    1. The crossbow has a minimum draw weight of one hundred twenty-five pounds;
    2. The crossbow has a working safety; and
    3. The crossbow is used with bolts and arrows not less than eighteen inches in length with a broad head having at least two sharp cutting edges, measuring at least three fourths of an inch in width.

§20-2-5h. Elk management area; elk damage fund; criminal penalties; rule-making.

a. Findings.—The Legislature finds that Eastern Elk were once a common, native species in the state prior to and following its formation, but historical records indicate native elk were extirpated from the state around 1875. Until recently, free roaming elk have not been present in the state. However, elk are now migrating to the state from Kentucky, which has an active elk restoration program. Therefore, the Division of Natural Resources has established an active elk restoration program in Southern West Virginia.
b. Elk management area.—The division has established an elk restoration management plan to reintroduce elk to all of Logan County, Mingo County, McDowell County and Wyoming County, and part of Boone County, Lincoln County and Wayne County. The director and the division may not expand the elk management area without statutory authorization.
c. Elk damage fund.—There is hereby created a special revenue account in the State Treasury to be known as the Elk Damage Fund to be administered by the division. Ten percent from all application fees for the hunting of elk are to be deposited into the Elk Damage Fund." Expenditures from the fund shall be for the payment of damages caused to agricultural crops, agricultural fences and personal gardens by elk.
d. Criminal penalties.—It shall be unlawful for any person to hunt, capture or kill any elk, or have in his or her possession elk or elk parts, except for elk lawfully taken, killed or obtained during an established open hunting season for elk or by permit.
    1. Any person who commits a violation of the provisions of this section is guilty of a misdemeanor and, upon conviction thereof, shall be fined not less than $1,000 nor more than $5,000, or confined in jail not less than thirty nor more than one hundred days, or both fined and confined.

2. Any person who commits a second violation of the provisions of this section is guilty of a misdemeanor and, upon conviction thereof, shall be fined not less than $2,000 nor more than $7,500, or confined in jail not less than thirty days nor more than one year, or both fined and confined.
3. Any person who commits a third or subsequent violation of the provisions of this section is guilty of a felony and, upon conviction thereof, shall be fined not less than $5,000 nor more than $10,000, or imprisoned in a state correctional facility not less than one year nor more than five years, or both fined and imprisoned.

e. Rulemaking. — The director shall propose rules for promulgation in accordance with the provisions of article three, chapter twenty-nine-a of this code to:
   1. Set forth the parameters of the elk management plan;
   2. Establish the procedures for the issuance of depredation permits to persons suffering damage from elk;
   3. Establish protocols for the control of elk outside the elk management area;
   4. Establish hunting application fees and procedures;
   5. Establish procedures for reimbursement from the elk damage fund to those with damage to agricultural crops, agricultural fences and personal gardens caused by elk; and (6) Establish protocols for ensuring elk imported to the state are healthy, tested for tuberculosis, brucellosis and other diseases of critical concern, and from an area where chronic wasting disease has not been detected.

§20-2-22a. Hunting, tagging and reporting bear; procedures applicable to property destruction by bear; penalties.

a. A person may not hunt, capture, or kill any bear, or have in his or her possession any bear or bear parts, except during the hunting season for bear in the manner designated by rule or law, and as provided in this section. For the purposes of this section, bear parts include, but are not limited to, the pelt, gallbladder, skull and claws of bear.
b. A person who kills a bear shall, within twenty-four hours after the killing, electronically register the bear. A game tag number shall be issued to the person and recorded in writing with the person's name and address, or on a field tag and shall remain on the skin until it is tanned or mounted. Any bear or bear parts not properly tagged shall be forfeited to the state for disposal to a charitable institution, school or as otherwise designated by the director.
c. It is unlawful:

1. To hunt bear without a bear damage stamp, as prescribed in section forty-four-b of this article, in addition to a hunting license as prescribed in this article;
2. To hunt a bear with:
   A. A shotgun using ammunition loaded with more than one solid ball; or
   B. A rifle of less than twenty-five caliber using rimfire ammunition;
3. To kill or attempt to kill any bear through the use of poison, explosives, snares, steel traps or deadfalls;
4. To shoot at or kill:
   A. A bear weighing less than seventy-five pounds live weight or fifty pounds field dressed weight, after removal of all internal organs;
   B. Any bear accompanied by a cub; or
   C. Any bear cub so accompanied, regardless of its weight;
5. To possess any part of a bear not tagged in accordance with the provisions of this section;
6. To enter a state game refuge with firearms for the purpose of pursuing or killing a bear except under the direct supervision of division personnel;
7. To hunt bear with dogs or to cause dogs to chase bear during seasons other than those designated by the division for the hunting of bear;
8. To pursue a bear with a pack of dogs other than the pack used at the beginning of the hunt once the bear is spotted and the chase has begun;
9. To possess, harvest, sell or purchase bear parts obtained from bear killed in violation of this section;
10. To organize for commercial purposes or to professionally outfit a bear hunt, or to give or receive any consideration whatsoever or any donation in money, goods or services in connection with a bear hunt, notwithstanding the provisions of sections twenty-three and twenty-four of this article; or
11. For any person who is not a resident of this state to hunt bear with dogs or to use dogs in any fashion for the purpose of hunting bear in this state except in legally authorized hunts.

d. The following provisions apply to bear destroying property:
   1. Any property owner or lessee who has suffered damage to real or personal property, including loss occasioned by the death or injury of livestock or the unborn issue of livestock, caused by an act of a bear may complain to any natural resources police officer of the division for protection against the bear.
   2. Upon receipt of the complaint, the officer shall immediately investigate the circumstances of the complaint. If the officer is unable to personally investigate the complaint, he or she shall designate a wildlife biologist to investigate on his or her behalf.

3. If the complaint is found to be justified, the officer or designated person may, together with the owner and other residents, proceed to hunt, destroy or capture the bear that caused the property damage: Provided, That only the natural resources police officer or the wildlife biologist may determine whether to destroy or capture the bear and whether to use dogs to capture or destroy the bear: Provided, however, That, if out-of-state dogs are used in the hunt, the owners of the dogs are the only nonresidents permitted to participate in hunting the bear.
4. A. When a property owner has suffered damage to real or personal property as the result of an act by a bear, the owner shall file a report with the director of the division. The report shall state whether or not the bear was hunted and destroyed and, if so, the sex, weight and estimated age of the bear. The report shall also include an appraisal of the property damage occasioned by the bear duly signed by three competent appraisers fixing the value of the property lost.
   B. The report shall be ruled upon and the alleged damages examined by a commission comprised of the complaining property owner, an officer of the division and a person to be jointly selected by the officer and the complaining property owner.
   C. The division shall establish the procedures to be followed in presenting and deciding claims under this section in accordance with article three, chapter twenty-nine-a of this code.
   D. All claims shall be paid in the first instance from the Bear Damage Fund provided in section forty-four-b of this article. In the event the fund is insufficient to pay all claims determined by the commission to be just and proper, the remainder due to owners of lost or destroyed property shall be paid from the special revenue account of the division.
5. In all cases where the act of the bear complained of by the property owner is the killing of livestock, the value to be established is the fair market value of the livestock at the date of death. In cases where the livestock killed is pregnant, the total value is the sum of the values of the mother and the unborn issue, with the value of the unborn issue to be determined on the basis of the fair market value of the issue had it been born.

e. Criminal penalties.
   1. Any person who commits a violation of the provisions of this section is guilty of a misdemeanor and, upon conviction thereof, shall be fined not less than $1,000 nor more than $5,000, which is not subject to suspension by the court, confined in jail not less than thirty nor more than one hundred days, or both fined and confined.

   Further, the person's hunting and fishing licenses shall be suspended for two years.

2. Any person who commits a second violation of the provisions of this section is guilty of a misdemeanor and, upon conviction thereof, shall be fined not less than $2,000 nor more than $7,500, which is not subject to suspension by the court, confined in jail not less than thirty days nor more than one year, or both fined and confined. The person's hunting and fishing licenses shall be suspended for life.
3. Any person who commits a third or subsequent violation of the provisions of this section is guilty of a felony and, upon conviction thereof, shall be fined not less than $5,000 nor more than $10,000, which is not subject to suspension by the court, imprisoned in a correctional facility not less than one year nor more than five years, or both fined and imprisoned.

§20-2-42w. Class Y special crossbow hunting permit for certain disabled persons.

a. A Class Y permit is a special statewide hunting permit 2 entitling a person to hunt all wildlife during established archery and firearm seasons if the person meets the following requirements:
   1. He or she holds a Class Q permit;
   2. He or she has a permanent and substantial loss of function in one or both hands while failing to meet the minimum standards of the upper extremity pinch, grip and nine-hole peg tests administered under the direction of a licensed physician; or
   3. He or she has a permanent and substantial loss of function in one or both shoulders while failing to meet the minimum standards of the shoulder strength test administered under the direction of a licensed physician.
b. The application form shall include a written statement or report prepared by the physician conducting the test no more than six months preceding the application and verifying that the applicant is physically disabled as described in this section. As part of the application, the applicant shall authorize, by written release, an examination of all medical records regarding his or her qualifying disability. When completed, the permit form constitutes a Class Y permit. The Class Y permit and a completed license application shall be submitted to the Division, which will issue a wallet-sized card to the permittee. The card and all other documents and identification required to be carried by this article shall be in the permittee's possession when hunting.
c. A Class Y permit must be accompanied by a valid statewide hunting license or the applicant must be exempt from hunting licenses as provided in this chapter.[467]

[467]HB 2515, West Virginia Legislature, http://www.legis.state.wv.us/Bill_Text_HTML/2015_SESSIONS/RS/pdf_bills/HB2515%20SUB%20ENR%20PRINTED.pdf.

WISCONSIN

There are three laws at this time, namely, AB 670,[468] SB 338,[469] and SB 196.[470] Wisconsin has introduced AB 433 to prohibit certain activities that interfere with hunting, fishing, and trapping;[471] AB 671 relating to the use of a drone to commit a crime and providing a criminal penalty;[472] SB 497 relating to the use of a drone to commit a crime and providing a criminal penalty;[473] and SB 498 relating to the operation of drones over state correctional institutions, the authority of political subdivisions to designate areas over which the operation of drones is limited and providing a penalty.[474]

Additionally, the City of Antigo has adopted Resolution 090-15 to support the efforts of Unmanned Systems Incorporated to establish UAV testing and training at the Langlade County Airport,[475] Chetek City has passed Code 118-89,[476] Green Bay has passed Section 27.310 relating to drone use at special events,[477] and Outagamie has passed Ordinance 10.195 limiting aeronautical activities of aerial devices,[478] and 10.35 subjecting certain entities to minimum standards of operation.[479]

Assembly Bill (AB) 670: Published March 31, 2016, this was an act to renumber and amend 114.105; to amend 114.04; and to create 114.045, 114.105(1) and 114.105(3) of the statutes relating to the operation of drones over state correctional institutions, the authority of political subdivisions to designate areas over which the operation of drones is limited, and providing a penalty.[480]

Senate Bill (SB) 338: Published April 3, 2016, this was an act to amend 29.083(1), 29.083(a)(intro.), 29.083(2)(a)5, 29.971(11r)(a); and to create 29.083(2)(a)6, 29.083(2)(a)7, 29.083(2)(a)8 and 29.971(11r)(am) of the statutes relating to interfering with hunting, fishing, and trapping and providing criminal penalties.[481]

Senate Bill (SB) 196: On May 23, 2013, the Wisconsin State Legislature enacted SB 196. It reads as follows:

The people of the state of Wisconsin, represented in senate and assembly, do enact as follows:

---

[468] AB 670, Wisconsin State Legislature, https://docs.legis.wisconsin.gov/2015/proposals/ab670.

[469] SB 338, Wisconsin State Legislature, https://docs.legis.wisconsin.gov/2015/proposals/sb338.

[470] SB 196, Wisconsin State Legislature, https://docs.legis.wisconsin.gov/2013/related/proposals/sb196.pdf.

[471] AB 433, Wisconsin State Legislature, http://docs.legis.wisconsin.gov/2015/proposals/ab433.

[472] AB 671, Wisconsin State Legislature, https://docs.legis.wisconsin.gov/2015/proposals/ab671.

[473] SB 497, Wisconsin State Legislature, https://docs.legis.wisconsin.gov/2015/proposals/reg/sen/bill/sb497.

[474] SB 498, Wisconsin State Legislature, https://docs.legis.wisconsin.gov/2015/proposals/reg/sen/bill/sb498.

[475] Resolution 090-15, City of Antigo, available at http://www.antigo-city.org/Council_8_12_15.pdf.

[476] Code of Ordinances Section 118-89, Chetek, http://uavs.insct.org/wp-content/uploads/2016/04/Chetek-WI-Code-of-Ordinances-118-89.pdf.

[477] Chapter 27.310, Green Bay, http://greenbaywi.gov/law/wp-content/uploads/2013/04/CHAPTER-27.pdf.

[478] Section 10-195, Outagamie County, http://uavs.insct.org/wp-content/uploads/2016/04/Outagamie-County-WI-Code-of-Ordinances-10.195.pdf.

[479] Section 10-35, Outagamie County, http://uavs.insct.org/wp-content/uploads/2016/04/Outagamie-County-WI-Code-of-Ordinances-10.35.pdf.

[480] AB 670, Wisconsin State Legislature, https://docs.legis.wisconsin.gov/2015/proposals/ab670.

[481] SB 338, Wisconsin State Legislature, https://docs.legis.wisconsin.gov/2015/proposals/sb338.

SECTION 1. 114.04 of the statutes is amended to read:
114.04 Flying and landing, limitations. Flight Subject to s. 175.55, and except as provided in s. 942.10, flight of or in aircraft or spacecraft over the lands and waters of this state is lawful, unless at such a low altitude as to interfere with the then existing use to which the land or water, or the space over the land or water, is put by the owner, or unless so conducted as to be imminently dangerous or damaging to persons or property lawfully on the land or water beneath. The landing of an aircraft or spacecraft on the lands or waters of another, without the person's consent, is unlawful, except in the case of a forced landing. For damages caused by a forced landing, however, the owner or lessee of the aircraft or spacecraft or the aeronaut or astronaut shall be liable, as provided in s. 114.05.

SECTION 2. 175.55 of the statutes is created to read:
175.55 Use of drones restricted. (1) In this section:

a. "Drone" means a powered, aerial vehicle that carries or is equipped with a recording device, as defined in s. 943.49 (1) (c), that does not carry a human operator, uses aerodynamic forces to provide vehicle lift, and can fly autonomously or be piloted remotely. A drone may be expendable or recoverable.
b. "Wisconsin law enforcement agency" has the meaning given in s. 165.77 (1) (c) and includes the department of justice and a tribal law enforcement agency.
    1. No Wisconsin law enforcement agency may use a drone to gather evidence or other information in a criminal investigation without first obtaining a search warrant under s. 968.12. This subsection does not apply to the use of a drone to assist in an active search and rescue operation, to locate an escaped prisoner, or if a law enforcement officer has reasonable suspicion to believe that the use of a drone is necessary to prevent imminent danger to an individual or to prevent imminent destruction of evidence.

SECTION 3. 941.292 of the statutes is created to read:
941.292 Possession of a weaponized drone.

1. In this section, "drone" means a powered, aerial vehicle that does not carry a human operator, uses aerodynamic forces to provide vehicle lift, and can fly autonomously or be piloted remotely. A drone may be expendable or recoverable.
2. Whoever sells, transports, manufactures, possesses, or operates any weaponized drone is guilty of a Class H felony.

SECTION 4. 942.10 of the statutes is created to read:
942.10 Use of a drone. Whoever uses a drone, as defined in s. 175.55 (1) (a), to photograph, record, or otherwise observe another individual in a place where the individual has

a reasonable expectation of privacy is guilty of Class A misdemeanor. This section does not apply to a law enforcement officer authorized to use a drone pursuant to s. 175.55 (2).

SECTION 5. 972.113 of the statutes is created to read:
972.113 Evidence obtained by a drone. Evidence or information obtained or collected in violation of s. 175.55 is not admissible in evidence in any criminal proceeding.[482]

WYOMING

Wyoming has introduced HB 18, also known as the Drone Protection Act, to specify requirements for the use of drones by law enforcement agencies, limit the use of information obtained with a drone by governmental entities, and require reports,[483] and HB 242, an act relating to criminal procedure, requiring a warrant for unmanned aerial surveillance, and prohibiting use of evidence obtained without a required warrant.[484]

---

[482] SB 196, Wisconsin State Legislature, https://docs.legis.wisconsin.gov/2013/related/proposals/sb196.pdf.
[483] HB 18, Wyoming State Legislature, http://legisweb.state.wy.us/2015/Engross/HB0018.pdf.
[484] HB 242, Wyoming State Legislature, http://legisweb.state.wy.us/2013/Introduced/HB0242.pdf.

CHAPTER 4
GOVERNMENT UAS RULES

1. Public or Governmental Regulations

Public aircraft operations are limited by federal statute to certain government operations within US airspace. Title 49 United States Code § 40102(a)(41) provides the definition of "Public Aircraft" and § 40125 provides the qualifications for public aircraft status. Whether an operation qualifies as a public aircraft operation is determined on a flight-by-flight basis, under the terms of the statute. The considerations when making this determination are aircraft ownership, the operator, the purpose of the flight, and the persons on board the aircraft.[1]

Where public use of UAS is concerned, the FAA developed guidance in a memorandum titled "Unmanned Aircraft Systems Operations in the US National Airspace System—Interim Operational Approval Guidance," which is also referred to as Policy Notice 05-01.[2] In this Policy Notice 05-01, the FAA set out guidance for public use of UAS by defining a process for evaluating applications for Certificate(s) of Waiver or Authorization (COAs) for UAS to operate in the NAS. This was to address the FAA's concern for safety, not only due to interference with commercial and general aviation aircraft but also with other airborne vehicles and persons or property on the ground.[3]

---

[1] *UAS Operations,* Federal Aviation Administration (FAA), http://www.faa.gov/uas/public_operations/.
[2] *UAS Operations,* Federal Aviation Administration (FAA), http://www.uavm.com/images/AFS-400_05-01_faa_uas_policy.pdf.
[3] *Id.*

At this time COAs are available to public entities that want to fly a UAS in civil airspace. These entities include law enforcement, firefighting, border patrol, disaster relief, search and rescue, military training, and many more. The COA allows the operator to use a defined block of airspace and includes special provisions unique to the proposed operation. COAs are issued for a specific period, typically up to two years. Most COAs require coordination with an appropriate air traffic control facility and may require a transponder on the UAS to operate in certain types of airspace. Because UAS technology cannot currently comply with "see and avoid" rules that apply to all aircraft, a visual observer or an accompanying "chase plane" must maintain visual contact with the UAS and serve as its "eyes" when operating outside airspace restricted from other users. As of September 2015, the FAA claims the average time to issue an authorization for non-emergency operations is less than 60 business days. The FAA has an expedited procedure in place to grant a one-time COA for time-sensitive emergency missions, such as disaster relief and humanitarian efforts.[4]

In order to ensure safety, the operator is required to establish the UAS's airworthiness either through FAA certification, DoD airworthiness statement, or by other approved means. Applicants also have to demonstrate that a collision with another aircraft or other airspace user is extremely improbable as well as complying with appropriate cloud and terrain clearances as required. This concept is tied to the roles of pilot in command and observer and includes minimum qualifications and currency requirements.[5]

The role of the observer is to observe the activity of the UAS and surrounding airspace, either through line of sight on the ground or in the air by means of a chase aircraft. Generally, this means that the pilot or the observer must be, typically, within one mile laterally and 3,000 feet vertically of the UAS. Furthermore, direct communication between the pilot in command and the observer must be maintained at all times. Unmanned aircraft flight above 18,000 feet must be conducted in accordance with Instrument Flight Rules (IFR), on an IFR flight plan, must obtain air traffic control clearance, must be equipped with at least a Mode C transponder (preferably a Mode S), operating lights and/or collision avoidance lights, and maintain communication between the pilot in command and air traffic control. It is worth noting that as of 2014 the FAA had issued in excess of 609 COAs.[6]

2. Advisory Circular (AC) 00-1.1A

It is important to consult Advisory Circular (AC) 00-1.1A entitled Public Aircraft Operations for additional information. This AC was published on February 12, 2014.[7]

---

[4] *Id.*
[5] *UAS Operations,* Federal Aviation Administration (FAA), http://www.faa.gov/regulations_policies/rulemaking/committees/documents/media/ARAC.Meeting.Minutes.12.06.12.pdf.
[6] *Id.*
[7] *Public Aircraft Operations,* FAA, https://www.faa.gov/documentLibrary/media/Advisory_Circular/AC_00-1_1A.pdf.

3. Title 10 for US Military UAS

Title 10 of the United States Code applies to the Armed Forces. Certain parts of it are pertinent to US military UAS pilots. Furthermore, in March 2011, the Department of Defense (DoD) UAS Task Force Airspace Integration Integrated Product Team issued version 2.0 of the UAS Airspace Integration Plan. This plan, as outlined by Deputy Secretary of Defense, William J. Lynn, III, recognizes the following:[8]

Over the past decade, UAS have become an integral part of the US Military and Government operations. Currently, there are many different systems and thousands of aircraft fielded and/or deployed within the DoD and Other Government Agencies (OGAs). They also support operations for Homeland Security (HLS), Homeland Defense (HD), and Defense Support of Civil Authorities (DSCA). In support of their operational, training, and test and evaluation missions, DoD UAS require safe and routine access to the US National Airspace System (NAS) similar to manned aircraft. Beyond the NAS, DoD's requirement to deploy globally into foreign and international airspace requires a high level of awareness of international law and conditions placed by foreign sovereigns on use of their territorial airspace.[9]

The challenges of UAS Airspace Integration (AI) are complex and multidimensional. They are influenced by the differences in UAS types, classes, and types of airspace, Programs of Record (POR) requirements, available technologies, and specific mission needs. Without a clear, regulatory path to integration, DoD intends to leverage new technologies, procedures, and policies to maximize regulatory compliance. This plan outlines a set of crucial UAS flight profiles to identify and focus resources to tailor solutions that enable near term routine operations. Near-term efforts can help increase UAS access to the NAS immediately, while a full set of regulations, policy and procedures, standards, and technology must be developed and considered to allow UAS appropriate access to the NAS in a safe and efficient manner.[10]

Some other material pertaining to US military UAS include:

- Unmanned Aircraft System Commander's Guide and Aircrew Training Manual (TC 3-04.61 – January 2014)
- Small Unmanned Aircraft System Aircrew Training Program (TC 3-04.62 – August 2013)
- MQ-1C Unmanned Aircraft System Commander's Aircrew Training Program and Aircrew Training Manual (TC 3-04.63 – August 2014)

[8]*UAS Airspace Integration Plan,* Department of Defense (DoD), http://www.acq.osd.mil/sts/docs/DoD_UAS_Airspace_Integ_Plan_v2_(signed).pdf.
[9]*Id.*
[10]*Id.*

CHAPTER 5
HOBBY OR RECREATIONAL UAS RULES

1. Recreational or Hobby Regulations

The new rules mentioned in the NPRM for small UAS would not apply to model aircraft. However, model aircraft operators must continue to satisfy all of the criteria specified in Sec. 336 of Public Law 112-95, including the stipulation that they be operated only for hobby or recreational purposes.[1]

2. Interpretation of Special Rule for Model Aircraft

Inundated with requests for clarification of Section 336 of the FAA Modernization and Reform Act of 2012, prompted the FAA, two years later, to release the interpretation of special rule for model aircraft. This action, the notice of interpretation with request for comment, provides interested persons with the opportunity to comment on the FAA's interpretation of the special rule for model aircraft established by Congress in the FAA Modernization and Reform Act of 2012. In this interpretation, the FAA clarifies that: model aircraft must satisfy the criteria in the Act to qualify as model aircraft and to be exempt from future FAA rulemaking action; and consistent with the Act, if a model aircraft operator endangers the safety of the National Airspace System, the FAA has the authority to take enforcement action against those operators for those safety violations.[2]

The FAA issued this interpretation because they had received many inquiries regarding the scope of the special rule for model aircraft in section 336 of the FAA Modernization

[1] *Small UAS Notice of Proposed Rulemaking,* Federal Aviation Administration (FAA), https://www.faa.gov/regulations_policies/rulemaking/recently_published/.

[2] FAA-2014-0396, Interpretation of the Special Rule for Model Aircraft, https://www.faa.gov/uas/media/model_aircraft_spec_rule.pdf.

and Reform Act of 2012 and the FAA's enforcement authority over model aircraft as affirmed by the statute. In this interpretation, they explain the history of FAA oversight of model aircraft operations and the new statutory requirements that apply to model aircraft operations, and then clarify how the FAA intends to apply its enforcement authority to model aircraft operations that endanger the safety of the National Airspace System (NAS).[3]

What follows below is the text of this interpretation.

Discussion of the Interpretation

I. Background of FAA Oversight of Model Aircraft Operations

Historically, the FAA has considered model aircraft to be aircraft that fall within the statutory and regulatory definitions of an aircraft, as they are contrivances or devices that are "invented, used, or designed to navigate, or fly in, the air." See 49 USC 40102 and 14 CFR 1.1. As aircraft, these devices generally are subject to FAA oversight and enforcement. However, consistent with FAA's enforcement philosophy, FAA's oversight of model aircraft has been guided by the risk that these operations present. The FAA first recognized in 1981 that "model aircraft can at times pose a hazard to full-scale aircraft in flight and to persons and property on the surface," and recommended a set of voluntary operating standards for model aircraft operators to follow to mitigate these safety risks. See Advisory Circular 91-57, Model Aircraft Operating Standards (June 9, 1981). These operating standards included restricting operations over populated areas, limiting use of the devices around spectators until after the devices had been flight tested and proven airworthy; restricting operations to 400 feet above the surface; requiring that the devices give right of way to, and avoid flying near manned aircraft, and using observers to assist in operations.

These guidelines were further clarified in 2007, when the FAA issued a policy statement regarding unmanned aircraft systems (UAS) operations in the NAS. See 72 Fed. Reg. 6689 (Feb. 13, 2007). In this policy statement, the FAA also recognized that UAS fall within the statutory and regulatory definition of "aircraft" as they are devices that are "used or [are] intended to be used for flight in the air with no onboard pilot." Id.; see also 49 U.S.C. 40102; 14 CFR 1.1. The FAA noted that they can be "as simple as a remotely controlled model aircraft used for recreational purposes or as complex as surveillance aircraft flying over hostile areas in warfare." The FAA then stated its current policy regarding UAS based on the following three categories: (1) UAS used as public aircraft; (2) UAS used as civil aircraft; and (3) UAS used as model aircraft.

---

[3] *Id.*

With respect to UAS used as model aircraft, the FAA reiterated the operating guidelines in AC 91-57, and further noted that to qualify as a model aircraft, the aircraft would need to be operated purely for recreational or hobby purposes, and within the visual line of sight of the operator. The policy statement also clarified that AC 91-57 applied only to modelers and "specifically excludes its use by persons or companies for business purposes." 72 FR at 6690.

II. Requirements to Qualify as a Model Aircraft under the FAA Modernization and Reform Act of 2012 (P.L. 112-95, section 336).

A. Statutory Requirements

On February 14, 2012, the President signed into law the FAA Modernization and Reform Act of 2012 (P.L. 112-95) (the Act), which established, in Section 336, a "special rule for model aircraft." In Section 336, Congress confirmed the FAA's long-standing position that model aircraft are aircraft. Under the terms of the Act, a model aircraft is defined as "an unmanned aircraft" that is "(1) capable of sustained flight in the atmosphere; (2) flown within visual line of sight of the person operating the aircraft; and (3) flown for hobby or recreational purposes." P.L. 112-95, section 336(c). Congress' intention to define model aircraft as aircraft is further established by section 331(8) of the Act, which defines an unmanned aircraft as "an aircraft that is operated without the possibility of direct human intervention from within or on the aircraft." Congress' definition of model aircraft is consistent with the FAA's existing definition of aircraft as "any contrivance invented, used, or designed to navigate, or fly in, the air." 49 U.S.C. 40102; see also 14 C.F.R. 1.1. Although model aircraft may take many forms, at a base level model aircraft are clearly "invented, used, or designed" to fly in the air. Id. Section 336 also prohibits the FAA from promulgating "any rule or regulation regarding a model aircraft, or an aircraft being developed as a model aircraft" if the following statutory requirements are met:

- the aircraft is flown strictly for hobby or recreational use;
- the aircraft is operated in accordance with a community-based set of safety guidelines and within the programming of a nationwide community-based organization;
- the aircraft is limited to not more than 55 pounds unless otherwise certified through a design, construction, inspection, flight test, and operational safety program administered by a community-based organization;
- the aircraft is operated in a manner that does not interfere with and gives way to any manned aircraft; and

- when flown within 5 miles of an airport, the operator of the aircraft provides the airport operator and the airport air traffic control tower . . . with prior notice of the operation. . . .
- P.L. 112-95, section 336(a)(1)-(5).

Thus, based on the language of the statute, we conclude that aircraft that meet the statutory definition and operational requirements, as described above, would be exempt from future FAA rulemaking action specifically regarding model aircraft. Model aircraft that do not meet these statutory requirements are nonetheless unmanned aircraft, and as such, are subject to all existing FAA regulations, as well as future rulemaking action, and the FAA intends to apply its regulations to such unmanned aircraft.

B. Model Aircraft Must Meet the Criteria in Section 336 to Be Exempt from Future Rulemaking

Congress directed that the FAA may not "promulgate any rule or regulation regarding a model aircraft, or an aircraft being developed as a model aircraft" if the aircraft is being operated, or being developed to be operated, pursuant to the five criteria enumerated in the statute as described above. P.L. 112-95, section 336(a). In other words, Congress has restricted the FAA from promulgating regulations, from the date when the statute was enacted, specifically regarding model aircraft that meet the terms of the statute.

However, the prohibition against future rulemaking is not a complete bar on rulemaking that may have an effect on model aircraft. As noted above, the rulemaking limitation applies only to rulemaking actions specifically "regarding a model aircraft or an aircraft being developed as a model aircraft." P.L. 112-95, section 336(a). Thus, the rulemaking prohibition would not apply in the case of general rules that the FAA may issue or modify that apply to all aircraft, such as rules addressing the use of airspace (e.g., the 2008 rule governing VFR operations in the Washington, DC area) for safety or security reasons. See 73 FR 46803. The statute does not require FAA to exempt model aircraft from those rules because those rules are not specifically regarding model aircraft. On the other hand, a model aircraft operated pursuant to the terms of section 336 would potentially be excepted from a UAS aircraft certification rule, for example, because of the limitation on future rulemaking specifically "regarding a model aircraft, or an aircraft being developed as a model aircraft." P.L. 112-95, section 336(a). The FAA interprets the section 336 rulemaking prohibition as one that must be evaluated on a rule-by-rule basis.

Although the FAA believes the statutory definition of a model aircraft is clear, the FAA provides the following explanation of the meanings of "visual line of sight" and "hobby or recreational purpose," terms used in the definition of model aircraft, because the FAA has received a number of questions in this area.

By definition, a model aircraft must be "flown within visual line of sight of the person operating the aircraft." P.L. 112-95, section 336(c)(2). (For purposes of the visual line of sight requirement, "operator" means the person manipulating the model aircraft's controls). Based on the plain language of the statute, the FAA interprets this requirement to mean that:

1. the aircraft must be visible at all times to the operator;
2. that the operator must use his or her own natural vision (which includes vision corrected by standard eyeglasses or contact lenses) to observe the aircraft; and
3. people other than the operator may not be used in lieu of the operator for maintaining visual line of sight. Under the criteria above, visual line of sight would mean that the operator has an unobstructed view of the model aircraft. To ensure that the operator has the best view of the aircraft, the statutory requirement would preclude the use of vision-enhancing devices, such as binoculars, night vision goggles, powered vision magnifying devices, and goggles designed to provide a "first-person view" from the model. (The FAA is aware that at least one community-based organization permits "first person view" (FPV) operations during which the hobbyist controls the aircraft while wearing goggles that display images transmitted from a camera mounted in front of the model aircraft. While the intent of FPV is to provide a simulation of what a pilot would see from the flight deck of a manned aircraft, the goggles may obstruct an operator's vision, thereby preventing the operator from keeping the model aircraft within his or her visual line of sight at all times). Such devices would limit the operator's field of view thereby reducing his or her ability to see-and-avoid other aircraft in the area. Additionally, some of these devices could dramatically increase the distance at which an operator could see the aircraft, rendering the statutory visual-line-of-sight requirements meaningless. Finally, based on the plain language of the statute, which says that aircraft must be "flown within the visual line of sight of the person operating the aircraft," an operator could not rely on another person to satisfy the visual line of sight requirement. See id. (emphasis added). While the statute would not preclude using an observer to augment the safety of the operation, the operator must be able to view the aircraft at all times.

The statute requires model aircraft to be flown strictly for hobby or recreational purposes. Because the statute and its legislative history do not elaborate on the intended meaning of "hobby or recreational purposes," we look to their ordinary meaning and also the FAA's previous interpretations to understand the direction provided by Congress. (In construing statutory language, agencies should assume that the ordinary meaning of the language accurately expresses the legislative purpose of Congress. Agencies are also permitted to presume that Congress was aware of the agencies' administrative or adjudicative interpretations of certain terms and intended to adopt those meanings. See *BedRoc Ltd. v. U.S.,* 541 U.S. 176, 183 (2004); *see also Haig v. Agee*, 453 U.S. 280, 300 (1981); *Lorillard v. Pons*, 434 U.S. 575, 580-81 (1978)). A definition of "hobby" is a "pursuit outside one's regular occupation engaged in especially for relaxation." Merriam-Webster Dictionary, *available at* www.merriam-webster.com (last accessed June 9, 2014). A definition of recreation is "refreshment of strength and spirits after work; a means of refreshment or diversion." Id. These uses are consistent with the FAA's 2007 policy on model aircraft in which the Agency stated model aircraft operating guidelines did not apply to "persons or companies for business purposes." See 72 FR at 6690. (The FAA has also addressed recreational use of aircraft by pilots in the Sport and Recreational Pilot Certificate rules, which prohibit those pilots from acting as pilot in command of an airplane carrying passengers or property for compensation or hire, or in furtherance of a business. 14 CFR 61.101(e), 61.315(c). As discussed in the Sport Pilot final rule, those prohibitions are designed to limit those pilots to "sport and recreational flying only." 69 FR 44722, 44839 (July 27, 2004)).

Any operation not conducted strictly for hobby or recreation purposes could not be operated under the special rule for model aircraft. Clearly, commercial operations would not be hobby or recreation flights. (A commercial operator is a "person, who, for compensation or hire, engages in the carriage by aircraft in air commerce of persons or property. . . . " See 14 CFR 1.1. The FAA would therefore not consider a commercial operation to be "flown strictly for hobby or recreational purposes" because it would be conducted for compensation or hire). Likewise, flights that are in furtherance of a business, or incidental to a person's business, would not be a hobby or recreation flight. Flights conducted incidental to, and within the scope of, a business where no common carriage is involved, generally may operate under FAA's general operating rules of part 91. See Legal Interpretation to Scott C. Burgess, from Rebecca B. MacPherson, Assistant Chief Counsel for Regulations (Nov. 25, 2008). Although they are not commercial operations conducted for compensation or hire, such operations do not qualify as a hobby or recreation flight because of the nexus between the operator's business and the operation of the aircraft. See, e.g., Legal Interpretation to BSTC Corporation, from Rebecca B. MacPherson, Assistant Chief Counsel for Regulations (June 22, 2009) (noting transportation of mining employees and guests appears to be incidental to and within scope of operator's geological business); Legal Interpretation to Scott C. Burgess (Nov. 25, 2008) (noting transportation of automotive dealership employees and guests must be incidental to and within scope of operator's real estate development business).

To provide guidance, the following are examples of flights that could be conducted as hobby or recreation flights and other types of flights that would not be hobby or recreation.

Hobby or Recreation	Not Hobby or Recreation
Flying a model aircraft at the local model aircraft club.	Receiving money for demonstrating aerobatics with a model aircraft.
Taking photographs with a model aircraft for personal use.	A realtor using a model aircraft to photograph a property that he is trying to sell and using the photos in the property's real estate listing. A person photographing a property or event and selling the photos to someone else.
Using a model aircraft to move a box from point to point without any kind of compensation.	Delivering packages to people for a fee. If an individual offers free shipping in association with a purchase or other offer, FAA would construe the shipping to be in furtherance of a business purpose, and thus, the operation would not fall within the statutory requirement of recreation or hobby purpose.
Viewing a field to determine whether crops need water when they are grown for personal enjoyment.	Determining whether crops need to be watered that are grown as part of commercial farming operation.

Operations that meet the section 336 definition of "model aircraft" must also meet the five additional criteria for model aircraft established in section 336(a) to be exempt from future rulemaking regarding model aircraft. These criteria, with the exception of the hobby and recreation standard that was previously addressed, are explained below.

Section 336(a)(2) requires model aircraft to be operated within a community-based set of safety guidelines and within the programming of a nationwide community-based organization. Congress explained that it intended "nationwide community-based organization" to mean, in part, a "membership based association that represents the aeromodeling community within the Unites States; [and] provides its members a comprehensive set of safety guidelines that underscores safe aeromodeling operations within the National Airspace System and the protection and safety of the general public on the ground . . . " U.S. House, FAA Modernization and Reform Act of 2012, Conference Report (to Accompany H.R. 658), 112 H. Rpt. 381 (Feb. 1, 2012) (discussion of special rule for model aircraft). Based on this language, which provides context to Congress' use of the term "nationwide community-based organization," the FAA expects that model aircraft operations conducted under section 336(a) will be operated according to those guidelines. ("[C]ommunity-based organizations," for example, would include groups such as the Academy of Model Aeronautics and others that meet the statutory definition).

Additionally, model aircraft are limited to 55 pounds or less. The statutory language does not specify whether it applies to 55 pounds unloaded or 55 pounds with other equipment, payload, or fuel, for example, on the aircraft. The FAA believes that Congress intended for the 55-pound limit to mean the weight of the aircraft at the time of the operation. If the weight of the aircraft, alone, was the determining factor then it could conceivably be loaded with equipment or payload increasing the weight of the aircraft at time of takeoff well in excess of 55 pounds, thereby increasing the risk of harm should the operation not proceed as planned. The weight at the time of operation is also consistent with the FAA's designation of small or large aircraft, which is determined by an aircraft's maximum certificated takeoff weight. See, e.g., 14 CFR 1.1 (defining small and large aircraft). Congress' recognition of the increased risk posed by heavier aircraft is demonstrated by the more stringent requirements for aircraft heavier than 55 pounds in the statute, which are discussed below. Accordingly, the

FAA interprets this provision to mean the weight of the aircraft at the time of the operation must not exceed 55 pounds, including the weight of any payload and fuel.

The statute creates an exception for model aircraft that exceed the 55-pound weight limit if the aircraft is "certified through a design, construction, inspection, flight test, and operational safety program administered by a community-based organization." P.L. 112-95, section 336(a)(3). If a nationwide community-based organization has provided its members with a set of safety guidelines that define a design, construction, inspection, flight test, and operational safety program then model aircraft constructed in accordance with that program may exceed 55 pounds and operate in accordance with section 336(a).

Model aircraft must not interfere with and must give way to any manned aircraft. This requirement needs no further explanation, and the FAA would expect that model aircraft operators abide by it. (This requirement is consistent with longstanding FAA guidance for model aircraft operators. See AC 91-57, para. 3 (advising model aircraft operators to "[g]ive right of way to, and avoid flying in the proximity of, full-scale aircraft."). We note that model aircraft interfering with, or that do not give way to, manned aircraft would be subject to enforcement action under section 336(b), as further explained in section III below.

Finally, the statute sets a requirement for model aircraft operating within 5 miles (for ease of determining distance, the FAA interprets the statute to mean 5 statute miles) of an airport to notify the airport operator and control tower, where applicable, prior to operating. (This requirement is consistent with longstanding FAA guidance for model aircraft operators. See AC 91-57, para. 3 (advising model aircraft operators to notify an airport operator, control tower, for flight service station when planning to operate within three miles of an airport). (If a group of modelers intends to operate in one area, one person could contact air traffic control on behalf of the group. Additionally, consistent with the statute, the FAA encourages operators who fly from a permanent location within 5 miles of an airport to "establish a mutually-agreed upon operating procedure with the airport operator and the airport air traffic control tower (when an air traffic facility is located at the airport)."). If the model aircraft operator provides notice of forthcoming operations which are then not authorized by air traffic or objected to by the airport operator, the FAA expects the model aircraft operator will not conduct the proposed flights. The FAA would consider flying model aircraft over the objections of FAA air traffic or airport operators to be endangering the safety of the NAS. Additionally, we note that following this 5-mile notification procedure would be read in conjunction with FAA rules governing airspace usage discussed below.

III. Scope of FAA's Enforcement Authority

As discussed above, if a model aircraft is operated consistently with the terms of section 336(a) and (c), then it would not be subject to future FAA regulations regarding model aircraft. However, Congress also recognized the potential for such operations to endanger other aircraft and systems of the NAS. Therefore, it specifically stated that "[n]othing in this section shall be construed to limit the authority of the Administrator to pursue enforcement action against persons operating model aircraft who endanger the safety of the national airspace system." P.L. 112-95, section 336(b).

Through this language, Congress specifically recognized the FAA's existing authority to take enforcement action to protect the safety of the NAS. (The NAS is broadly described as "the common network of U.S. airspace; air navigation facilities, equipment and services, airports or landing areas; aeronautical charts, information and services; rules, regulations and procedures, technical information, and manpower and material. Included are system components shared jointly with the military." See FAA Aeronautical Information Manual (Apr. 3, 2014), available at http://www.faa.gov/air_traffic/publications/media/AIM_Basic_4-03-14.pdf). Moreover, it did not limit the FAA's authority to take enforcement action where a violation of a regulation results in the endangerment of the NAS. As demonstrated by the FAA's statutory and regulatory authorities, our charge to protect the safety of the NAS is not only intended to protect users of the airspace, but is also intended to protect persons and property on the ground (See 49 U.S.C. 40103(b)(2) (authorizing the FAA to prescribe air traffic regulations to protect people and property on the ground); Adm'r v. Johnson, NTSB Order No. EA-1008, 1977 WL 22279 at *2 (May 10, 1977) (recognizing FAA authority to promulgate regulations to protect persons and property on the ground); Adm'r v. Page, NTSB Order No. EA-2786, 1988 WL 250725 at *3 (July 19, 1988) (finding FAA's rulemaking and enforcement authority extends to areas away from runways and taxiways – in this case the ramp of a fixed base operator).

For example, the FAA regulates low-altitude operations to protect people and property on the ground. The FAA permits aircraft operations below 500 feet when flown over open water and in sparsely populated areas. 14 CFR 91.119(c). Such operations may not be conducted "closer than 500 feet to any person, vessel, vehicle, or structure." Id. Therefore, although such low-altitude operations may pose a lower risk to aircraft flying much higher, the operation may still pose a risk to persons and property on the ground warranting enforcement action when conducted unsafely. See, e.g., Adm'r v. Kachalsky, NTSB Order No. EA-4847, 2000 WL 1072332 (July 24, 2000) (affirming a violation of § 91.119(c) for operating within 500 feet of a dwelling in a sparsely populated area); Adm'r v. Beissel, NTSB Docket No. SE-19436, 2013 WL 7809754 (Dec. 11, 2014) (ordering suspension of a pilot certificate when pilot flew a helicopter less than 40 feet above the surface of a lake).

Reading the broad reference to the NAS, along with Congress' clear interest in ensuring that model aircraft are safely operated, we conclude that Congress intended for the FAA to be able to rely on a range of our existing regulations to protect users of the airspace and people and property on the ground. Therefore, regardless of whether a model aircraft satisfies the statutory definition and operational requirements described above, if the model aircraft is operated in such a manner that endangers the safety of the NAS, the FAA may take enforcement action consistent with Congress' mandate.

IV. Examples of Regulations That Apply to Model Aircraft

The FAA could apply several regulations in part 91 when determining whether to take enforcement action against a model aircraft operator for endangering the NAS. The FAA's general operating and flight rules are housed in part 91 of the FAA's regulations. These rules are the baseline rules that apply to all aircraft operated in the United States with limited exceptions (Part 91 does not apply to moored balloons, kites, unmanned rockets, and unmanned free balloons, and ultralights vehicles operated under14 CFR parts 101 and 103) and are the appropriate rules to apply when evaluating model aircraft operations. See 14 CFR 91.1.

Rules relevant to these operations fall generally into three categories: (1) how the aircraft is operated; (2) operating rules for designated airspace; and, (3) special restrictions such as temporary flight restrictions (TFRs) and notices to airmen (NOTAMs). These rules are discussed in greater detail below.

Rules addressing operation of the aircraft may include prohibitions on careless or reckless operation and dropping objects so as to create a hazard to persons or property. See 14 CFR 91.13 through 91.15. Additionally, § 91.113 establishes right-of-way rules for converging aircraft. (Additionally, model aircraft must not interfere with and must always give way to any manned aircraft. Section 336(a)(4)). Model aircraft that do not comply with those rules could be subject to FAA enforcement action.

Rules governing operations in designated airspace are found in §§ 91.126 through 91.135. In general, those rules establish requirements for operating in the various classes of airspace, and near airports in non-designated airspace to minimize risk of collision in higher traffic airspace. Generally, if an operator is unable to comply with the regulatory requirements for operating in a particular class of airspace, the operator would need authorization from air traffic control to operate in that area. See, e.g., 14 CFR 91.127(a), 91.129(a). Operations within restricted areas designated in part 73 would be prohibited without permission from the using or controlling agency. Accordingly, as part of the requirements for model aircraft operations within 5 miles of an airport set forth in section 336(a)(4) of P.L. 112-95, the FAA would expect modelers operating model aircraft in airspace covered by §§ 91.126 through 91.135 and part 73 to obtain authorization from air traffic control prior to operating.

The third category of rules relevant to model aircraft operations are rules relating to operations in areas covered by temporary flight restrictions and NOTAMs found in §§ 91.137 through 91.145. The FAA would expect that model aircraft operations comply with restrictions on airspace when established under these rules.

Other rules in part 91, or other parts of the regulations, may apply to model aircraft operations, depending on the particular circumstances of the operation. The regulations cited above are not intended to be an exhaustive list of rules that could apply to model aircraft operations. The FAA anticipates that the cited regulations are the ones that would most commonly apply to model aircraft operations.

Issued in Washington, DC, on June 18, 2014.[4]

3. Memorandum Related to Educational Use of UAS

On May 4, 2016, after almost two years of public requests for clarification, the FAA released a memorandum to address the educational use of UAS, as it pertains to hobby or recreational use.[5] What follows is the entire memo:

This interpretation addresses: (1) use of unmanned aircraft for hobby or recreational purposes at educational institutions and community-sponsored events; and (2) student use of unmanned aircraft in furtherance of receiving instruction at accredited educational institutions.

There is uncertainty in the model aircraft community about when an unmanned aircraft is a model aircraft operated for hobby or recreation or is an operation requiring FAA authorization. The FAA has received many inquiries from students and educational institutions offering coursework in the design, construction, and operation of small, unmanned aircraft with respect to the types of activities in which students and faculty lawfully may engage pursuant to the existing legal framework.

In light of these questions, we are issuing this interpretation to clarify that:

---

[4]FAA-2014-0396, Interpretation of the Special Rule for Model Aircraft, https://www.faa.gov/uas/media/model_aircraft_spec_rule.pdf.

[5]*Memorandum Related to Educational Use of UAS,* FAA, http://www.faa.gov/uas/regulations_policies/media/Interpretation-Educational-Use-of-UAS.pdf.

- A person may operate an unmanned aircraft for hobby or recreation in accordance with section 336 of the FAA Modernization and Reform Act of 2012 (FMRA) at educational institutions and community-sponsored events (to include demonstrations at schools, Boy or Girl Scout meetings, science clubs, etc.) provided that person is (1) not compensated, or (2) any compensation received is neither directly nor incidentally related to that person's operation of the aircraft at such events.
- A student may conduct model aircraft operations in accordance with section 336 of the FMRA in furtherance of his or her aviation-related education at an accredited educational institution.
- Faculty teaching aviation-related courses at accredited educational institutions may assist students who are operating a model aircraft under section 336 and in connection with a course that requires such operations, provided the student maintains operational control of the model aircraft such that the faculty member's manipulation of the model aircraft's controls is incidental and secondary to the student's (e.g., the faculty member steps in to regain control in the event the student begins to lose control, to terminate the flight, etc.).

On June 25, 2014, the FAA published in the Federal Register its interpretation of the Special Rules for Model Aircraft, section 336 of the FMRA. 79 Fed. Reg. 36172 (June 25, 2014). A model aircraft means an unmanned aircraft that is: (1) capable of sustained flight in the atmosphere; (2) flown within visual line of sight of the person operating the aircraft; and (3) flown for hobby or recreational purposes. Currently, the FAA is reviewing the more than 33,500 comments to that Special Rule. In addition, on February 23, 2015, the FAA published in the Federal Register its Notice of Proposed Rulemaking (NPRM) on the Operation and Certification of Small Unmanned Aircraft Systems (UAS). 80 Fed. Reg. 9544.

Separate from these two actions, the FAA continues to receive a number of questions on the use of model aircraft to conduct demonstrations and on student use of model aircraft tin connection with participation in coursework at educational institutions. The FAA finds it necessary to clarify the applicability of section 336 of the FMRA and of the FAA's operating requirements for UAS. The FAA recognizes that UAS increasingly are being used in education, including science, technology, education, and math (STEM) education, which is the focus of President Obama's Educate to Innovate Initiative.

Hobbyist Use of UAS to Conduct Demonstrations

Section 336(a) of the FMRA provides special rules for model aircraft that require the aircraft to be:

1. Flown strictly for hobby or recreational use;
2. Operated in accordance with a community-based set of safety guidelines and within the programming of a nationwide community-based organization;

3. Limited to not more than 55 pounds unless otherwise certified through a design, construction, inspection, flight test, and operational safety program administered by a community-based organization;
4. Operated in a manner that does not interfere with and gives way to any manned aircraft; and
5. When flown within 5 miles of an airport, the operator of the aircraft provides the airport operator and the airport air traffic control tower (when an air traffic control facility is located at the airport) with prior notice of the operations (model aircraft operators flying from a permanent location within 5 miles of an airport should establish a mutually-agreed upon operating procedure with the airport operator and the airport air traffic control tower).

If an unmanned aircraft is operated as a model aircraft in accordance with the above, then it does not require FAA authorization. A key element in determining whether an operation may qualify as a model aircraft operation is that it must be flown for "hobby or recreational" purposes. The FAA's interpretation of section 336 relies on common definitions of the terms "hobby" and "recreational." Previous agency guidance addressed the parameters of hobby or recreational use:

Any operation not conducted strictly for hobby or recreation purposes could not be operated under the special rule for model aircraft. Clearly, commercial operations would not be hobby or recreation flights. Likewise, flights that are in furtherance of a business, or incidental to a person's business, would not be a hobby or recreation flight.

The FAA interprets "hobby or recreational" use to include operation of UAS to conduct demonstrations at accredited educational institutions or at other community-sponsored events provided the aircraft is not being operated for compensation, in furtherance of a business or incidental to a business. Therefore, a model aircraft hobbyist or enthusiast lawfully may fly UAS at accredited educational institutions or other community-sponsored events to promote the safe use of UAS and encourage student interest in aviation as a hobby or for recreational purposes provided the hobbyist receives no compensation of any form (including honorarium or reimbursement of costs), or any such compensation neither directly nor indirectly furthers the hobbyist's business or operation of the UAS and he or she follows the provisions of section 336.

Student Operation of Model Aircraft for Educational Purposes

If not operated as "model aircraft" under section 336 of the FMRA, currently there are three ways to lawfully conduct unmanned aircraft operations in the United States: (1) as public aircraft operations pursuant to the requirements of the public aircraft statute and under a Certificate of Waiver or Authorization (COA) from the FAA; (2) as limited commercial operations by type certificated UAS, provided the operator obtains a COA from the FAA;

or (3) pursuant to a Section 333 of the FMRA grant of exemption based on the Secretary of Transportation's determination that a certificate of airworthiness is not required, and provided the operator obtains a COA from the FAA.

Each of these three methods is available to educational institutions (including their faculty and students) that want to operate UAS, including for commercial, research and development, and any other non-hobby or non-recreational purpose. Each, however, requires the educational institution or its faculty and students to meet statutory prerequisites and obtain from the FAA the requisite approvals (in the form of exemptions and COAs).

Many educational institutions are keenly interested in having students operate unmanned aircraft as model aircraft under section 336 of the FMRA in connection with their academic coursework at those schools. The educational community contends that these operations not only meet the definition of model aircraft but also meet the unique need of students, which is learning how to design, construct, and operate small unmanned aircraft as a component of a variety of science, technology, and aviation-related educational curricula. Students also are interested in operating small unmanned aircraft for other educational purposes such as in connection with television, film, or photography courses.

The FAA has considered whether a student's coursework of learning how to operate and use a UAS constitutes a hobby or recreational activity within the meaning of section 336's definition of model aircraft. The FAA believes students operating UAS as one component of a curricula pertaining to principles of flight, aerodynamics, and airplane design and construction promotes UAS safe use and advances UAS-related knowledge, understanding, and skills. UAS also may provide students a useful tool in other academic curricula such as television, film production, or arts generally. Although it may be argued that the student's knowledge and skills obtained through such coursework are necessary for a diploma or degree, which subsequently can lead to an aviation-related job or increased earning potential, the FAA finds this link simply too attenuated to transform student UAS use, as a component of an accredited educational curriculum, into a non-hobby or non-recreational use within the meaning of section 336. A person that operates a UAS strictly for hobby or recreation learns about principles of flight, aerodynamics, and airplane construction may subsequently use such knowledge when gainfully employed, but that does not transform what is otherwise a hobby or recreational activity into a non-hobby or non-recreational pursuit.

Therefore, we find that the use of small, unmanned aircraft by students at accredited educational institutions as a component of science, technology, and aviation-related educational curricula or other coursework such as television and film production or the arts more closely reflects and embodies the purposes of "hobby or recreational" use of model aircraft and is consistent with the intent of section 336 of the FMRA. Accordingly, the FAA concludes that student use of UAS at accredited educational institutions as a component of their science, technology and aviation-related educational curricula, or other coursework such as television and film production or the arts, is "hobby or recreational use" within the meaning of the FMRA. The student is, however, responsible for meeting and complying with all other elements required for lawful model aircraft operations pursuant to section 336 of the FMRA, including the student not receiving any form of compensation (including reimbursement of costs, honorarium, etc.) directly or incidentally to his or her operation of the model aircraft. The prohibition on receiving compensation, while broad, does not preclude a student from operating UAS in connection with fulfilling a specific course's requirement while also receiving financial aid, participating in work-study programs or being a paid research assistant to a faculty member teaching such course.

Faculty Use of Model Aircraft

The FAA recognizes that faculty participation in the student's learning experience often is an integral component of the student's educational experience and that faculty should be able to participate in and contribute to the unmanned aircraft activities in which students can engage as hobbyists. However, a faculty member engaging in the operation of an unmanned aircraft, as part of professional duties for which he or she is paid, would not be engaging in a hobby or recreational activity. Rather, the faculty member is being compensated for his or her teaching or research activity, including any UAS operation arising from or related to the faculty member's teaching a course or conducting research.

Likewise, a student operating UAS for research on behalf of a faculty member is associated with the faculty member's professional duties and compensation and, thus, is not hobby or recreational use by the student pursuant to section 336. Student operation of UAS for the professional research objectives of faculty renders the operation non-hobby or non-recreational. Accordingly, a faculty member conducting research may not rely on section 336's concept of "hobby or recreational use" to either operate a UAS or direct student UAS operations in connection with such research.

Nevertheless, faculty teaching a course or curricula that uses unmanned aircraft as a component of that course may provide limited assistance to students operating unmanned aircraft as part of that course without changing the character of the student's operation a s a hobby or recreational activity or requiring FAA authorization for the faculty member to operate. The FAA finds that de minimis limited instructor participation in student operation of UAS as part of coursework does not rise to the level of faculty conducting an operation outside of the hobby or recreation construct.

This limited circumstance would apply to courses at accredited institutions where the operation of the unmanned aircraft is secondary to the design and construction of the aircraft, such that the primary purpose of the course is not operating an unmanned aircraft. For example, an instructor teaching an engineering course in which construction and operation of UAS are one part of the curriculum would be able to conduct limited UAS operations. In that case, students would fly UAS to test the validity of design or construction methods to show mastery of the principles of the course. The faculty member's UAS operation would be secondary to the purpose of instructing engineering courses. In contrast, this limited circumstance would not apply to a course related to UAS flight instruction. In that case, the student's primary purpose for taking the course is to learn to fly a UAS and flight would be expected to be demonstrated on a regular basis. In that case, the faculty member's UAS operation is closely tied to his or her purpose of instructing how to fly a UAS.

Conclusion

UAS may be used to conduct demonstrations at schools or other community-sponsored events provided the person operating the aircraft is (1) not compensated, or (2) any compensation received is neither directly not indirectly related to that person's operation of the aircraft at such events.

Students that operate model aircraft in connection with fulfilling an accredited educational institution's curricula lawfully may conduct model aircraft operations for hobby and recreational purposes pursuant to section 336 of the FMRA, provided they do not receive compensation, directly or incidentally, arising from or related to such operations. Faculty at these educational institutions teaching such curricula may assist students with their model aircraft operations under section 336, provided that the operations are used to teach such curricula to students enrolled in those courses and the faculty member's participation is limited to de minimis participation in the student's UAS operations. We emphasize that these operations must be conducted under the provisions of section 336.

The FAA emphasizes that faculty members who wish to operate UAS outside of these parameters must seek authorization through one of the three methods discussed above. We also note that this interpretation was drafted prior to issuance of the final rule for Operation and Certification of Small UAS Rule and this interpretation may need to be revisited depending on its provisions.

4. Advisory Circular (AC) 91-57A

On September 2, 2015, the FAA updated its 1981 Advisory Circular (AC) 91-57 with AC91-57A, which outlined the obligations of model aircraft/UAS operators.[6] On January 11, 2016, the FAA published the first change to this advisory circular, to correct a typographical error translating form Section 336 of Public Law 112-95.[7] This is listed below with additional author commentary added in *italics* where deemed necessary to elaborate or clarify.[8]

---

[6]AC 91-57A, http://www.faa.gov/documentLibrary/media/Advisory_Circular/AC_91-57A.pdf.
[7]AC 91-57A with change 1, http://www.faa.gov/documentLibrary/media/Advisory_Circular/AC_91-57A_Ch_1.pdf.
[8]*Id.*

Subject: Model Aircraft Operating Standards
Date: January 11, 2016
Initiated by: AJV-115
AC No. 91-57A Change 1

1. PURPOSE. *This advisory circular (AC) provides guidance to persons operating Unmanned Aircraft (UA) for hobby or recreation purposes meeting the statutory definition of "model aircraft" contained in Section 336 of Public Law 112-95, the FAA Modernization and Reform Act of 2012.*

   Typically the purpose of an Advisory Circular is guidance as opposed to a regulation such as those in 14 Code of Federal Regulations (CFR). In the past it was taken to mean, "encourages voluntary compliance with."[9] Note the term "persons operating Unmanned Aircraft" because in the past, in the 1981 AC, it was applied to "model aircraft operators."[10] "Model aircraft" now "means an unmanned aircraft that is –

   1. Capable of sustained flight in the atmosphere;
   2. Flown within visual line of sight of the person operating the aircraft; and
   3. Flown for hobby or recreational purposes.[11]

   Section 336 of Public Law 112-95 starts out by saying that the Administrator of the FAA MAY NOT promulgate any rule or regulation regarding a model aircraft, or an aircraft being developed as a model aircraft. . . ."[12]

   This AC describes means by which model aircraft may be operated safely in the National Airspace System (NAS).

   Note the term "may" suggesting guidance as opposed to the term "must" denoting a regulatory behavior.

   Nothing in this AC changes the requirement to comply with the statute or any applicable regulations.

   What statutes or regulations were model aircraft operators required to "comply with" before? The author posits that it was just AC91-57 of 1981 that applied, and that was not a statute, nor a regulation but only an advisory circular! And what statutes or regulations apply now? Well, again the author posits, an updated version of an advisory circular, AC91-57A! Still there is no statute or regulation because PL 112-95 prohibits the FAA from promulgating rules or regulations for model aircraft.

---

[9]Model Aircraft Operating Standards of 1981, https://www.faa.gov/regulations_policies/advisory_circulars/index.cfm/go/document.information/documentid/22425.
[10]*Id.*
[11]Public Law 112-95 (February 14, 2012), http://www.gpo.gov/fdsys/pkg/PLAW-112publ95/content-detail.html.
[12]*Id.*

2. APPLICABILITY. This AC provides information for any person who engages in model aircraft operations using model aircraft as defined by statute. (See paragraph 6.)

Again note the term "information" suggesting guidance of a voluntary nature and not a regulation. Note too the use of "a person who engages in model aircraft operations" as opposed to earlier in the AC where it says "persons operating Unmanned Aircraft."

If you look at paragraph 6 in the AC you will notice that the definition is identical to the one in PL112-96 EXCEPT for the addition of the word "only" before the terms "hobby or recreational purposes."[13]

3. REFERENCES. Title 49 U.S.C. § 40101; P.L. 112-95 (126 Stat. 77 *et seq.*).

Title 49 U.S.C. § 40101 by definition is policy as defined below:

§40101. Policy

a. Economic Regulation. – In carrying out subpart II of this part and those provisions of subpart IV applicable in carrying out subpart II, the Secretary of Transportation shall consider the following matters, among others, as being in the public interest and consistent with public convenience and necessity:

1. assigning and maintaining safety as the highest priority in air commerce.[14]

4. RELATED MATERIAL (current editions).

This list of related materials assumes a person flying a model aircraft for fun is going to spend an inordinate amount of time reading, and updating editions, of all these documents, which are typically applicable to pilots of manned aircraft.

- *Subtitle VII of Title 49, United States Code (49 USC)*[15]
- Title 14 of the Code of Federal Regulations (14 CFR)[16]
- Subtitle B of Public Law 112-95 (Feb. 14, 2012)[17]
- Aeronautical Information Manual (AIM)[18]
- Temporary Flight Restriction (TFR) listing: http://tfr.faa.gov/tfr2/list.html

This listing for the entire nation can be found online but thankfully you can search just by state.[19]

---

[13]Public Law 112-95 (February 14, 2012), http://www.gpo.gov/fdsys/pkg/PLAW-112publ95/content-detail.html.
[14]Title 49 U.S.C. § 40101, http://www.gpo.gov/fdsys/granule/USCODE-2011-title49/USCODE-2011-title49-subtitleVII-partA-subparti-chap401-sec40101.
[15]Subtitle VII of Title 49, United States Code, http://www.gpo.gov/fdsys/pkg/USCODE-2011-title49/html/USCODE-2011-title49.htm.
[16]Title 14 of the Code of Federal Regulations, http://www.ecfr.gov/cgi-bin/text-idx?tpl=/ecfrbrowse/Title14/14tab_02.tpl.
[17]Public Law 112-95 (February 14, 2012), http://www.gpo.gov/fdsys/pkg/PLAW-112publ95/content-detail.html.
[18]Aeronautical Information Manual, http://www.faa.gov/air_traffic/publications/media/aim_basic_4-03-14.pdf.
[19]Temporary Flight Restriction, http://tfr.faa.gov/tfr2/list.html.

- Aeronautical Navigation Products (Charts): http://www.faa.gov/air_traffic/flight_info/aeronav/[20]
- Notices to Airman: https://www.faa.gov/air_traffic/publications/notices/[21]

5. EXPLANATION OF CHANGE. This AC has been changed to correct a typographical error translating from Section 336 of Public Law 112-95.
6. MODEL AIRCRAFT OPERATIONS.
   a. Terminology.
   Note that what follows are four definitions. The first two are for an aircraft, the third is for an unmanned aircraft, and the fourth is for model aircraft. The assumption is that since model aircraft are "aircraft" as further explained in the Pirker case, and model aircraft are also "unmanned aircraft" then all four definitions apply.
      1. 49 USC § 40102 defines an aircraft as "any contrivance invented, used, or designed to navigate, or fly in, the air."[22]
      14 CFR § 1.1 defines an aircraft as "a device that is used or intended to be used for flight in the air."[23]
      2. Public Law 112-95 defines unmanned aircraft as an aircraft that is operated without the possibility of direct human intervention from within or on the aircraft.[24]
      3. Section 336 of P.L. 112-95 defines a model aircraft as an unmanned aircraft that is capable of sustained flight in the atmosphere, flown within visual line of sight of the person operating the aircraft, and flown only for hobby or recreational purposes.[25] *As mentioned above, you will notice that the definition is identical to the one in PL 112-96 EXCEPT for the addition of the word "only" before the terms "hobby or recreational purposes."*
   b. Model Aircraft Hazards in the NAS. While aero-modelers generally are concerned about safety and exercise good judgment when flying model aircraft for the hobby and recreational purposes for which they are intended, they may share the airspace in which manned aircraft are operating.
   The term "may" denotes that they are allowed to be there but do not have to be as if they do not actually have a right to be there.

[20]Aeronautical Navigation Products, http://www.faa.gov/air_traffic/flight_info/aeronav/.
[21]Notices to Airman, https://www.faa.gov/air_traffic/publications/notices/.
[22]49 USC § 40102, http://www.gpo.gov/fdsys/pkg/USCODE-2011-title49/html/USCODE-2011-title49-subtitleVII-partA-subparti-chap401-sec40102.htm.
[23]14 CFR § 1.1, http://www.ecfr.gov/cgi-bin/text-idx?rgn=div8&node=14:1.0.1.1.1.0.1.1.
[24]Public Law 112-95, http://www.gpo.gov/fdsys/pkg/PLAW-112publ95/pdf/PLAW-112publ95.pdf.
[25]*Id.*

Unmanned aircraft, including model aircraft, may pose a hazard to manned aircraft in flight and to persons and property on the surface if not operated safely.

Model aircraft are now a subset of unmanned aircraft.

Model aircraft operations that endanger the safety of the National Airspace System, particularly careless or reckless operations or those that interfere with or fail to give way to any manned aircraft may be subject to FAA enforcement action.

This advisory circular is now beginning to sound like a regulation, namely, 14 CFR § 91.13, which reads as follows:

a. Aircraft operations for the purpose of air navigation. No person may operate an aircraft in a careless or reckless manner so as to endanger the life or property of another.

b. *Aircraft operations other than for the purpose of air navigation. No person may operate an aircraft, other than for the purpose of air navigation, on any part of the surface of an airport used by aircraft for air commerce (including areas used by those aircraft for receiving or discharging persons or cargo), in a careless or reckless manner so as to endanger the life or property of another. 26*

This advisory circular is now beginning to sound like another regulation, namely, 14 CFR § 91.113, which defines the right-of-way rules for manned aircraft.[27]

This "may" language appears again, but does not define exactly what kind of enforcement an operator would face. However, if the Pirker case is taken as precedent then it would be found in 49 USC § 46301. Civil Penalties.[28] *More on that in another part of the book.*

c. Determination of "Model Aircraft" Status. Whether a given unmanned aircraft operation may be considered a "model aircraft operation" is determined with reference to section 336 of Public Law 112-95:

This would follow the separation as mentioned in Public Law 112-95 of Feb 14, 2012, also known as the FAA Modernization and Reform Act, where the operation is based on public use, civil use, or hobby and recreational use.

---

[26]14 CFR § 91.13, http://www.ecfr.gov/cgi-bin/text-idx?SID=b743e7ea0c6fca930b5d2c86a983e5f2&mc=true&node=se14.2.91_113&rgn=div8.

[27]14 CFR § 91.113, http://www.ecfr.gov/cgi-bin/text-idx?SID=b743e7ea0c6fca930b5d2c86a983e5f2&mc=true&node=se14.2.91_1113&rgn=div8.

[28]49 USC § 46301, http://www.gpo.gov/fdsys/granule/USCODE-2011-title49/USCODE-2011-title49-subtitleVII-partA-subpartiv-chap463-sec46301.

1. The aircraft is flown strictly for hobby or recreational use;
2. The aircraft operates in accordance with a community-based set of safety guidelines and within the programming of a nationwide community-based organization (CBO);
   This differs from PL 112-95 only in the use of "operates" as opposed to "is operated."
   As to a definition for what constitutes "community-based set of safety guidelines" one must look to FAA-2014-0396 (Interpretation of the Special Rule for Model Aircraft)[29] *to see that the FAA meant groups like Academy of Model Aeronautics (AMA) and others like them who offer their members a safety code for flying their aircraft.*[30] *More on this in another part of the book.*
3. The aircraft is limited to not more than 55 pounds, unless otherwise certified through a design, construction, inspection, flight test, and operational safety program administered by a CBO;
   This is because the NPRM that is currently in committee is making the rule for UAS that weigh 55 pounds or less.
4. The aircraft operates in a manner that does not interfere with, and gives way to, any manned aircraft; and
   This again sounds like the regulation for right-of way rules for manned aircraft per 14 CFR § 91.113, mentioned above.
5. When flown within 5 miles of an airport, the operator of the model aircraft provides the airport operator and the airport air traffic control tower (when an air traffic facility is located at the airport) with prior notice of the operation. Model aircraft operators flying from a permanent location within 5 miles of an airport should establish a mutually agreed upon operating procedure with the airport operator and the airport air traffic control tower (when an air traffic facility is located at the airport).
   Is the 5 miles measurement in nautical miles or statute miles? A statute mile is 5,280 feet in length. A nautical mile is 6,076 feet in length. If Section 333 Exemptions were to be used as an example then it would be nautical miles. However, if one were to use the interpretation of special rule for model aircraft then it would be statute miles. And what about airport? Would that be public or private or even abandoned? In the case of private unlisted airports, where is one to find

---

[29]FAA-2014-0396, Interpretation of the Special Rule for Model Aircraft, https://www.faa.gov/uas/media/model_aircraft_spec_rule.pdf.
[30]Academy of Model Aeronautics (AMA) safety code, https://www.modelaircraft.org/files/105.pdf.

that number to the airport owner/operator? So in the case of an air traffic control tower that operates part-time, what is one to do? Does one not fly the UAS when the tower is closed, or assume it is safe to fly as there may likely be less manned aircraft operations? How much time is prior notice? A day? An hour? Are the operators or controllers to give a "clearance" or thumbs up of some kind? Do they have the authority to deny your request? To date most of these questions have not been clarified. In November 2015 the FAA issued JO7210.891 which clarified that the tower controller may in fact deny your request. A "mutually agreed upon operating procedure" sounds like a memorandum of understanding or some such similar contract, which the author would recommend you get in writing signed by the person with authority to do so. And again who has the authority to sign such a document? The tower manager for a controlled airport? The owner listed on the deed to the land whereupon the private airport resides? In November 2015 the FAA issued JO7210.891 which clarified that the tower controller should issue a formal letter of agreement with the UA operator.

d. Public Law 112-95 recognizes the authority of the Administrator to pursue enforcement action against persons operating model aircraft who endanger the safety of the National Airspace System.

This is somewhat true because this is not a regulation for people to follow so much as Public Law 112-95 is a document telling the FAA to reform and modernize and come up with regulations for UAS. Hence the NPRM was initiated, which, according to the Administrative Procedure Act, is how regulations come into being. Case law, if you see NTSB Administrative Law Judges docket, for evidence that this is indeed occurring, Pirker being a prime example.

Accordingly, model aircraft operators must comply with any Temporary Flight Restrictions (TFR).

Note the "must" language which sounds very much like a regulation!

TFRs are issued over specific locations due to disasters, or for reasons of national security; or when determined necessary for the management of air traffic in the vicinity of aerial demonstrations or major sporting events. Do not operate model aircraft in designated areas until the TFR is no longer in force.

What follows here is a definition of a TFR, assumedly because not all model aircraft operators are pilots and hence may not be aware of such knowledge! "DO NOT" sounds like pretty regulatory language! But wait! Isn't this an

AC? Which implies guidance. And didn't the FAA Modernization and Reform Act of 2012 prohibit the FAA from regulating hobby or recreational use? Model aircraft must not operate in Prohibited Areas, Special Flight Rule Areas or, the Washington National Capital Region Flight Restricted Zone, without specific authorization.

"MUST NOT" sounds pretty regulatory too! Would a hobbyist know where to look to get "specific authorization"? Pilots would know the controlling center is listed on the edge of the aeronautical chart, but not all model aircraft operators are pilots!

Such areas are depicted on charts available at http://www.faa.gov/air_traffic/flight_info/aeronav/. Additionally, model aircraft operators should be aware of other Notices to Airmen (NOTAMS) which address operations near locations such as military or other federal facilities, certain stadiums, power plants, electric substations, dams, oil refineries, national parks, emergency, services and other industrial complexes. In addition to the previously mentioned link, information regarding published NOTAMS can be found at: https://www.faa.gov/air_traffic /publications/notices/.

It looks like the FAA realized that not all model aircraft operators are pilots, hence the brief explanation.

The requirement to not fly within TFRs, or other circumstances where prohibited, would apply to operation of model aircraft that would otherwise comply with section 336 of Public Law 112-95.

The "requirement to not fly" means essentially "do not fly." This looks like a regulation dressed up in the clothing of an AC. Section 336 in other words means hobby or recreational operators.

e. Model aircraft operators should follow best practices including limiting operations to 400 feet above ground level (AGL).

"Should follow" again implies voluntary compliance. "Best practices" is a somewhat vague term. Again are these as defined by organizations like AMA and what if an operator does not belong to such organizations? Does the term apply to the "Know before you fly" campaign?[31]

f. All other operators and for additional information on Unmanned Aircraft Systems please visit: http://www.faa.gov/uas/

"All other operators" are who specifically? Are they public use or civil use operators?

Elizabeth L. Ray
Vice President, Mission Support Services
Air Traffic Organization

[31] Know Before You Fly, http://knowbeforeyoufly.org.

5. Registration

On October 19, 2015, US Transportation Secretary Anthony Foxx and FAA Administrator Michael Huerta announced the creation of a task force to develop recommendations for a registration process for UAS. The task force would be composed of 25 to 30 diverse representatives from the UAS and manned aviation industries, the federal government, and other stakeholders. The group would advise the Department on which aircraft should be exempt from registration due to a low safety risk, including toys and certain other small UAS. The task force also would explore options for a streamlined system that would make registration less burdensome for commercial UAS operators. The task force could make additional safety recommendations as it deemed appropriate. Secretary Foxx directed the group to deliver its report by November 20.[32]

According to Secretary Foxx, registering unmanned aircraft would help build a culture of accountability and responsibility, especially with new users who have no experience operating in the US aviation system, and it would help protect public safety in the air and on the ground. Every day, the FAA receives reports of potentially unsafe UAS operations. Pilot sightings of UAS doubled between 2014 and 2015. The reports ranged from incidents at major sporting events and flights near manned aircraft, to interference with wildfire operations. According to FAA Administrator Huerta, these reports signal a troubling trend, and registration would help make sure that operators know the rules and remain accountable to the public for flying their unmanned aircraft responsibly. While the task force does its work, the FAA would continue its aggressive education and outreach efforts, including the Know Before You Fly Campaign (see section 10 below) and No Drone Zone initiatives with the nation's busiest airports. The agency also would continue to take strong enforcement action against egregious violators. At the same time, it would continue working with stakeholders to improve safety to ensure further integration and innovation in this promising segment of aviation.[33]

Secretary Foxx was joined by representatives from the following stakeholder groups: (1) The Association for Unmanned Vehicle Systems International; (2) Academy of Model Aircraft; (3) Air Line Pilots Association; (4) American Association of Airport Executives; (5) Helicopter Association International; (6) PrecisionHawk; (7) AirMap/ Small UAV Coalition; and (8) Consumer Electronics Association.[34]

On October 29, 2015, FAA Administrator Michael Huerta announced the membership of the UAS Registration Task Force. Earl Lawrence, director of the FAA's UAS Integration Office, and Dave Vos of Google X would co-chair the group. Department of Transportation (DOT) Secretary Anthony Foxx and Administrator Huerta announced the formation

---

[32]UAS registration requirement press release, US Department of Transportation, available at https://www.transportation.gov/briefing-room/us-transportation-secretary-anthony-foxx-announces-unmanned-aircraft-registration.
[33]*Id.*
[34]*Id.*

of the Task Force the prior week. The Task Force membership represented a range of stakeholder viewpoints, interests and knowledge of the objectives and scope. Task Force membership was by invitation only and participation was voluntary. Interested parties who were not members of the Task Force could submit comments to the public docket.[35]

Secretary Foxx set a deadline of November 20 for the Task Force to complete its recommendations. The group met formally from November 3–5 before developing recommendations on a streamlined registration process and minimum requirements on which unmanned aircraft should be registered. Given the urgency of this issue, the DOT and FAA would move expeditiously to consider the Task Force's recommendations.

Task Force Members included: Nancy Egan – 3D Robotics; Richard Hanson – Academy of Model Aeronautics; George Novak – Aerospace Industries Association; Chuck Hogeman and Randy Kenagy – Air Line Pilots Association; Jim Coon – Aircraft Owners and Pilots Association; Sean Cassidy – Amazon Prime Air; Ben Gielow – Amazon Retail; Justin Towles – American Association of Airport Executives; Brian Wynne – Association of Unmanned Vehicle Systems International; Parker Brugge – Best Buy; Douglas Johnson – Consumer Electronics Association; Brendan Schulman – DJI; Paul Feldman – General Aviation Manufacturers Association; Dave Vos – GoogleX (Co-Chair); Tony Bates – GoPro; Matt Zuccaro – Helicopter Association International; Mike Fergus – International Association of Chiefs of Police; John Perry – Management Association for Private Photogrammetric Surveyors; Brandon Declet – Measure; Randall Burdett – National Association of State Aviation Officials; Sarah Wolf – National Business Aviation Association; Baptiste Tripard – Parrot; Tyler Collins – PrecisionHawk; Gregory McNeal – Small UAV Coalition; and Thomas Head – Walmart.[36]

Along with the FAA and DOT, the following federal agencies would provide expert support to the Task Force: Department of Commerce, Department of Defense, Department of Homeland Security, Department of the Interior, Office of Management and Budget, National Aeronautics and Space Administration and the Department of State.[37]

In November 2015, the FAA chartered the UAS Registration Task Force (RTF) Aviation Rulemaking Committee (ARC) to provide recommendations to the FAA on registration requirements and process for small UAS (those weighing under 55 lbs.), including those used for commercial purposes, and all model aircraft. Although the FAA did not currently enforce the requirement for small UAS used for hobby or recreational purposes to be registered, the rapid proliferation of these aircraft in the national airspace has caused the FAA to reevaluate this policy in the interests of public safety and the safety of the National Airspace System (NAS). The stated objective of the Task Force was to develop

---

[35] *UAS Registration Task Force,* FAA, http://www.faa.gov/news/updates/?newsId=84125&omniRss=news_updatesAoc&cid=101_N_U.

[36] *Id.*

[37] *Id.*

recommendations for the creation of a registration process, which ultimately would contribute to an enforceable rule imposed by the FAA. The intent of establishing this registration framework, according to the FAA, was to promote a culture of accountability while achieving a maximum level of compliance.[38]

Recommendations from this Task Force are summarized as follows: (1) Owners fill out an electronic registration form through the web or through an application (app) to include name and street address, for any and all small UAS that the registrant owns prior to operation of their UAS in the NAS; (2) Immediately receive an electronic certificate of registration and a personal universal registration number for use on all small UAS owned by that person; and (3) Mark the registration number (or registered serial number) on all applicable small UAS prior to their operation in the NAS. Based primarily upon an assessment of available safety studies and risk probability calculations, and notwithstanding determinations in other countries with differing weight thresholds, the Task Force recommended an exclusion from the registration requirement for any small unmanned aircraft weighing a total of 250 grams or less.[39] Information on US citizenship or residence status would not be required. There would be a minimum age requirement of 13 years to register to comply with the Children's Online Privacy Protection Act, 15 United States Code § 6501-6505. Children under that age operate a small UAS under a parent or guardian's registration.[40]

On December 14, 2015, the FAA announced its Small UAS Rule, which can be summarized as follows:

- Registration is a statutory requirement that applies to all aircraft
- Applies to small UAS weighing 0.55lb (250g) up to less than 55lb (25kg) including payload (e.g., camera)
- Registration begins Dec 21, 2015
- First 30 days are free (till Jan 20, 2016) then $5
- Must register no later than Feb 19, 2016 if UAS was previously operated prior to Dec 21, 2015
- Must register before first flight outdoors if UAS purchased after Dec 21, 2015
- Must be 13 years of age to register (parent/guardian must register for younger children)
- Streamlined and user-friendly web-based aircraft registration process
- Link for registration at www.faa.gov/uas/registration
- Provide NAME, HOME ADDRESS, EMAIL

---

[38] *Id.*
[39] *UAS RTF ARC*, Federal Aviation Administration (FAA), http://www.faa.gov/uas/publications/media/RTFARCFinalReport_11-21-15.pdf.
[40] *Id.*

- Certificate of Aircraft Registration/Proof of Ownership includes ID Number for UAS owner and this ID number must be marked on UAS
- Hobby/recreational users must only register once for all their models and use the same ID number that will be valid for 3 years[41]

On December 21, 2015, 14 CFR Part 48 became effective. This is the regulation pertaining to the registration and marking requirements for small, unmanned aircraft. It reads as follows:

Subpart A – General

§48.1 Applicability.

a. This part provides registration and identification requirements for small unmanned aircraft that are part of a small unmanned aircraft system as defined in §1.1 of this chapter.
b. Small unmanned aircraft eligible for registration in the United States must be registered and identified in accordance with either:
    1. The registration and identification requirements in this part; or
    2. The registration requirements in part 47 and the identification and registration marking requirements in subparts A and C of part 45.
c. Small unmanned aircraft intended to be operated outside of the territorial airspace of the United States, or registered through a trust or voting trust, must be registered in accordance with subparts A and B of part 47 and satisfy the identification and registration marking requirements of subparts A and C of part 45.

§48.5 Compliance dates.

a. *Small unmanned aircraft used exclusively as model aircraft.* For small unmanned aircraft operated by the current owner prior to December 21, 2015, compliance with the requirements of this part or part 47 is required no later than February 19, 2016. For all other small unmanned aircraft, compliance with this part is required prior to operation of the small unmanned aircraft.
b. *Small unmanned aircraft used as other than model aircraft.* Small unmanned aircraft owners authorized to conduct operations other than model aircraft operations must register the small unmanned aircraft in accordance with part 47 of this chapter. Beginning March 31, 2016, small unmanned aircraft operated as other than model aircraft may complete aircraft registration in accordance with this part.

---

[41] *Small UAS Rule,* FAA, http://www.faa.gov/news/press_releases/news_story.cfm?newsId=19856&amp%3Bcid=TW378.

§48.10 Definitions.

For purposes of this part, the following definitions apply:

Citizen of the United States or U.S. citizen means one of the following:

1. An individual who is a citizen of the United States or one of its possessions.
2. A partnership each of whose partners is an individual who is a citizen of the United States.
3. A corporation or association organized under the laws of the United States or a State, the District of Columbia, or a territory or possession of the United States, of which the president and at least two-thirds of the board of directors and other managing officers are citizens of the United States, which is under the actual control of citizens of the United States, and in which at least 75 percent of the voting interest is owned or controlled by persons that are citizens of the United States.

Registry means the FAA, Civil Aviation Registry, Aircraft Registration Branch.

Resident alien means an individual citizen of a foreign country lawfully admitted for permanent residence in the United States as an immigrant in conformity with the regulations of the Department of Homeland Security (8 CFR Chapter 1).

§48.15 Requirement to register.

No person may operate a small unmanned aircraft that is eligible for registration under 49 U.S.C. 44101-44103 unless one of the following criteria has been satisfied:

a. The owner has registered and marked the aircraft in accordance with this part;
b. The aircraft weighs 0.55 pounds or less on takeoff, including everything that is on board or otherwise attached to the aircraft; or
c. The aircraft is an aircraft of the Armed Forces of the United States.

§48.20 Eligibility for registration.

A small unmanned aircraft may be registered under 49 U.S.C. 44103 and under this part only when the aircraft is not registered under the laws of a foreign country and is—

a. Owned by a U.S. citizen;
b. Owned by an individual citizen of a foreign country lawfully admitted for permanent residence in the United States;
c. Owned by a corporation not a citizen of the United States when the corporation is organized and doing business under the laws of the United States or a State within the United States, and the aircraft is based and primarily used in the United States; or
d. An aircraft of—

1. The United States Government; or
2. A State, the District of Columbia, a territory or possession of the United States, or a political subdivision of a State, territory, or possession.

§48.25 Applicants.

a. To register a small unmanned aircraft in the United States under this part, a person must provide the information required by §48.100 to the Registry in the form and manner prescribed by the Administrator. Upon submission of this information, the FAA issues a Certificate of Aircraft Registration to that person.
b. A small unmanned aircraft must be registered by its owner using the legal name of its owner, unless the owner is less than 13 years of age. If the owner is less than 13 years of age, then the small unmanned aircraft must be registered by a person who is at least 13 years of age.
c. In accordance with 49 U.S.C. 44103(c), registration is not evidence of aircraft ownership in any proceeding in which ownership of an unmanned aircraft by a particular person is in issue.
d. In this part, "owner" includes a buyer in possession, a bailee, a lessee of a small unmanned aircraft under a contract of conditional sale, and the assignee of that person.

§48.30 Fees.

a. The fee for issuing or renewing a Certificate of Aircraft Registration for aircraft registered in accordance with §48.100(a) is $5.00 per aircraft.
b. The fee for issuing or renewing a Certificate of Aircraft Registration for aircraft registered in accordance with §48.100(b) is $5.00 per certificate.
c. Each application for and renewal of a Certificate of Aircraft Registration must be accompanied by the fee described in paragraphs (a) and (b), as applicable, paid to the Federal Aviation Administration through the web-based aircraft registration system, or in another manner if prescribed by the Administrator.

Subpart B – Certificates of Aircraft Registration for Small Unmanned Aircraft

§48.100 Application.

a. *Required information: Persons intending to use the small unmanned aircraft as other than a model aircraft.* Each applicant for a Certificate of Aircraft Registration issued under this part must submit all of the following information to the Registry:

1. Applicant name and, for an applicant other than an individual, the name of the authorized representative applying for a Certificate of Aircraft Registration.
2. Applicant's physical address and, for an applicant other than an individual, the physical address for the authorized representative. If the applicant or authorized representative does not receive mail at their physical address, a mailing address must also be provided.
3. Applicant's email address or, for applicants other than individuals, the email address of the authorized representative.
4. The aircraft manufacturer and model name.
5. The aircraft serial number, if available.
6. Other information as required by the Administrator.

b. *Required information: Individuals intending to use the small unmanned aircraft exclusively as a model aircraft.* Each applicant for a Certificate of Aircraft Registration issued under this part must submit all of the following information to the Registry:
   1. Applicant name.
   2. Applicant's physical address and if the applicant does not receive mail at their physical address, a mailing address must also be provided.
   3. Applicant's email address.
   4. Other information as required by the Administrator.

c. *Provision of information.* The information identified in paragraphs (a) and (b) of this section must be submitted to the Registry through the Web-based small unmanned aircraft registration system in a form and manner prescribed by the Administrator.

d. *Issuance of Certificate of Aircraft registration.* The FAA will issue a Certificate of Aircraft Registration upon completion of the application requirements provided in paragraph (a) or (b) of this section as applicable.

§48.105 Requirement to maintain current information.

a. The holder of a Certificate of Aircraft Registration must ensure that the information provided under §48.100 remains accurate.

b. The holder of a Certificate of Aircraft Registration must update the information using the web-based small unmanned aircraft registration system within 14 calendar days of the following:
   1. A change in the information provided under §48.100.
   2. When aircraft registration requires cancellation for any reason including sale or transfer, destruction, or export.

§48.110 Registration: Persons intending to use small unmanned aircraft for purposes other than as model aircraft.

a. *Certificate of Aircraft Registration.* A Certificate of Aircraft Registration issued in accordance with §48.100 for aircraft used for purposes other than as model aircraft constitutes registration only for the small unmanned aircraft identified on the application.

b. *Effective date of registration.* An aircraft is registered when the applicant receives a Certificate of Aircraft Registration for the specific aircraft. The effective date of registration is shown by the date of issue on the Certificate of Aircraft Registration issued for the aircraft.

c. *Registration renewal.* A Certificate of Aircraft registration issued under this part expires 3 years after the date of issue unless it is renewed.

    1. The holder of a Certificate of Aircraft Registration must renew the Certificate by verifying, in a form and manner prescribed by the Administrator, that the information provided in accordance with §48.100 of this subpart is accurate and if it is not, provide updated information. The verification may take place at any time within the six months preceding the month in which the Certificate of Aircraft registration expires.
    2. A certificate issued under this paragraph expires three years from the expiration date of the previous certificate.

d. *Other events affecting effectiveness of Certificate.* Each Certificate of Aircraft Registration issued by the FAA under this subpart is effective, unless registration has ended by reason of having been revoked, canceled, expired, or the ownership is transferred, until the date upon which one of the following events occurs:

    1. Subject to the Convention on the International Recognition of Rights in Aircraft when applicable, the aircraft is registered under the laws of a foreign country.
    2. The small unmanned aircraft is totally destroyed or scrapped.
    3. The holder of the Certificate of Aircraft Registration loses U.S. citizenship.
    4. Thirty days have elapsed since the death of the holder of the Certificate of Aircraft Registration.
    5. The owner, if an individual who is not a citizen of the United States, loses status as a resident alien, unless that person becomes a citizen of the United States at the same time.
    6. The owner is a corporation other than a corporation which is a citizen of the United States and one of the following events occurs:
        i. The corporation ceases to be lawfully organized and doing business under the laws of the United States or any State thereof; or
        ii. The aircraft was not operated exclusively within the United States during the period of registration under this part.

§48.115 Registration: Individuals intending to use small unmanned aircraft exclusively as a model aircraft.

a. *Certificate of Aircraft Registration.* A Certificate of Aircraft Registration issued in accordance with §48.100 for small unmanned aircraft used exclusively as model aircraft constitutes registration for all small unmanned aircraft used exclusively as model aircraft owned by the individual identified on the application.
b. *Effective date of registration.* An aircraft is registered when the applicant receives a Certificate of Aircraft Registration. The effective date of registration is shown by the date of issue on the Certificate of Aircraft Registration issued under this part.
c. *Registration renewal.* A Certificate of Aircraft registration issued under this part expires 3 years after the date of issue unless it is renewed.
    1. The holder of a Certificate of Aircraft Registration must renew the Certificate by verifying, in a form and manner prescribed by the Administrator, that the information provided in accordance with §48.100(b) and (c) of this part is accurate and if it is not, provide updated information. The verification may take place at any time within the six months preceding the month in which the Certificate of Aircraft registration expires.
    2. A certificate issued under this paragraph expires three years from the expiration date of the previous certificate.
d. *Other events affecting effectiveness of Certificate.* Each Certificate of Aircraft Registration issued by the FAA under this part is effective, unless registration has ended by reason of having been revoked, canceled or expired, or until the date upon which one of the following events occurs:
    1. The holder of the Certificate of Aircraft Registration loses U.S. citizenship.
    2. Thirty days have elapsed since the death of the holder of the Certificate of Aircraft Registration.
    3. The owner, if an individual who is not a citizen of the United States, loses status as a resident alien, unless that person becomes a citizen of the United States at the same time.

§48.120 Invalid registration.

The registration of a small unmanned aircraft is invalid if, at the time it is made—

a. The aircraft is registered in a foreign country;
b. The applicant is not the owner, except when the applicant registers on behalf of an owner who is under 13 years of age;
c. The applicant is not eligible to submit an application under this part; or
d. The interest of the applicant in the aircraft was created by a transaction that was not entered into in good faith, but rather was made to avoid (with or without the owner's knowledge) compliance with 49 U.S.C. 44101-44103.

§48.125 Foreign civil aircraft.

Except for corporations eligible to register under §48.20(c), the FAA will issue a recognition of ownership to persons required to comply with the provisions of this part pursuant to an authorization to operate issued under part 375 of this title. The recognition of ownership does not have the effect of U.S. aircraft registration.

Subpart C – Aircraft Marking

§48.200 General.

a. No person may operate a small unmanned aircraft registered in accordance with this part unless the aircraft displays a unique identifier in accordance with the requirements of §48.205 of this subpart.
b. A unique identifier is one of the following:
    1. The registration number issued to an individual or the registration number issued to the aircraft by the Registry upon completion of the registration process provided by this part; or
    2. If authorized by the Administrator and provided with the application for Certificate of Aircraft Registration under §48.100 of this part, the small unmanned aircraft serial number.

§48.205 Display and location of unique identifier.

a. The unique identifier must be maintained in a condition that is legible.
b. The unique identifier must be affixed to the small unmanned aircraft by any means necessary to ensure that it will remain affixed for the duration of each operation.
c. The unique identifier must be readily accessible and visible upon inspection of the small, unmanned aircraft. A unique identifier enclosed in a compartment is readily accessible if it can be accessed without the use of any tool.[42]

6. FAA Order 8900.338

On December 22, 2015, the FAA released Order 8900.338 entitled New Requirements for Registering and Marking Small Unmanned Aircraft. The effective date was 12/22/15 and the cancellation date is 12/22/16. It is reproduced below in its entirety:

1. Purpose of This Notice. This notice is intended to inform Flight Standards Service (AFS) field employees of new requirements for the registration and marking

[42]14 CFR 48, ECFR, http://www.ecfr.gov/cgi-bin/text-idx?SID=36ce0c51390f31ff80788437c8f20703&mc=true&node=20151216y1.36.

of small unmanned aircraft found in Title 14 of the Code of Federal Regulations (14 CFR) parts 47 and 48. These requirements may supersede the current guidance in Federal Aviation Administration (FAA) Order 8900.1, Flight Standards Information Management System (FSIMS), relative to surveillance, incident/accident investigation, and reporting of activity involving small Unmanned Aircraft Systems (sUAS). This notice contains guidance that is pertinent to part 48. A list of Frequently Asked Questions (FAQ) can be found in Appendix A.

2. Audience. The primary audience for this notice is Flight Standards District Office (FSDO) aviation safety inspectors (ASI) and assigned sUAS focal points. The secondary audience includes AFS divisions and branches at headquarters (HQ) and the regional offices.
3. Where You Can Find This Notice. You can find this notice on the MyFAA employee Web site at https://employees.faa.gov/tools_resources/orders_notices. Inspectors can access this notice through FSIMS at http://fsims.avs.faa.gov. Operators can find this notice on the FAA's Web site at http://fsims.faa.gov. This notice is available to the public at http://www.faa.gov/regulations_policies/orders_notices and http://rgl.faa.gov.
4. Applicability. This notice applies to sUAS owners and operators subject to the new registration requirements under 14 CFR parts 47 and 48. Under the new rule, unmanned aircraft weighing less than 55 pounds and more than 0.55 pounds (250 grams) on takeoff, including everything that is on board or otherwise attached to the aircraft and operated outdoors in the National Airspace System (NAS), must be registered and marked. All unmanned aircraft are considered aircraft, as defined by Title 49 of the United States Code (49 U.S.C.) § 44101(a).
5. Background.
    a. Request For Information (RFI). In October 2015, the Department of Transportation (DOT) and the FAA published an RFI, titled "Clarification of the Applicability of Aircraft Registration Requirements for Unmanned Aircraft Systems (UAS) and Request for Information Regarding Electronic Registration for UAS." The RFI was published under Federal Register (FR) Docket No. FAA-2015-4378.
    b. Unmanned Aircraft Systems (UAS) Registration Task Force Aviation Rulemaking Committee (ARC). On October 20, 2015, the FAA chartered the UAS Registration Task Force ARC. The ARC provided a forum for industry to discuss and provide recommendations to the FAA. The ARC was tasked specifically with developing recommendations for the registration of sUAS, including developing criteria for which UAS would need to be registered, a registration process, and methods for proving registration and marking. The ARC submitted a final report with their recommendations to the FAA on November 21, 2015.

c. Final Rule. The FAA published the Registration and Marking Requirements for Small Unmanned Aircraft final rule on December 16, 2015. The rule contains the registration and marking requirements for small unmanned aircraft. The requirements of the rule became effective on December 21, 2015. The rule creates a new part 48 that provides an alternative to the paper-based part 47 registration process. Part 48 provides a streamlined, Web-based aircraft registration process for the registration of small unmanned aircraft. This simplified registration process will facilitate compliance with the statutory requirement for small unmanned aircraft, including model aircraft, to register prior to operation. It also provides a method for marking small unmanned aircraft that is more appropriate. This rule does not address issues related to operations, certification, or training. These issues will be covered in a separate rule. This rule does not prohibit registration of a small unmanned aircraft using the part 47 process. sUAS owners can register their unmanned aircraft under part 48 at https://www.registermyuas.faa.gov/.

6. Summaries of Requirements and Rule Provisions.

a. Registration and Marking. sUAS owners and/or operators intending to fly a small unmanned aircraft for hobby or recreation; for civil and commercial purposes; and for conducting public aircraft operations will be required to register and mark the small unmanned aircraft in accordance with part 47 or part 48 prior to operating. The FAA is strongly encouraging sUAS owners to register under part 48, which was designed solely for the registration of small unmanned aircraft and offers a faster, more streamlined registration process than what is available under part 47.

b. Weight Requirement and Registration Number. Small unmanned aircraft weighing less than 55 pounds and more than 0.55 pounds (250 grams) on takeoff, including everything that is on board or otherwise attached to the aircraft and operated outdoors in the NAS, will be required to display a registration number. The registration number will be composed of multiple alphanumeric characters and will not be a traditional N-number.

c. Registration Certificate. Registration certificates must be readily available to law enforcement or any representative of the Administrator. This means that the owner must have a printed or electronic copy available to show upon request.

d. Readable and Easily Located Markings. The rule requires registration markings to be readable and easily located on the small unmanned aircraft. Only the aircraft must be marked with the registration number. This may include placing a marking in a compartment on the unmanned aircraft that can be opened without the use of tools. The rule requires the registration of the unmanned aircraft, not the associated system equipment (e.g., controller).

e. Registration System Access. Law enforcement access to the registration system must be coordinated through the Office of Security and Hazardous Materials Safety (ASH), Law Enforcement Assistance Program (LEAP). Law enforcement entities should contact the local LEAP special agent (SA). Currently, there is no public search function.

f. Registration Under Part 47 and Part 48. Unlike registration under part 47, which provides for registration of individual aircraft described by manufacturer name, model, and serial number, registration under part 48 provides a new option for individual persons who intend to operate their sUAS. For individuals intending to use their sUAS for hobby or recreation, part 48 does not require individual aircraft to be described and issues a single registration number to the owner for use on all of their recreational sUAS. sUAS intended to be used nonrecreationally will be registered individually with each aircraft being assigned a unique sUAS registration number. Part 48 should offer the vast majority of sUAS owners a system that is preferable to the existing registration system under part 47. Part 48 offers an online registration system that will provide a Certificate of Aircraft Registration almost immediately. Part 48 will allow an owner to register multiple small unmanned aircraft for a single $5 fee, provided the owner is an individual and the intended use of the small unmanned aircraft is noncommercial (i.e., hobby or recreation). sUAS intended for other uses will require a $5 fee for each aircraft registered. sUAS owners can register their unmanned aircraft under part 48 at https://www.registermyuas.faa.gov/.

7. Surveillance, Investigation and Enforcement. ASIs may observe sUAS in performing surveillance or responding to accidents, incidents, occurrences, or complaints reported to the FSDO.

a. Surveillance. ASIs should be prepared to support sUAS activity within their area of geographic responsibility. Familiarity with sUAS local (FSDO), regional, and HQ representatives, and focal points will help with these tasks. In an effort to provide consistency across the agency, ASIs are encouraged to contact and notify their management and sUAS focal points of any questions or concerns regarding this rule or sUAS in general. The use of Risk-Based Decision Making and compliance philosophy, along with current practice and procedures, should be used to support proper surveillance and vigilance over sUAS operations and the NAS. The sUAS focal points have access to a list of current exemption holders (i.e., persons authorized to operate for purposes other than as model aircraft pursuant to an exemption issued in accordance with section 333 of Pub. L. 112-95) and should receive notification of any new exemption holders within the FSDO area of geographic responsibility. A list of exemption holders can also be found at http://www.faa.gov/uas/legislative_programs/section_333/333_authorizations/.

b. Impact on Accident, Incident, Occurrence, and Complaint Investigation. During the conduct of accident, incident, occurrence, and complaint investigations involving sUAS, ASIs will ensure that the unmanned aircraft meet the compliance requirements for registration and marking. In addition, ASIs will inform their local sUAS focal point and coordinate as necessary to ensure sUAS events are channeled up appropriately through the chain of command and to HQ divisions.

c. Education, Compliance, and Enforcement Philosophy. ASIs shall use the guidance published in FAA Notice 8900.313, Education, Compliance, and Enforcement of Unauthorized Unmanned Aircraft Systems Operators, and Order 8000.373, Federal Aviation Administration Compliance Philosophy, on the process of contact, education, and enforcement generally to be provided to individuals who are the subject of investigations involving sUAS aircraft. Both ASH and AFS are delegated authorities and have responsibilities regarding FAA's aviation security and safety missions. Some authorities and responsibilities are unique to each organization, while others are interrelated. Information sharing and collaboration between the organizations is critical to the achievement of the mission objectives. If a registration issue is discovered in the course of a sUAS investigation wherein a law enforcement, national security, or national defense nexus exists, the ASI must coordinate with the local ASH Servicing Security Element (SSE).

d. ASI Access to sUAS Registrant Records. Access to the Web-based sUAS registration system will provide ASIs with a searchable database of sUAS registrants. The FAA has determined that it is in the public interest to publish 14 CFR Part 48 as expeditiously as possible. As a result, the searchable database functionality of the system is not yet operational and will become available in future releases of the registration platform.

e. Documenting Activities. Individuals with sUAS compliance responsibilities will document their activities into the Program Tracking and Reporting Subsystem (PTRS) database using the guidance in Order 8900.1, Volume 16. In addition, use of the following codes contained below in Table 1, PTRS Codes, will ensure the highest Flight Standards Automation System (FSAS) data quality.

8. Disposition. We will incorporate the information in this notice into Order 8900.1 before this notice expires. Direct questions concerning the information in this notice to the General Aviation and Commercial Division (AFS-800) at (202) 267-1100.

Table 1. PTRS Codes

PTRS FIELD	DATA
Activity Field	Various
14 CFR Field	Enter "107" or "91" as appropriate
Registration Number Field	Enter the registration number if registered under part 47; if registered under part 48, enter information in the Comment field
Make, Model, and Series (M/M/S) Field	Populate M/M/S field if sUAS is assigned an M/M/S in the Enhanced Flight Standards Automation System (eFSAS) database; leave blank if not in database and enter the information in Section III – Equipment Ancillary
National Use Field	Enter "UAS-P-S" (if sUAS is Public Use*), "UAS-H-S" (if sUAS is Hobby Use**), or "UAS-C-S" (if sUAS is Civil Use***)
Comment Field	If sUAS is registered under part 48, enter "H" in the Primary Area field, enter "399" in the Keyword field, enter "I" in the Opinion Code field, enter registration number in Comment field. ASIs should also note if a sUAS was not registered but should have been under the requirements of parts 47 and 48.

* Public Aircraft: Aircraft operated by a government entity (including Federal, State, or local governments, and U.S. Department of Defense (DOD) and its military branches) for certain purposes as described in 49 U.S.C. §§ 40102(a)(41) and 40125. A public aircraft status determination is based on the operation. Refer to 14 CFR part 1, § 1.1 for a complete definition and to the current edition of Advisory Circular (AC) 00-1.1, Public Aircraft Operations.
** Hobby Aircraft: sUAS used for hobby or recreational purposes only.
*** Civil Aircraft: Aircraft other than public aircraft.

Appendix A. Frequently Asked Questions (FAQ)

1. What unmanned aircraft are covered by the new registration requirements?

   Unmanned aircraft operated outdoors and weighing less than 55 pounds and more than 0.55 pounds (250 grams) on takeoff, including everything that is on board or otherwise attached to the aircraft. (Refer to part 48, § 48.15.)

2. When must sUAS owners who purchased and operated their sUAS before December 21, 2015, register their aircraft?

   For small unmanned aircraft operated by the current owner prior to December 21, 2015, compliance with the requirements of part 47 or part 48 is required no later than February 19, 2016. (Refer to § 48.5.)

3. If a person received a sUAS as a holiday gift when must it be registered by?

The rule is effective on December 21, 2015. Any sUAS acquired as a gift or otherwise, on or after December 21, 2015, and intended to be used exclusively for noncommercial use (e.g., recreation, hobby, and model aircraft) must be registered before outdoor use. (Refer to § 48.5.)

4. Will there be a unique registration number for each small unmanned aircraft that is required to register?

   An individual owner may be able to use the same registration number and Certificate of Registration for multiple aircraft. As provided under § 48.115(a), a Certificate of Aircraft Registration issued in accordance with § 48.100 for small unmanned aircraft used exclusively as model aircraft constitutes registration for all small unmanned aircraft used exclusively as model aircraft owned by the individual identified on the application. (Refer to § 48.115(a).) Each small unmanned aircraft intended to be used other than as a model aircraft and owned by either a corporation or individual will be issued a Certificate of Aircraft Registration with a unique registration number. (Refer to § 48.110(a).)

5. Is there a minimum age that a sUAS owner must be before they can register their small unmanned aircraft online?

   Yes. A person must be 13 years of age or older before they are permitted to register a small unmanned aircraft using part 48. If the owner is less than 13 years of age, then the small unmanned aircraft must be registered by a person who is at least 13 years of age. If a person turns 13 years of age or older after the aircraft has been registered, the FAA does not expect the registration to be transferred during the period of registration (i.e., prior to the expiration date of the registration). Acceptable example: A parent purchases an sUAS for a child who is under 13 and registers the aircraft for their use. When the child turns 13, the aircraft may remain registered to the parent or may be transferred to the now 13-year-old owner. (Refer to § 48.25.)

6. Can you register a small unmanned aircraft under part 48 using a paper form?

   No. Registration under part 48 will occur through an online Web-based system only. (Refer to § 48.100(c).)

7. How long does it take to get a Certificate of Aircraft Registration under part 48?

   The owner should receive a Certificate of Aircraft Registration immediately when registering under part 48 using the online system. (Refer to § 48.100(c).)

8. When registering under part 48, how will the Certificate of Aircraft Registration be delivered to the sUAS owner?

   Upon completion of the registration process, the Certificate of Aircraft Registration will be delivered electronically to the owner through the online registration system. (Refer to § 48.100(d).)

9. What information is displayed on the Certificate of Aircraft Registration?

The small unmanned aircraft owner's name, issue date, expiration date, and registration number are displayed on the Certificate of Aircraft Registration.

10. Is there a fee for registration under part 48?

Yes. The fee is $5. (Refer to § 48.30.)

11. Is it $5 to register each aircraft?

Not necessarily. For persons intending to use the small unmanned aircraft other than as a model aircraft, it is $5 to register each aircraft. However, for individuals intending to use the small unmanned aircraft exclusively as model aircraft, it is only $5 to register that individual's fleet of small unmanned aircraft under part 48. (Refer to § 48.30.)

12. Is there a registration renewal requirement for sUAS like there is for manned aircraft?

Yes. A Certificate of Aircraft Registration issued in accordance with part 48 is effective once the registration process is complete and must be renewed every three years. Each Certificate of Aircraft Registration will display an effective date and an expiration date. (Refer to § 48.115(c).)

13. Is there a fee for registration renewal under part 48?

Yes. The fee is the same as initial registration, $5. (Refer to § 48.115(c).)

14. Are small unmanned aircraft subject to the new registration requirements also required to mark their small unmanned aircraft?

Yes. Small unmanned aircraft subject to the registration requirements must display a unique identifier. The identifier will be either an FAA-issued registration number or the Administrator may authorize the use of the small unmanned aircraft serial number. The registration number will be composed of multiple alphanumeric characters and will not be a traditional N-number. The registration number can be affixed by permanent marker, label, engraving, or other means, as long as the number is readily accessible and maintained in a condition that is readable and legible upon close visual inspection. (Refer to §§ 48.200 and 48.205.)

For the latest information on FAQs, visit https://www.faa.gov/uas/registration/faqs.[43]

7. Taylor v. FAA

On January 4, 2016, the FAA was hauled into Federal Court over UAS Registration rules for the first time. The lawsuit, challenging the FAA registration of small UAS that applies to hobby and recreational users as being illegal, was filed in the Court of Appeals for the

[43] FAA Order 8900.338, FAA, http://www.faa.gov/regulations_policies/orders_notices/index.cfm/go/document.information/documentID/1028709.

District of Columbia on December 24, 2015, by a model aircraft enthusiast, Maryland resident, and insurance attorney, John A. Taylor on his own behalf. Taylor is requesting that the court issue an order declaring that the [FAA's registration rule] is void and prohibited by Section 336 of the FAA Modernization and Reform Act of 2012. He alleges that that section specifically prohibits the FAA from promulgating any new rules or regulations regarding model aircraft if they are flown for hobby or recreational purposes.[44]

Taylor requested an emergency stay of the FAA's registration requirement while the case winds its way through the legal process. That request was denied by the Court of Appeals on December 24, 2015, stating that Taylor "has not satisfied the stringent requirements for a stay pending court review."[45] Both parties filed their briefs, and oral arguments were scheduled for some time after August 2016.[46]

8. TechFreedom v. FAA

On February 17, 2016, TechFreedom filed suit against the FAA in the U.S. Court of Appeals for the District of Columbia arguing that the UAS registration rules put in place in December 2015 violate a federal law that prohibits the FAA from regulating recreational UAS.[47] This is currently the second suit of its kind challenging this law. The petition is reproduced below:

PETITION FOR REVIEW

Pursuant to 49 U.S.C. § 46110, 5 U.S.C. § 706, and Rule 15(a) of the Federal Rules of Appellate Procedure, TechFreedom respectfully petitions this Court for review of an interim final rule of the Federal Aviation Administration ("FAA") setting forth registration and marking requirements for small unmanned aircraft, 80 Fed. Reg. 78593 (Dec. 16, 2015). A copy of the interim final rule is attached as Exhibit 1 (the "IFR"). This Petition is being timely filed with the Court within sixty days of the issuance of the IFR, pursuant to 49 U.S.C. § 46 110(a) and calculated in accordance with Federal Rule of Appellate Procedure 26(a)(1). Venue is proper in this Court pursuant to 49 U.S.C. § 46110(a).

Founded in 2010, TechFreedom is a non-profit, non-partisan 501(c)(3) tax-exempt think tank dedicated to educating policymakers, the media, and the public about Internet policy. We defend the freedom to tinker and innovate, as well as the freedom of expression using new technologies such as "drones," i.e., small unmanned aircraft systems.

---

[44] *FAA sued in Federal Court over UAS Registration Rules*, Forbes, http://www.forbes.com/sites/johngoglia/2016/01/04/faa-sued-in-federal-court-over-drone-registration-rules/#6757548d2c79.

[45] *Id.*

[46] The Lawsuit over Drone Registration: Taylor vs. Huerta & FAA, http://dronelife.com/2016/06/14/lawsuit-drone-registration-taylor-vs-huerta-faa/

[47] *TechFreedom sues the FAA on Drone Regulations*, TechFreedom, http://techfreedom.org/post/139547849574/techfreedom-sues-the-faa-on-drone-regulations.

TechFreedom has purchased a drone that it wishes to operate in the national airspace system. However, because this drone weighs more than 250 grams, TechFreedom is prohibited from operating this drone unless it abides by the IFR's registration requirement and pays a $5.00 fee. See 80 Fed. Reg. 78593, 78645—648 (to be codified at 14 C.F.R. § 48. 1—.205). The IFR is contrary to law, as Section 336 of the FAA Modernization and Reform Act of 2012 prohibits the agency from "promulgat[ing] any rule or regulation regarding a model aircraft . . . notwithstanding any other provision of law relating to the incorporation of unmanned aircraft systems into Federal Aviation Administration plans and policies." Pub. L. No. 112-95, § 336(c), 126 Stat. 11.

Moreover, even if Congress had not enacted this provision, the IFR exceeds the FAA's authority to require the registration of aircraft under 49 U.S.C. § 44101—106, as it purports to require the registration of persons who own model aircraft—not of the aircraft themselves. See 80 Fed. Reg. 78593, 78647 (to be codified at 14 C.F.R. § 48.115). Finally, the agency promulgated the IFR without observance of procedure required by law, as the FAA failed to show good cause for dispensing with the Administrative Procedure Act's notice-and-comment rulemaking process. See 5 U.S.C. § 553(b)(3)(B). Thus, the FAA's actions are unlawful, as they were "arbitrary, capricious, an abuse of discretion, [and done without observance of procedure required by law[.]" 5 U.S.C. § 706].

Accordingly, TechFreedom respectfully requests that this court hold unlawful, vacate, enjoin, and set aside the IFR, and that it provide such additional relief as may be appropriate.

Dated: February 16, 2016[48]

In April 2016, FAA lawyers convinced the U.S. Court of Appeals for the District of Columbia Circuit to dismiss the case, arguing TechFreedom and Taylor had missed the deadline to request a judicial review of the regulation. As a result of the court's decision, Taylor and all other U.S. drone owners will continue to be required to obey the FAA regulations.[49]

9. Academy of Model Aeronautics (AMA)

The Academy of Model Aeronautics (AMA) is the world's largest model aviation association, representing a membership of more than 175,000 from every walk of life, income level, and age group. It is a self-supporting, nonprofit organization whose purpose is to promote the development of model aviation as a recognized sport and worthwhile recreation activity. The organization is open to anyone interested in model aviation. The AMA is the official national body for model aviation in the United States. AMA sanctions more than 2,000 model competitions throughout the country each year, and certifies official model flying records on a national and international level. Additionally, the AMA is the organizer

---

[48] *Id.*

[49] *Federal Court Dismisses Lawsuit against Drone Regulations,* Heartland News, available at http://news.heartland.org/newspaper-article/2016/04/19/federal-court-dismisses-lawsuit-against-drone-regulations.

of the annual National Aeromodeling Championships, the world's largest model airplane competition. The AMA is the chartering organization for more than 2,500 model airplane clubs across the country. AMA offers its chartered clubs official contest sanction, insurance, and assistance in getting and keeping flying sites. The AMA is the voice of its membership, providing liaison with the FAA, the Federal Communications Commission (FCC), and other government agencies through their national headquarters in Muncie, Indiana. AMA also works with local governments, zoning boards, and parks departments to promote the interests of local chartered clubs. The AMA is an associate member of the National Aeronautic Association (NAA). Through NAA, AMA is recognized by the Fédération Aéronautique Internationale (FAI), the world governing body of all aviation activity, as the only organization, which may direct US participation in international aeromodeling activities.[50]

The AMA promotes safety with the following Academy of Model Aeronautics National Model Aircraft Safety Code, which it distributes to its members and became effective on January 1, 2014. It reads as follows:

A. GENERAL: A model aircraft is a non-human-carrying aircraft capable of sustained flight in the atmosphere. It may not exceed limitations of this code and is intended exclusively for sport, recreation, education and/or competition. All model flights must be conducted in accordance with this safety code and any additional rules specific to the flying site.
   1. Model aircraft will not be flown:
      a. In a careless or reckless manner.
      b. (b) At a location where model aircraft activities are prohibited.
   2. Model aircraft pilots will:
      a. Yield the right of way to all human-carrying aircraft.
      b. See and avoid all aircraft and a spotter must be used when appropriate. (AMA Document #540-D.)
      c. Not fly higher than approximately 400 feet above ground level within three (3) miles of an airport without notifying the airport operator.
      d. Not interfere with operations and traffic patterns at any airport, heliport or seaplane base except where there is a mixed-use agreement.
      e. Not exceed a takeoff weight, including fuel, of 55 pounds unless in compliance with the AMA Large Model Airplane program. (AMA Document 520-A.)
      f. Ensure the aircraft is identified with the name and address or AMA number of the owner on the inside or affixed to the outside of the model aircraft. (This does not apply to model aircraft flown indoors.)

[50]*Model Aircraft,* Academy of Model Aeronautics (AMA), http://www.modelaircraft.org/aboutama/whatisama.aspx.

g. Not operate aircraft with metal-blade propellers or with gaseous boosts except for helicopters operated under the provisions of AMA Document #555.

h. Not operate model aircraft while under the influence of alcohol or while using any drug that could adversely affect the pilot's ability to safely control the model.

i. Not operate model aircraft carrying pyrotechnic devices that explode or burn, or any device which propels a projectile or drops any object that creates a hazard to persons or property.

Exceptions:

- ❑ Free Flight fuses or devices that burn producing smoke and are securely attached to the model aircraft during flight.
- ❑ Rocket motors (using solid propellant) up to a G-series size may be used provided they remain attached to the model during flight. Model rockets may be flown in accordance with the National Model Rocketry Safety Code but may not be launched from model aircraft.
- ❑ Officially designated AMA Air Show Teams (AST) are authorized to use devices and practices as defined within the Team AMA Program Document. (AMA Document #718.)

j. Not operate a turbine-powered aircraft, unless in compliance with the AMA turbine regulations. (AMA Document #510-A.)

3. Model aircraft will not be flown in AMA sanctioned events, air shows or model demonstrations unless:
   a. The aircraft, control system and pilot skills have successfully demonstrated all maneuvers intended or anticipated prior to the specific event.
   b. An inexperienced pilot is assisted by an experienced pilot.
4. When and where required by rule, helmets must be properly worn and fastened. They must be OSHA, DOT, ANSI, SNELL or NOCSAE approved or comply with comparable standards.

B. RADIO CONTROL (RC)

1. All pilots shall avoid flying directly over unprotected people, vessels, vehicles or structures and shall avoid endangerment of life and property of others.
2. A successful radio equipment ground-range check in accordance with manufacturer's recommendations will be completed before the first flight of a new or repaired model aircraft.
3. At all flying sites a safety line(s) must be established in front of which all flying takes place. (AMA Document #706.)

a. Only personnel associated with flying the model aircraft are allowed at or in front of the safety line.
b. At air shows or demonstrations, a straight safety line must be established.
c. An area away from the safety line must be maintained for spectators.
d. Intentional flying behind the safety line is prohibited.

4. RC model aircraft must use the radio-control frequencies currently allowed by the Federal Communications Commission (FCC). Only individuals properly licensed by the FCC are authorized to operate equipment on Amateur Band frequencies.
5. RC model aircraft will not knowingly operate within three (3) miles of any pre-existing flying site without a frequency-management agreement. (AMA Documents #922 and #923.)
6. With the exception of events flown under official AMA Competition Regulations, excluding takeoff and landing, no powered model may be flown outdoors closer than 25 feet to any individual, except for the pilot and the pilot's helper(s) located at the flightline.
7. Under no circumstances may a pilot or other person touch an outdoor model aircraft in flight while it is still under power, except to divert it from striking an individual.
8. RC night flying requires a lighting system providing the pilot with a clear view of the model's attitude and orientation at all times. Hand-held illumination systems are inadequate for night flying operations.
9. The pilot of an RC model aircraft shall:
    a. Maintain control during the entire flight, maintaining visual contact without enhancement other than by corrective lenses prescribed for the pilot.
    b. Fly using the assistance of a camera or First-Person View (FPV) only in accordance with the procedures outlined in AMA Document #550.
    c. Fly using the assistance of autopilot or stabilization system only in accordance with the procedures outlined in AMA Document #560.

C. FREE FLIGHT

1. Must be at least 100 feet downwind of spectators and automobile parking when the model aircraft is launched.
2. Launch area must be clear of all individuals except mechanics, officials, and other fliers.
3. An effective device will be used to extinguish any fuse on the model aircraft after the fuse has completed its function.

D. CONTROL LINE

1. The complete control system (including the safety thong where applicable) must have an inspection and pull test prior to flying.
2. The pull test will be in accordance with the current Competition Regulations for the applicable model aircraft category.
3. Model aircraft not fitting a specific category shall use those pull-test requirements as indicated for Control Line Precision Aerobatics.
4. The flying area must be clear of all utility wires or poles and a model aircraft will not be flown closer than 50 feet to any above-ground electric utility lines.
5. The flying area must be clear of all nonessential participants and spectators before the engine is started.[51]

10. Association for Unmanned Vehicle Systems International (AUVSI)

The AUVSI Foundation is a 501(c)(3) charitable organization that was established to support the educational initiatives of the Association for Unmanned Vehicle Systems International (AUVSI). The AUVSI Foundation focuses on the future of the robotics industry by developing programs that will attract and equip students for a career in this rapidly growing field. Through a variety of efforts, the AUVSI Foundation provides students with the opportunity to experience fun, hands-on robotics activities that promote STEM education (science, technology, engineering, and math). At the K-12 level, the Foundation introduces students to the wide range of robotics programs available to them, fueling their imaginations while enhancing their education. As students' skill levels advance, the AUVSI Foundation hosts robotics competitions that challenge students to apply their engineering skills in the development of autonomous ground, air and maritime vehicles. To date, the AUVSI Foundation has awarded nearly $1 million in prize money to participating schools since the competitions began in 1991.[52]

11. Know Before You Fly

Know Before You Fly is an education campaign founded by the Association for Unmanned Vehicle Systems International (AUVSI), the Academy of Model Aeronautics (AMA), and the Small UAV Coalition in partnership with the FAA to educate prospective users about the safe and responsible operation of UAS. As excitement and enthusiasm continues to grow around UAS, and the regulatory framework continues to take shape, more consumers are looking to buy UAS for personal use and more businesses are looking to use UAS too. These prospective operators want to fly, and fly safely, but any do not realize that just because you can buy a UAS does not mean you can fly it anywhere or for any purpose.

---

[51] *Safety Code,* Academy of Model Aeronautics (AMA), https://www.modelaircraft.org/files/105.pdf.

[52] Association for Unmanned Vehicle Systems International (AUVSI), http://www.auvsi.org/home.

Know Before You Fly provides prospective users with the information and guidance they need to fly safely and responsibly.[53]

12. B4UFLY

B4UFLY is an easy-to-use smartphone app that helps unmanned aircraft operators determine whether there are any restrictions or requirements in effect at the location where they want to fly. On August 28, 2015, the FAA released B4UFLY to approximately 1,000 beta testers, including members of industry, government, and the public. This limited beta test is expected to run for several months, after which the FAA plans to make a final version of B4UFLY available for the general public. The beta version of the app is for iOS devices only, but the FAA intends to release a future version for both iOS and Android devices.[54]

Key features of the B4UFLY app include: (1) A clear "status" indicator that immediately informs the operator about their current or planned location. For example, it shows flying in the Special Flight Rules Area around Washington, DC is prohibited; (2) Information on the parameters that drive the status indicator; (3) A "Planner Mode" for future flights in different locations; (4) Informative, interactive maps with filtering options; and (5) Links to other FAA UAS resources and regulatory information.[55]

On January 6, 2016, FAA Administrator Michael Huerta announced the public release of the B4UFLY mobile application following an initial beta testing period. B4UFLY tells users about current or upcoming requirements and restrictions in areas of the National Airspace System (NAS) where they may want to operate their unmanned aircraft system (UAS). The app is now available for Apple devices and can be downloaded from the APP Store. The FAA also is releasing a beta version of B4UFLY for Android devices, which can be downloaded from Google. According to Huerta, the FAA expects that B4UFLY will help raise public awareness about what it means to operate unmanned aircraft safely and it is another important part of the FAA's education and awareness efforts to foster a culture of safety and accountability for the UAS community.[56]

The B4UFLY app includes a number of enhancements the FAA developed as a result of user feedback during the beta testing announced in May 2015. Within two taps, users know if it is safe to fly at their current location. The app provides a status indicator that tells users: "Proceed with Caution," "Warning – Action Required," or "Flight Prohibited." The app also features a planner mode that allows users to select a different time and location for an upcoming flight and determine if there are any restrictions at that place and time. By law, hobbyists who want to fly within five miles of an airport must notify the airport operator and the air traffic control facility (if there is one) prior to flying. For now,

---

[53]Know Before You Fly, http://knowbeforeyoufly.org.

[54]B4UFLY, Federal Aviation Administration (FAA), https://www.faa.gov/uas/b4ufly/.

[55]*Id.*

[56]*FAA Releases B4UFLY Smartphone App*, FAA, https://www.faa.gov/news/updates/?newsId=84508.

B4UFLY will ask users who are supposed to notify the airport before flying for voluntary information about their planned flight. This will not meet the statutory requirement to notify the airport and air traffic control facility, but the data will help the agency make informed policy decisions related to notification. This information will not be publicly available.[57]

13. Fly Safe with Your Drone

On November 27, 2015, just in time for Black Friday shopping, the FAA released their new safety brochure, Fly Safe with Your Drone. It contains the following information:[58]

I FLY SAFE

All drones are aircraft – even the ones at the toy store. So when I fly a drone I am a pilot. Before I fly I always go through my pre-flight checklist. I regularly check the safety guidelines at faa.gov/uas. Fly smart, fly safe, and have fun!

I fly below 400 feet.

I always fly within visual line of sight.

I'm aware of FAA airspace requirements: faa.gov/go/uastfr

I never fly over groups of people.

I never fly over stadiums and sports events.

I never fly within 5 miles of an airport without first contacting air traffic control and airport authorities.

I never fly near emergency response efforts such as fires.

I never fly near other aircraft.

I never fly under the influence.[59]

14. NOTAM for TFR

This section is identical to that in section III Commercial UAS Rules – Part 4.

15. Washington, DC Flight Restricted Zone (FRZ)

This section is identical to that in section III Commercial UAS Rules – Part 5.

16. 2009 Special Notice

This section is identical to that in section III Commercial UAS Rules – Part 6.

---

[57] *Id.*

[58] *Fly Safe with Your Drone*, FAA, http://www.faa.gov/news/updates/media/2015-FAA-383-UAS_Holiday_Pre-flight-checklist_1200x627_ae05.pdf.

[59] *Id.*

ABOUT THE AUTHOR

Dr. Sarah Nilsson has both an aviation and a legal background. She began her career as a pilot in 1994 and is a Federal Aviation Administration (FAA)–licensed airline transport pilot in multi- and single-engine fixed-wing airplanes. Since 2000 she has been a gold seal flight instructor and has trained hundreds of pilots. She also has flown air cargo and private business jets and holds a type-rating in the CE-500 series.

Currently, Sarah is founder and managing attorney of Nilsson Law, PLLC, as well as a certified mediator. Since January 2015, Sarah has served as full-time faculty at Embry-Riddle Aeronautical University (ERAU) in Prescott, Arizona, teaching Aviation Law, Global UAS Regulations (a course she designed to fulfill a need within the UAS degree program), Business Law, and Business Ethics. She also teaches Global UAS Risk Management, another course she designed, for the ERAU Worldwide Campus. Sarah volunteers with the FAA as a FAASTeam Safety Representative.

In addition to her own research on Drone Law and Space Law and Policy, she coauthored with Scott Hamilton the 6th edition of *Practical Aviation and Aerospace Law*, a national aviation law textbook. She has made multiple media appearances as an expert in UAS regulation, law, and implications for the public.

Dr. Nilsson holds a PhD in business administration from Northcentral University, conferred in 2011. In 2014, she also graduated with honors with a juris doctorate from Arizona Summit Law School. In 2014, Sarah became a licensed attorney in the State of Arizona. She previously graduated with honors summa cum laude with a bachelor of science in professional aeronautics from Embry-Riddle Aeronautical University, and minors in Aviation Business Administration and Aviation Safety. In 2006, Sarah also obtained a Master of Aeronautical Science from Embry-Riddle with concentrations in Aviation Safety, Aerospace Operations, and Human Factors.

INDEX